Polycrystalline Materials–from Design to (Micro)Structural Characterization and Applications

Polycrystalline Materials–from Design to (Micro)Structural Characterization and Applications

Guest Editors

Sanja Burazer
Lidija Androš Dubraja

Basel • Beijing • Wuhan • Barcelona • Belgrade • Novi Sad • Cluj • Manchester

Guest Editors

Sanja Burazer
Ruđer Bošković Institute
Zagreb
Croatia

Lidija Androš Dubraja
Ruđer Bošković Institute
Zagreb
Croatia

Editorial Office
MDPI AG
Grosspeteranlage 5
4052 Basel, Switzerland

This is a reprint of the Special Issue, published open access by the journal *Crystals* (ISSN 2073-4352), freely accessible at: https://www.mdpi.com/journal/crystals/special_issues/C0Q608C87Z.

For citation purposes, cite each article independently as indicated on the article page online and as indicated below:

Lastname, A.A.; Lastname, B.B. Article Title. *Journal Name* **Year**, *Volume Number*, Page Range.

ISBN 978-3-7258-2725-1 (Hbk)
ISBN 978-3-7258-2726-8 (PDF)
https://doi.org/10.3390/books978-3-7258-2726-8

Cover image courtesy of Sanja Burazer and Lidija Androš Dubraja

© 2025 by the authors. Articles in this book are Open Access and distributed under the Creative Commons Attribution (CC BY) license. The book as a whole is distributed by MDPI under the terms and conditions of the Creative Commons Attribution-NonCommercial-NoDerivs (CC BY-NC-ND) license (https://creativecommons.org/licenses/by-nc-nd/4.0/).

Contents

About the Editors vii

Preface ix

Vassilis Psycharis, Manolis Chatzigeorgiou, Dimitra Koumpouri, Margarita Beazi-Katsioti and Marios Katsiotis
Structure–Superstructure Inter-Relations in Ca_2SiO_4 Belite Phase
Reprinted from: *Crystals* **2022**, *12*, 1692, https://doi.org/10.3390/cryst12121692 1

Alberto Ubaldini, Flavio Cicconi, Sara Calistri, Stefano Salvi, Chiara Telloli, Giuseppe Marghella, et al.
Removal of Organic Materials from Mytilus Shells and Their Morphological and Chemical-Physical Characterisation
Reprinted from: *Crystals* **2024**, *14*, 464, https://doi.org/10.3390/cryst14050464 16

Peishen Ni, Yongxuan Chen, Wenxin Yang, Zijian Hu and Xin Deng
Research on Microstructure, Synthesis Mechanisms, and Residual Stress Evolution of Polycrystalline Diamond Compacts
Reprinted from: *Crystals* **2023**, *13*, 1286, https://doi.org/10.3390/cryst13081286 34

Glorija Medak, Andreas Puškarić and Josip Bronić
The Influence of Inserted Metal Ions on Acid Strength of OH Groups in Faujasite
Reprinted from: *Crystals* **2023**, *13*, 332, https://doi.org/10.3390/cryst13020332 51

Ivana Landripet, Andreas Puškarić, Marko Robić and Josip Bronić
Fine Tuning of Hierarchical Zeolite Beta Acid Sites Strength
Reprinted from: *Crystals* **2024**, *14*, 53, https://doi.org/10.3390/cryst14010053 63

Maria Poienar, Paula Svera, Bogdan-Ovidiu Taranu, Catalin Ianasi, Paula Sfirloaga, Gabriel Buse, et al.
Electrochemical Investigation of the OER Activity for Nickel Phosphite-Based Compositions and Its Morphology-Dependent Fluorescence Properties
Reprinted from: *Crystals* **2022**, *12*, 1803, https://doi.org/10.3390/cryst12121803 79

Jana Chrappová, Yogeswara Rao Pateda, Lenka Bartošová and Erik Rakovský
Investigating the Formation of Different $(NH_4)_2[M(H_2O)_5(NH_3CH_2CH_2COO)]_2[V_{10}O_{28}] \cdot nH_2O$ ($M = Co^{II}$, Ni^{II}, Zn^{II}, $n = 4$, $M = Cd^{II}$, Mn^{II}, $n = 2$) Crystallohydrates
Reprinted from: *Crystals* **2024**, *14*, 685, https://doi.org/10.3390/cryst14080685 96

Essam A. Ali, Rim Bechaieb, Rashad Al-Salahi, Ahmed S. M. Al-Janabi, Mohamed W. Attwa and Gamal A. E. Mostafa
Supramolecular Structure, Hirshfeld Surface Analysis, Morphological Study and DFT Calculations of the Triphenyltetrazolium Cobalt Thiocyanate Complex
Reprinted from: *Crystals* **2023**, *13*, 1598, https://doi.org/10.3390/cryst13111598 109

Marko Očić and Lidija Androš Dubraja
Intermolecular Interactions in Molecular Ferroelectric Zinc Complexes of Cinchonine
Reprinted from: *Crystals* **2024**, *14*, 978, https://doi.org/10.3390/cryst14110978 127

Lixin Hou, Dingding Jing, Yanfeng Wang and Ying Bao
CO_2 Promoting Polymorphic Transformation of Clarithromycin: Polymorph Characterization, Pathway Design, and Mechanism Study
Reprinted from: *Crystals* **2024**, *14*, 394, https://doi.org/10.3390/cryst14050394 139

About the Editors

Sanja Burazer

Sanja Burazer is a research associate in the Division of Materials Physics at the Ruđer Bošković Institute in Zagreb, Croatia. She holds a PhD in Chemistry from the Faculty of Science, University of Zagreb. She spent a year working in the Department of Condensed Matter Physics, Faculty of Mathematics and Physics, Charles University in Prague, the Czech Republic, as a postdoctoral researcher. Sanja has participated in four bilateral projects as a collaborator. Her current research interests include structure determination from PXRD data and the investigation of structure–property relationships, especially in batteries but also in hydrogen storage, sensors, and magnetoelectric materials.

Lidija Androš Dubraja

Lidija Androš Dubraja is a senior research associate in the Division of Materials Chemistry at the Ruđer Bošković Institute in Zagreb, Croatia. She holds a PhD in Chemistry from the Faculty of Science, University of Zagreb. She has worked as a postdoctoral researcher at the Karlsruhe Institute of Technology, Germany. In 2019, she received the installation research grant from the Croatian Science Foundation to study molecular ferroelectrics. Her research focuses on coordination chemistry, crystallography, and the electrical properties of materials in various forms (crystal, bulk, thin film).

Preface

This Special Issue focuses on the use of powder X-ray diffraction data and complementary analytical techniques in the study of various polycrystalline materials. It highlights the performance and potential applications of polycrystalline materials and provides insights into structure–property correlation and the evaluation of various properties. This Reprint covers a range of structure–property relationships and contains 10 papers. One of these is a review paper by Psycharis et al. that discusses the structure–superstructure relationships in the belite phase, an important component of Portland cement. Ubaldini et al. discussed the removal of organic materials from Mytilus shells and their morphological and physico-chemical characterization. Microstructure, synthesis mechanisms and residual stress development of polycrystalline diamond compacts were published in a research paper by Ni et al. The influence of inserted metal ions on the acid strength of hydroxyl groups in faujasite zeolite is described in the work of Medak et al., while Landripet et al. reported on the fine-tuning of the acid strength of hierarchical beta zeolites. The work of Poienar et al. presents electrochemical studies of the oxygen evolution reaction for nickel phosphite materials and their morphology-dependent fluorescence properties. Chrappová et al. reported on the study of various decavanadates crystallizing as different crystal hydrates. Ali et al. investigated intermolecular interactions in triphenyltetrazolium cobalt thiocyanate complexes, and Očić and Androš Dubraja in a molecular ferroelectric zinc complexes with natural alkaloids. Hou et al. published a study on the CO_2-promoted polymorphic transformation of clarithromycin, including the characterization of polymorphs, development of reaction pathways and investigation of the mechanism.

Sanja Burazer and Lidija Androš Dubraja
Guest Editors

Review

Structure–Superstructure Inter-Relations in Ca$_2$SiO$_4$ Belite Phase

Vassilis Psycharis [1,*], Manolis Chatzigeorgiou [1,2], Dimitra Koumpouri [3], Margarita Beazi-Katsioti [2] and Marios Katsiotis [4]

[1] Institute of Nanoscience and Nanotechnology, National Center for Scientific Research "Demokritos", 15310 Athens, Greece
[2] School of Chemical Engineering, National Technical University of Athens, 15772 Athens, Greece
[3] Independent Researcher, 15451 Athens, Greece
[4] Group Innovation & Technology, TITAN Cement S.A., 11143 Athens, Greece
* Correspondence: v.psycharis@inn.demokritos.gr; Tel.: +30-210-6503346

Abstract: Belite, the second most abundant mineralogical phase in Portland cement, presents five polymorphs which are formed at different temperatures. The increased interest in belite cement-based products is due to the lower environmental impact associated with the lower energy consumption. The importance of belite polymorphs formed at higher temperatures for cement industry applications is high, because they present better hydraulic properties. Thus, any study that helps to explore the structure relations of all belite polymorphs is of interest for both scientific and practical points of view. In the present work, a systematic structure–superstructure relation study is presented for all polymorphs, and it is based on the work of O'Keefe and Hyde (1985). In this pioneering work, generally, the structures of oxides are considered as having common characteristics with prototype structures of alloys. The basic result of the present work is the fact that all the polymorphs adopt a common architecture which is based on capped trigonal prisms of Ca cations, which host the Si one, and the oxygen anions occupy interstitial sites, i.e., an architecture in conformity with the model which considers the oxide structures as stuffed alloys. This result supports the displacive character of the transformation structural mechanism that links the five polymorphs based on the cation sites in their structures. However, based on the sites of oxygen anions, it could be considered as of diffusion character. The study of belite polymorphs is also of interest to products obtained by doping dicalcium silicate compounds, which present interesting luminescent properties.

Keywords: belite; structure superstructure; stuffed alloys

1. Introduction

Alite (~50–70%) and belite (~20–30%) are the main mineralogical phases of Portland cement (OPC), accompanied by the presence of tricalcium aluminate and ferrite phases. Using the compact mineralogical code names, these four phases are represented by C$_3$S for Ca$_3$SiO$_5$ (alite), C$_2$S for Ca$_2$SiO$_4$ (belite), C$_3$A for Ca$_3$Al$_2$O$_6$ or (CaO)$_3$(Al$_2$O$_3$) (tricalcium aluminate), and C$_4$AF for Ca$_4$Al$_2$Fe$_2$O$_{10}$ or (CaO)$_4$(Al$_2$O$_3$)(Fe$_2$O$_3$), where C = CaO, S = SiO$_2$, A = Al$_2$O$_3$ and F = Fe$_2$O$_3$ [1]. All the mineral phases of OPC cement are available in different polymorphic forms. The polymorphism of the mineral phases of cement depends on several parameters, such as the temperature and duration of heat treatment, cooling rate, the composition of the raw mixture and the presence of "stabilizer oxides" in the raw mixture. The formation of different polymorphic mineral phases in cement products is critical for their performance (hydraulic properties), as their presence directly affects the physical (fineness, setting time and volume stability) and mechanical (strength development) properties of cement.

For belite, tricalcium aluminate and ferrite phases, the structures of their different polymorphs are well studied and known, i.e., C$_2$S [2–4], C$_3$A [5,6] and C$_4$AF [7,8]. Alite and belite present many polymorphs. Their structure–superstructure relations are used to explore the structural connection among them and also to understand the mechanism governing the transformation from one polymorph to the other [9,10]. In a series of papers [9,11–14], a thorough analysis of the structure–superstructure relations for alite is given. In the framework of this analysis [13], new structural models for alite polymorphs have been proposed and are used in the analysis of powder diffraction diagrams with the Rietveld method of industrial cement products.

The increased interest in the use of belite-based cement products is due to the need for lower energy consumption and their positive environmental impact compared to OPC [15]. The recent increased research interest and study of luminescent belite-based doped compounds [16] make the understanding of belite polymorph formation and stabilization very attractive and the need for accurate belite polymorphism identification essential. The belite phase presents five polymorphs, and these are listed below based on the decreasing formation temperature, i.e., α-Ca_2SiO_4 (α-C_2S, 1545 °C), α'_H-Ca_2SiO_4 (α'_H-C_2S, 1250 °C), α'_L-Ca_2SiO_4 (α'_L-C_2S, 1060 °C), β-Ca_2SiO_4 (β-C_2S, <500 °C) and γ-Ca_2SiO_4 (γ-C_2S, stable at room temperature). The "activity" (better hydraulic properties) of the belite polymorphs is higher at high formation temperatures (belite polymorph activity: $\alpha > \alpha'_H > \alpha'_L > \beta$) [1,17], while γ-C_2S does not exhibit hydraulic properties. β-C_2S is the high-temperature monoclinic polymorph of the Calcio-Olivine γ-C_2S polymorph but does not belong to the olivine group. The structure relations of the family of belite polymorphs are discussed in detail in [10]. It is well known that the first four of them, α-, α'_H-, α'_L- and β-C_2S, are inter-related with reversible phase transitions [4,18–20]. The structural basis for these transformations is explained by their similarity to the structure of K, Na-glaserite sulfate, $K_3Na[SO_4]_2$ and its derivative β-$K_2[SO_4]$ Arcanite (α'_H-, α'_L- and β-C_2S) [10]. The main result of these studies is the coordination polyhedra formed by oxygen atoms around the cations, and the framework build by the polyhedra arrangement within the structures of these polymorphic phases. The relation of the framework formed by the coordination polyhedra in the structure of glaserite and that of β-C_2S is also discussed in [21].

The description of structures based on polyhedra of anions is the traditional point of view introduced by Pauling [22] and is based on the idea that the "small" cations occupy positions in the voids left in their crystal structure by the larger atoms of anions such as oxygens [23,24]. A new approach has been introduced by O'Keefe and Hyde [23], where the role of ions is reversed, i.e., the structure description is based on the packing of cations and the anions are inserted at interstitial sites among them. In the beginning, this new approach was considered purely geometric, and no bonding type considerations were implied for the atoms involved in the packing within a crystal structure. However, recently, in addition to the achieved simplifications in the presentation of many structures, the role of cations was reconsidered as the packing of cations observed within the ionic structures resamples those observed in their parent metal structures [25]. The geometric characteristics, arrangement and distances in the parent alloy structures are retained in the ionic one [24]. More interestingly, based on the pioneering work in this field by O'Keeffe and Hyde [23], it is found that in ternary oxides with two different metals the packing of cations corresponds to the packing of the binary alloys of the metals. Therefore, in the case of β-C_2S, the cations' packing resembles that of Ni_2In prototype structure ($B8_b$, Strukturbericht designations). These observations have made the authors O'Keefe and Hyde characterize the oxides as "stuffed alloys". With a lack of data from the other b elite polymorphs, the discussion by Barbier and Hyde (1985) [26] using the cation-staffed model of oxides is restricted on the structure inter-relations of β- and γ-Ca_2SiO_4 dicalcium silicates. In addition, they extended their study to α'- and β-Sr_2SiO_4 polymorphs which resemble the structures of the corresponding polymorphs of belite. The ideas developed in the works of Barbier and Hyde [26] and O'Keeffe and Hyde [23] are applied systematically in the present work for the study of all polymorphs of belite. This analysis of all the belite polymorphic structures, which is based on the observation of common building blocks of prototype alloy structures, revealed first the geometric basis of the structure–superstructure inter-relations and second possible transformation mechanisms.

2. Methodology

In this section, the methodology of the present study is developed, which essentially consists at a first step of the description of the arrangement of cations in prototype alloy structures and their relation to the structures of β-C_2S and γ-C_2S belite polymorphs, as have been discussed in previous studies [23,26]. Then, in the next section this discussion is extended to the other polymorphs as well. According to Mumme [4], the Ca_2SiO_4 compound presents five polymorphs (Scheme 1):

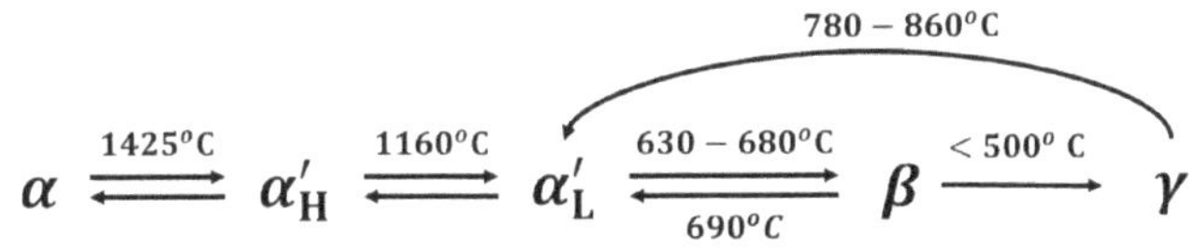

$$\alpha \underset{}{\overset{1425°C}{\rightleftarrows}} \alpha'_H \underset{}{\overset{1160°C}{\rightleftarrows}} \alpha'_L \underset{690°C}{\overset{630-680°C}{\rightleftarrows}} \beta \overset{<500°C}{\longrightarrow} \gamma$$

(with 780 – 860°C arrow from α'_L to γ)

Scheme 1. Belite polymorphs formed at different temperatures.

The decreasing order of formation/stabilization temperature of the belite polymorphs is α-, α'_H-, α'_L-, β- and γ-C_2S. β-C_2S is metastable and can be formed during cooling but cannot be produced from γ-C_2S on heating. Unless β-C_2S is stabilized during cooling, the α- and α'-C_2S polymorphs revert to the stable form of γ-C_2S. The subscript H stands for the high- and L for the low-temperature form of α' polymorphs. Stabilization of the "active" belite polymorphs and also prevention of the transition from β to γ-C_2S can be achieved by rapid cooling and/or by incorporation of a "stabilizer" (substitution of Ca and Si cations by cations of Ti, B, S, K, etc.) in the belite structure [19,20,27–29].

In their original articles, Barbier and Hyde [26] and O' Keefe and Hyde [23] use two prototype metal alloy structures to describe β-C_2S and γ-C_2S belite polymorphs. These are the prototype structure of $PbCl_2$ (Strukturbericht designations C23) and Ni_2In ($B8_b$ type), respectively. The β-C_2S polymorph is also compared to the Ca_2Si alloy, which is similarly characterized by the C23 Strukturbericht structure type. The β- and γ-C_2S polymorphs are also discussed in comparison to the structures of low- and high-temperature K_2SO_4 compounds, which belong to the Strukturbericht C23 and B8b types of compounds, respectively.

In Table 1, we provide general information and ICSD codes for all the compounds that are discussed in the present work. Detailed crystallographic data for the structures of all the polymorphs and prototype phases are provided in Table S2, together will all references from where all the data were obtained. In the present work, a systematic structural study of all the polymorphs of belite and their inter-relations is attempted using the approach of O'Keefe and Hyde. To ensure that the structural models were obtained by analyzing data from materials synthesized in similar ways, we decided to use the models given in reference Mumme (1995) [3] for β-C_2S and Mumme (1996) [4] for the rest of belite polymorphs. Details for the structures of all polymorphs are given in Table S2. Plots of the structures were drawn using the Diamond-3.1 program package [30].

The prototype structure of Ni_2In ($B8_b$) crystallizes in the $P6_3/mmc$ Space Group (S.G.) [31]. Atoms in the unit cell are arranged as in Figure 1A. Following the original description [23], Ni atoms are packed in trigonal Ni_6 prisms that host the In atoms and, by sharing their trigonal faces, form columns parallel to the direction [110]. The trigonal prisms, as they are the basic polyhedra that characterize the structures studied in this work, are demarcated by orange lines in all the pictures. These columns are arranged in a zig-zag fashion above and below (1–100) planes, and by edge-sharing they form "walls" parallel to these planes. In Figure 1B, the arrangement of atoms is shown in projection on the (11–20) planes, and in Figure 1C the same view is shown tilted to make clear the relative arrangement of crystallographic planes. It is mentioned that the atoms on the "wall" in the middle of Figure 1B,C, those indicated with the yellow arrows, are shifted along the direction normal to the (11–20) planes by a distance half the height of the trigonal prisms, which is equal to the d_{11-20} plane distance. For the description of all the structures discussed in the present work on a common basis, it is useful to use in addition to the hexagonal cell the ortho-hexagonal one [32]. The relative orientation of ortho-hexagonal cell axes is shown in Figure 1B and their relation with the hexagonal axes in Figure 2D.

Table 1. Structural information and parameters for the prototype 8B2 and C23 prototype structures and the five polymorphs of belite (Ca_2SiO_4).

Phase Code Name	ICSD #	Space Group	V/Z
$8B_b$ ($InNi_2$)	640098 [31]	$P6_3/mmc$	-
C23 (Ca_2Si)	158275 [32]	$Pnma$	-
α_H-C_2S	82998 [4]	$P6_3/mmc$	97.10 Å^3
α_{Tr}-C_2S	82999 [4]	P-$3m1$	97.10 Å^3
α'_H-C_2S	82997 [4]	$Pnma$	91.94 Å^3
α'_L-C_2S	82996 [4]	$Pna2_1$	90.79 Å^3
β-C_2S	81096 [3]	$P2_1/n$	86.45 Å^3
γ-C_2S	82994 [4]	$Pbnm$	96.18 Å^3

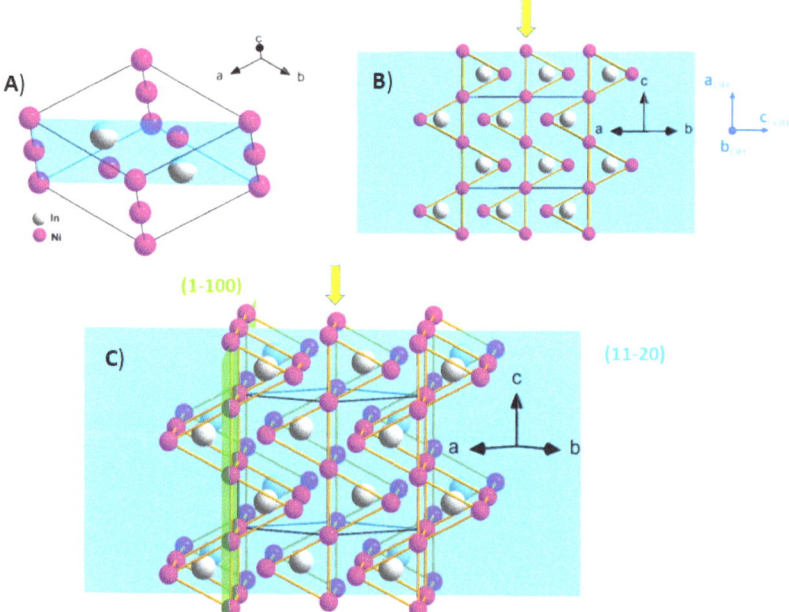

Figure 1. (**A**) Arrangement of atoms in the cell of the Ni_2In structure. (**B**) "Walls" of trigonal prisms parallel to the (1–100) planes projected on the (11–20) planes. The relative orientation of the ortho-hexagonal reference system is also indicated. The former plane is normal, and the second one is parallel to the plane of the figure and (**C**) tilted view. The planes (11–20) and (1–100) are shown in cyan and light green color, respectively. The orange lines are delimitating the trigonal prisms. The yellow arrows indicate the middle "wall" arrangement of trigonal prisms and the reference system, with axes drawn in blue indicating the orientation of the ortho-hexagonal axes system.

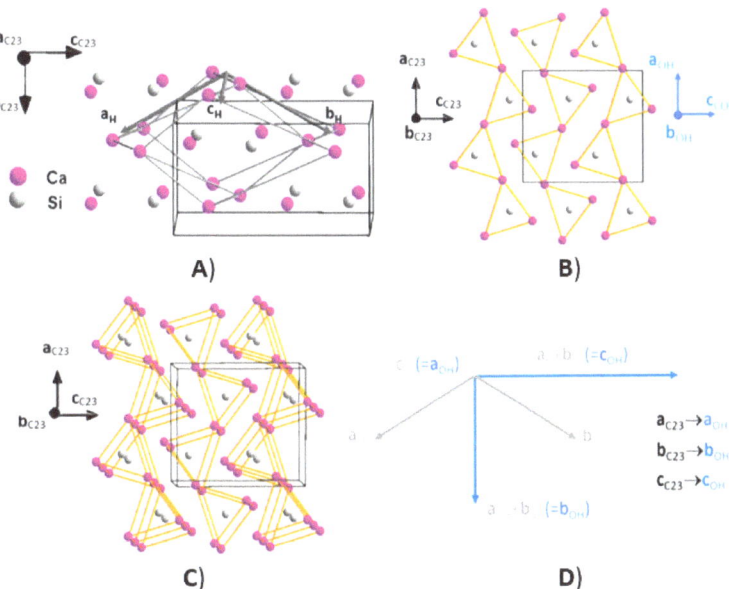

Figure 2. (**A**) Arrangement of atoms in the cell of the Ca$_2$Si structure cell. Light grey lines joining Ca cations indicate the orientation of the hexagonal part of the structure within the prototype structure C23. With dark grey vectors the orientation of the reference hexagonal cell is also shown. (**B**) "Walls" of trigonal prisms parallel to the planes (001) projected on the (010) planes. The former planes are normal, and the latter ones are parallel to the plane of the figure. The relative orientation of the ortho-hexagonal reference system is also indicated. (**C**) Presents a tilted view of (**B**). (**D**) Schematic presentation of the unit cell axes relations for the hexagonal and ortho-hexagonal cells. The correspondences of the cell axes of the orthorhombic C23 type structure with those of ortho-hexagonal cell are also indicated.

The arrangement of atoms In the Ca$_2$Si alloy, which crystallizes in the *Pnma* S.G. [32], with the C23 Strukturbericht structure type, is shown in Figure 2. The atoms in the structure of the Ca$_2$Si alloy are packed again in trigonal Ca$_6$ prisms which host the Si atoms and by sharing their trigonal faces form columns parallel to the b crystallographic axis. These columns are arranged in a zig-zag fashion above and below the (001) planes, and by edge-sharing they form "walls" parallel to these planes. In Figure 2B, the arrangement of atoms Is shown in projection on (010) planes, and in Figure 2C the same view is shown tilted. It is mentioned that also in the present case, the atoms on the "wall" in the middle of Figure 2B,C are shifted along the a-axis direction by a distance half the high of the trigonal prisms, which is equal to the d$_{100}$ plane distance. The main difference in the present arrangement from the one presented in Figure 1A,C is that in the second case the "walls" are puckered. In Figure 2A, in addition to atom arrangement within the unit cell the relative position of the hexagonal axes system is shown, which helps to derive the relationship which joins the hexagonal with the orthorhombic cell. The relative orientation of the cell axes of the two systems is shown in Figure 2D. The relative orientation of the ortho-hexagonal cell axes with those of the orthogonal ones is also indicated (Figure 2B,D), and they are identical. Thus, there is a one-to-one correspondence among a$_{C23}$, b$_{C23}$ and c$_{C23}$ and a$_{OH}$, b$_{OH}$ and c$_{OH}$ axes, respectively, which is expressed with the relations a$_{C23}$ → a$_{OH}$ = c$_H$, b$_{C23}$ → b$_{OH}$ = a$_H$ + b$_H$ and c$_{C23}$ → c$_{OH}$ = −a$_H$ + b$_H$, where the C23, OH and H subscripts stand for orthorhombic, ortho-hexagonal and hexagonal cells, respectively. The prototype structures of Ni$_2$In and Ca$_2$Si, which are discussed in the present section, are used as a basis to describe the structure inter-relation for all belite polymorphs. It worth to notice that the c_{OH}-axis is normal to the planes of the "walls" and the a_{OH} and b_{OH} axes lie within these planes.

The transformation of β-C$_2$S to γ-C$_2$S has been discussed as a transformation of the C23 to B8b prototype arrangement of the Ca and Si cations [23,26]. The relative arrangements of Ca^{+2} cations and SiO_4^{-4} anions in the structures of these two polymorphs of belite are shown in Figures 3 and 4, respectively. Ca and Si atoms in the β-C$_2$S polymorph of belite occupy the corresponding Ca and Si sites of the Ca$_2$Si (C23) prototype structure. In the γ-C$_2$S polymorph, Ca and Si atoms occupy the sites of Ni and In atoms in the Ni$_2$In B8$_b$ prototype structure, respectively. The oxygen atoms in both cases are inserted at interstitial sites (Figures 3 and 4). In reference [23], the β-C$_2$S to γ-C$_2$S transformation is discussed in relation to the observed 11.2% of unit cell volume increase when the transformation takes place. This is also mentioned in reference [33], where ternary alloys undergo a C23 to B8b type structural transformation. The same transformation is also discussed on the basis of unit cell axes changes in both works. In Table 2, the unit cell axes relations for all studied structures with those of the ortho-hexagonal system are given, and through these relations all the structure inter-relations and changes are discussed. For ternary alloys, the transformation of 8Bb to C23 results in changes within the **a**, **b** planes, i.e., within the (001) "wall" planes of the ortho-hexagonal cell [33]. In references [26,34], the emphasis is given to the presentation of changes that take place at a local level and more specifically to the atomic displacements of the atoms that contribute to the formation of trigonal prisms and their surroundings. An attempt has been made in the past to extend the description based on the cation-staffed model of oxides to other belite polymorphs, using standard alloy structures, but so far, the discussion has been limited either to data from analogous structures [26] or to the interpretation of results from TEM studies [35]. The availability of structural data for all belite polymorphs [3,4] is exploited in the next section of the present work for the examination of all the structures of the corresponding polymorphs with the O'Keefe and Hyde model.

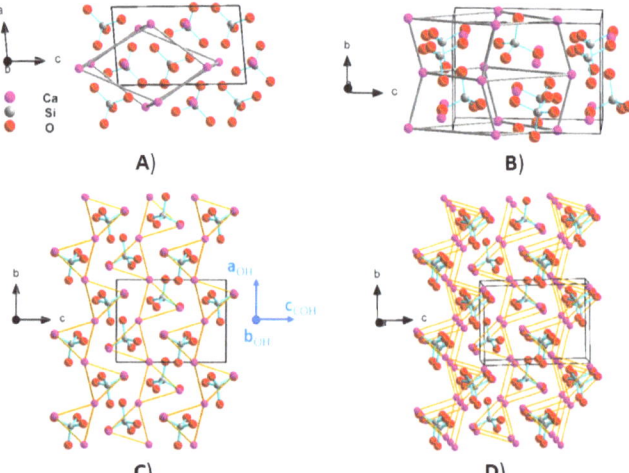

Figure 3. (**A**) The unit cell content of β-Ca$_2$SiO$_4$ polymorph viewed along the b-axis. Light grey lines joining Ca cations indicate the orientation of the hexagonal part of the structure and (**B**) side view of unit cell content along the a-axis is shown. (**C**) View along a-axis and the packing arrangement of trigonal prisms is demarcated with orange lines. The relative orientation of the ortho-hexagonal reference system is also indicated. (**D**) A tilted view of (**C**). The Si-O bonds are indicated by cyan lines.

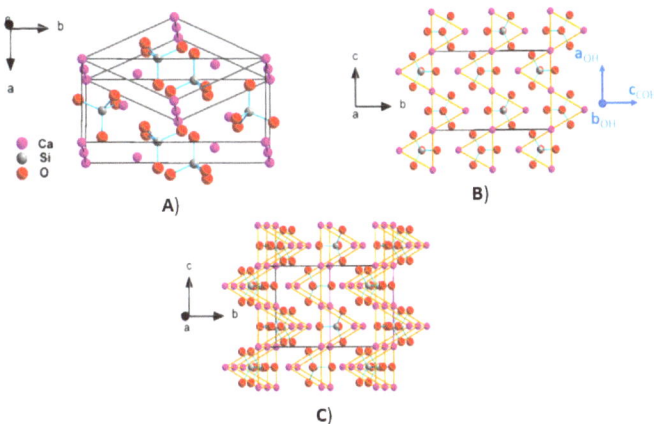

Figure 4. (**A**) The unit cell content of γ-Ca$_2$SiO$_4$ polymorph viewed along c-axis. Light grey lines joining Ca cations indicate the orientation of the hexagonal part of the structure. (**B**) View along the a-axis and the packing arrangement of trigonal prisms is demarcated with orange lines. The relative orientation of the ortho-hexagonal reference system is also indicated. (**C**) Tilted view of (**B**). The Si-O bonds are indicated by cyan lines.

Table 2. Unit cell relation of the C23 prototype structure and the unit cells of the five polymorphs of belite with the basic hexagonal unit cell of the 8B2 structure.

Phase Code Name	Unit Cell Axes	Cell Volume Relations
8B$_b$(InNi$_2$)	a_H, b_H and c_H	V_H
C23(Ca$_2$Si)	$a_{c23} = c_H$, $b_{c23} = a_H + b_H$ and $c_{c23} = -a_H + b_H$	$2 \times V_H$
α$_H$-C$_2$S	$a_{αH} \to a_H$, $b_{αH} \to b_H$ and $c_{αH} \to c_H$ $a_{αH} = 5.532$ Å, $b_{αH} = 5.532$ and $c_{αH} = 7.327$	$V_{αH} \to V_H$
α$_{Tr}$-C$_2$S	$a_{αTr} = a_{αH}$, $b_{αTr} = b_{αH}$ and $c_{αTr} = c_{αH}$	$V_{αTr} = V_{αH}$
α'$_H$-C$_2$S	$a_{α'H} = c_{αH} (\to a_{oH})$, $b_{α'H} = a_{αH} + b_{αH} (\to b_{oH})$ and $c_{α'H} = -a_{αH} + b_{αH} (\to c_{oH})$ $a_{α'H} = 6.871$ Å $\to a_{oH} = 7.327$ Å, $b_{α'H} = 5.601$ Å $\to b_{oH} = 5.532$ Å and $c_{α'H} = 9.556$ Å $\to c_{oH} = 9.581$ Å	$2 \times V_{αH}$
α'$_L$-C$_2$S	$a_{α'L} = 3c_{αH} (\to 3a_{oH})$, $b_{α'L} = -a_{αH} + b_{αH} (\to c_{oH})$, $c_{α'L} = -a_{αH} - b_{αH} (\to -b_{oH})$, $a_{α'L} = 20.527$ Å $\to 3a_{oH} = 21.981$ Å, $b_{α'L} = 9.496$ Å $\to c_{oH} = 9.581$ Å and $c_{α'L} = 5.590$ Å $\to b_{oH} = 5.532$ Å	$6 \times V_{αH}$
β-C$_2$S	$a_β = a_{αH} - b_{αH} (\to -b_{oH})$, $b_β = c_{αH} (\to a_{oH})$, $c_β = -a_{αH} + b_{αH} (\to c_{oH})$, $a_β = 5.512$ Å $\to (b_{oH}) = 5.532$ Å, $b_β = 6.758$ Å $\to a_{oH} = 7.327$ Å, $c_β \sin β = 9.284$ Å $\to c_{oH} = 9.581$ Å and ($β_β = 94.581° \to α_{oH} = 90.0°$)	$2 \times V_{αH}$
γ-C$_2$S	$a_γ = a_{αH} + b_{αH} (\to b_{oH})$, $b_γ = -a_{αH} + b_{αH} (\to c_{oH})$ and $c_γ = c_H (\to a_{oH})$ $a_γ = 5.076$ Å $\to b_{oH} = 5.532$ Å, $b_γ = 11.214$ Å $\to c_{oH} = 9.581$ Å and $c_γ = 6.758$ Å $\to a_{oH} = 7.327$ Å	$2 \times V_{αH}$

3. Structure Inter-Relations for Belite Polymorphs

In this section, the structure inter-relation of all belite polymorphs are studied based on common patterns of the arrangement of cations using ideas which have been introduced previously. In Figure 5, the two models for α-C$_2$S, the high-temperature polymorph of belite, are presented, which have been proposed by Mumme [4]. The difference between the proposed models, a trigonal and a hexagonal one, concerns the arrangements of oxygen atoms of the SiO$_4$ tetrahedra, but in both models, the arrangement of Ca and Si cations resembles that of the B8b prototype structure, and these are shown in Figures 5B and 5D, respectively. The Rietveld refinement has promoted the hexagonal model for the α-C$_2$S polymorph [4], and this is the reason why the hexagonal cell axes are used to express these relations. As has been already mentioned above, the best way to track the changes which are observed in this family of compounds is by using the ortho-hexagonal reference system which is derived from the hexagonal cell of the α-C$_2$S polymorph. The relative orientation of the ortho-hexagonal cell axes with respect to those of the reference hexagonal cell is shown in Figure 5B,D for the α-C$_2$S polymorph. In the first model, where the structure is described in a trigonal S.G. (Table 1), deviation of one oxygen atom from the c-axis is observed (Figure 5A). In the second model, where the structure is described in a hexagonal S.G. (Table 1), the oxygen atoms show severe disorder, and half of the SiO$_4$ tetrahedra point in one direction and half of them in the opposite one (Figure 5C). In Figure 6A,C, the unit cell content for the α'$_H$- and α'$_L$-C$_2$S polymorphs are shown. In the same figure, their projection along the b-axis for the former (Figure 6B) and also along the c-axis for the latter (Figure 6D) are shown, revealing the resemblance to the C23 prototype structure for both polymorphs. The relative orientation of the ortho-hexagonal reference system with the crystal systems of α'$_H$-Ca$_2$SiO$_4$ and α'$_L$-Ca$_2$SiO4 polymorphs is also indicated in Figure 6B,C, respectively. Concerning the SiO$_4$ tetrahedra, in the case of the α'$_H$ polymorph, all oxygen atoms are disordered, in contrast to the α'$_L$ where all the oxygens are ordered (Figure 6). In Figure 6A,B, the gray lines that join calcium atoms outline the part of the structure that corresponds to the basic hexagonal cell. A direct comparison of Figure 5A,B (or Figure 5C,D) with Figure 1A (or Figure 1B) makes clear that the arrangement of cations in the α-C$_2$S polymorph of belite resembles that in the Ni$_2$In 8B$_b$ prototype, irrespective of the S.G. used for the structure description of this polymorph. Thus, for the study of the structure inter-relations of belite polymorphs, the structure of the α-C$_2$S polymorph is used for the rest of the text, as a prototype structure. The inter-relation of all the polymorphs of belite is concluded from the relation of the unit cell axes of each polymorph with the hexagonal cell axes of the α-C$_2$S polymorph and more specifically through their relation with the ortho-hexagonal ones. In Table 2, the inter-relation of the cells for the C23 and 8B$_b$ prototype structures are also listed, and their relation is presented schematically in Figure 2D. The relation among the structures of these two prototype structures is concluded on the fact that there is a one-to-one correspondence of the structures at an atomic level. This relation extends also to the reference systems of both structures, as there is also a one-to-one correspondence of the axes of the C23 structure with those of the ortho-hexagonal cell axes deduced from the hexagonal cell of the 8B$_b$ structure (Figure 2D). These relations are deduced concerning Figures 1A and 2A for the prototype C23 and 8B$_b$ structures. From the previous discussion and the presentations given so far for all the structures, and more specifically for those of belite polymorphs, it is clear that the arrangements of Ca and Si cations resemble that of the prototype structures, and the ortho-hexagonal reference system of the axes of the α-C$_2$S polymorph could be used for the expression of the structure inter-relations. The relations given in Table 2 for the unit cell axes of the polymorphs α-C$_2$S, α'$_H$-C$_2$S, α'$_L$-C$_2$S, β-C$_2$S and γ-C$_2$S with those of ortho-hexagonal ones were derived based on Figure 5D (or Figure 5B), Figure 6B,D, Figures 3C and 4B, respectively. In Table 2, in addition to the vector relations of the cell axes the values of the corresponding axes are given in order to facilitate their comparisons. In Figure 7A, the vector relations for all belite polymorphs with the ortho-hexagonal cell derived from the hexagonal cell of the α-C$_2$S polymorph are schematically presented. In Figure 7B, a diagram for the variation in the unit cell values is given for all the polymorphs as compared to the values of the ortho-hexagonal cell. Based on the variation in the cell values given in Figure 7B and Table 2, during the transformation of α-C$_2$S (8Bb) → α'$_H$-C$_2$S (C23) the variation in the cell axes concerns the $a_{α'H}$, $b_{α'H}$ axes (a,b ortho-hexagonal plane) with $c_{α'H}$ remaining almost constant in agreement with the corresponding variation observed for 8Bb → C23 transformation for ternary alloys [33]. For the rest of the polymorphs which have the C23 prototype structures (α'$_L$-C$_2$S and β-C$_2$S), a slight decrease is observed for all axes. The highest one is related to the equivalent to the c_{oH}, $c_β$-axis of the beta polymorph. With the transformation β-C$_2$S (C23) → γ-C$_2$S (8Bb), the $c_γ$ (equivalent to $a_{αH}$-axis) value remains constant ($b_β$ = 6.758Å → $c_γ$ = 6.758 Å) and the $a_γ$ value (equivalent to $b_{αH}$-axis) is reduced ($a_β$ = 5.512 Å → $a_γ$ = 5.076 Å) but the $b_γ$ value (equivalent to $c_{αH}$-axis) increases significantly ($c_\beta \sin\beta$ = 9.284 Å → b_γ = 11.214 Å). It is worth noticing that this axis is normal in the planes of "walls" of trigonal prisms. In all the studies of β-C_2S (C23) → γ-C_2S (8Bb) transformation [23,26,33,35], the high increase in the volume of the cells of the corresponding cells is mentioned. In the study [35], which concerns ternary alloys, this variation is expressed by reference to the V_{cell}/Z (cell volume per formula unit) parameter. By using this value (last column of Table 1), this parameter starts from the value 97.10 Å^3 for the α_H-C_2S polymorph then decreases to the value of 86.45 Å^3 for the β-C_2S polymorph and then increases again to the value of 96.18 Å^3 for the γ-C_2S polymorph.

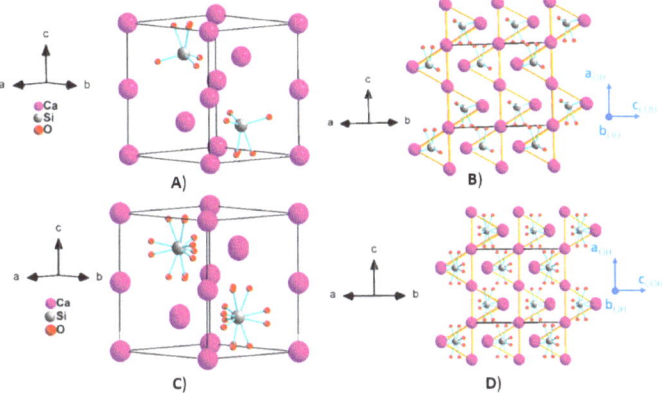

Figure 5. (**A**) The unit cell content and (**B**) the packing arrangement of trigonal prisms for the α_{Tr}-Ca_2SiO_4 polymorph structural model is shown. (**C**) The unit cell content and (**D**) the packing arrangement of the trigonal prisms for the α_H-Ca_2SiO_4 polymorph structural model are shown. Si-O bonds are indicated with cyan lines. In (**B,D**), the structures are projected on the (11–20) planes, the trigonal prisms are demarcated with orange lines and the relative orientation of the ortho-hexagonal reference system is also indicated.

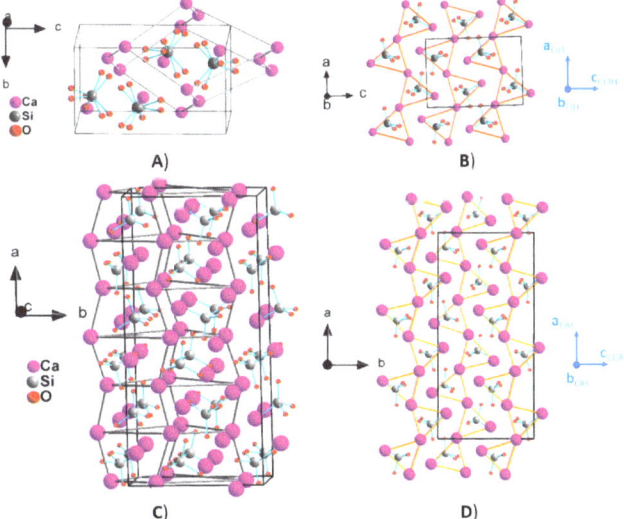

Figure 6. (**A**) The unit cell content and (**B**) the packing arrangement of trigonal prisms for α'_H-Ca_2SiO_4 polymorph is shown. (**C**) The unit cell content and (**D**) the packing arrangement of trigonal prisms for α'_L-Ca_2SiO4 polymorph structural model are shown. Si-O bonds are indicated with cyan lines. In (**B,D**), the structures are projected on the (010) planes, the trigonal prisms are demarcated with orange lines and the relative orientation of the ortho-hexagonal reference system is also indicated.

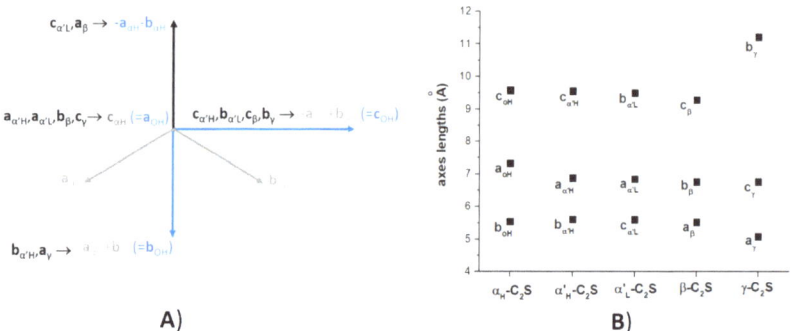

Figure 7. (**A**) The unit cell vector relations for the ortho-hexagonal axes (a_{oH}, b_{oH}, $c_{oH} = -a_{\alpha H} + b_{\alpha H}$) with those of the hexagonal axes of α_H-C_2S polymorph ($a_{\alpha H}$, $b_{\alpha H}$ and $c_{\alpha H}$) are shown: $a_{oH} = c_{\alpha H}$, $b_{oH} = a_{\alpha H} + b_{\alpha H}$ and $c_{oH} = -a_{\alpha H} + b_{\alpha H}$. The correspondence of the unit cell axes for the different polymorphs are also indicated. (**B**) Variation in unit cell values for all polymorphs of belite. The correspondences of the values for the axes of all the polymorphs with those of ortho-hexagonal cell of α_H-C_2S polymorph are given in Table 2.

Based on the original description in Barbier and Hyde [26], the basic building blocks for β- and γ-C_2S polymorphs are the trigonal prisms of the Ca_6Si type, shown in Figure 8. According to the previous discussion of the present work, these are also the building units of all the other belite polymorphs, resulting in the formation of "walls" consisting of edge-sharing columns of trigonal prisms. Adjacent walls are shifted along the stacking axis of the trigonal prisms, and this shift results in the capping of the slightly distorted orthorhombic faces of trigonal prisms belonging to one "wall" with Ca atoms belonging to neighboring "walls". The number of cupping Ca^{+2} cations that lie at the equator plane is five. Figures S1–S6 give the detailed arrangement of calcium and silicon cations and also those of oxygen anions in the structure of all the polymorphs. In Figure 7, the arrangements of Ca^{+2} cations around the Si^{+4} one is presented with a coding name presentation but with direct correspondence with the detailed presentations of all the polymorphs in Figures S1–S6. Table S1 lists the different geometric parameters of all the trigonal prisms for all polymorphs. The average plane of the capped positions (B1, ... , B5 sites in Figure 7) are shifted by half of the high of the trigonal prism in the α-C_2S trigonal/hexagonal (2.766/2.766 Å) and α'_H-C_2S (2.801/2.801Å) cases of the α-C_2S trigonal/hexagonal (2.766/2.766) and α'_H-C_2S (2.801/2.801) and are approximately equal in the case of γ-C_2S (2.548/2.534 Å). For α'_L-C_2S and for the prisms which host Si1 and Si2*** anions (Figure S4), the distances from the bases at the top (2.846 and 2.850 Å, respectively) are longer than those from the bottom of the prisms (2.737 and 2.738), and the opposite holds for the corresponding distances for the equator plane for the prism that hosts the Si3 anion (2.960/2.615). The average distances of Si from Ca cations which lie at both bases are equal, as is the distance from the corresponding planes of both trigonal bases in α-C_2S trigonal/hexagonal polymorphs (3.522/3.522 Å and 2.766/2.766 Å for trigonal 3.429/3.429 Å and 2.766/2.766 Å for the hexagonal model). Pairs of values with longer distances correspond to average distances of Si from Ca anions at the bases of trigonal prisms and the pairs of shorter ones correspond to distances to the base planes of the prisms. Both types of distances are equal in the case of α'_H-C_2S (3.533/3.533 Å and 2.801/2.801) and are approximately equal in the case of the β-C_2S polymorph (3.472/3.485 and 2.756/2.736 Å). For α'_L-C_2S and for the prisms which host Si1 and S2*** anions (Figure S4), the distances of Si cations from the Ca ones that lie at the bottom are longer than those from the top, and the opposite holds for the corresponding distances for the Si3 anion in the respective prism. The same trend is observed for the corresponding distances of Si anions from the planes at the bottom and the top of the prisms (3.595/3.537 Å and 2.828/2.762 Å for S1; 3.649/3.407 Å and 2.985/2.576Å for S2***; 3.535/3.640 Å and 2.537/3.006 Å for S3, i.e., longer distances correspond to average distances of Si from Ca anions at the bases of trigonal prisms). Although in the description given in [25] the number of anions on the equator plane is five, only the case of the α-C_2S trigonal/hexagonal polymorph seems to be satisfied based on the distances of Si cation from the Ca cations which occupy the B1, ..., B5 sites. For trigonal polymorphs, the Si-B1, ..., Si-B5 distances fall in the range of the values 3.198–3.782 Å, and those for hexagonal polymorphs in the range of 3.195–3.756 Å. In all the other cases, there are always two Ca cations that occupy sites at longer distances from the other three (Table S1). These are B3 and B5 (Si-B3: 4.226 Å and Si-B5: 3.778 Å) for $α'_H$-C_2S; B2 and B4 for Si1 (Si1-B2: 3.728 and Si1-B4: 3.898 Å); B2 and B5 for Si2*** (Si2***-B3: 4.167 and Si2***-B5: 3.999 Å); B2 and B4 for Si3 (Si3-B3: 3.889 and Si3-B4: 4.340 Å) for $α'_L$-C_2S; B2 and B4 (Si-B2: 3.812 and Si-B4: 4.009 Å) for β-C_2S; and B3 and B4 (Si-B3: 4.834 and Si-B4: 4.834 Å) for γ-C_2S. This result indicates that the nearest neighbors on the equator plane are five only for the high-temperature polymorphs (trigonal or hexagonal α-C_2S phase), and for the rest of the polymorphs the nearest neighbors are three, and thus the coordination polyhedron is a tricapped trigonal prism. Another factor that is important for the description of the structures of belite polymorphs is the sites that are occupied by the oxygen atoms. These are marked as T1, T2 and T3 and D2 and D3 in Figure 7. The T1, T2 and T3 sites are occupied in all polymorphs except γ-C_2S (Figures S1–S6). The D1 sites are occupied in the case of α-C_2S trigonal polymorphs (Figure S1) and $α'_H$-C_2S polymorphs (Figure S3) and D1 and D2 are occupied in the case of α-C_2S hexagonal polymorphs (Figure S2) and $α'_H$-C_2S polymorphs (Figure S4). The oxygen atoms in the γ-C_2S polymorph occupy unique sites that do not have a relationship with those occupied in the C_2S polymorphs. In the same polymorph, the silicon polyhedron is a tricapped trigonal prism and the O1, O2, O3 and O3ii atoms occupy interstitial sites within the $SiCa_3$ tetrahedral type polyhedral, as presented in Figure S6 (Si1-Ca1*-Ca1''-Ca2 for O1, S1-Ca2**-Ca1-Ca1' for O2i, Si1-Ca2#-Ca1-Ca2' for O3 and Si1-Ca2***-Ca1'-Ca2'' for O3ii; for symmetry codes, see Figure S6 caption). Another geometric parameter that has been discussed [26,36] as characterizing the relationship between the number of cupped positions and the dimensions of the trigonal prism is the ratio of the height (h of the prism) divided by the average length (<l>) of the triangular base edges. This parameter is h/<l> = 1.507, 1.555, 1.554, 1.561, 1582, 1.572 and 1.444 for the α-C_2S trigonal/hexagonal, $α'_H$-C_2S, $α'_L$-C_2S(Si1), $α'_L$-C_2S(Si2***), $α'_L$-C_2S(Si3), β-C_2S and γ-C_2S polymorphs, respectively, i.e., it increases upon transforming to a polymorph stabilized at lower temperature and takes the lowest value for the RT stabilized gamma polymorph. The similarities and characteristics of the different polymorphs of belite presented in this paragraph could support the general statement of Vegas in work [24], that cations in a structure could be considered as big molecules, and the anions act as an external factor such as pressure or temperature and thus could change their bonding patterns, which finally results in phase transformations. In addition, the present study supports the suggestion that the transformation mechanism, based on the relative disposition of cations for the different polymorphs, is of displacive character [26], but if we consider the sites occupied by oxygens at different polymorphs it could be considered as diffusion.

The present study could be of interest for the recent application of belite rare-earth doped compounds with the general formula $Ca_{2-x}Sr_xSiO_4$:Ce^{3+} [16], which present interesting luminescent properties. Depending on the Sr content, the compound crystallizes in the β- or $α'_H$-C_2S polymorph. This result is consistent with the observation that the larger cations conform with an elongation of the height of trigonal prisms and thus to higher h/<l> values [36]. According to the values given in Table S1, the highest values for the h/<l> parameter are observed for the polymorphs $α'_H$-C_2S, $α'_L$-C_2S and β-C_2S. The samples with composition $Ca_{1.65}Sr_{0.25}SiO_4$: 0.10Eu^{2+} and $Ca_{1.45}Sr_{0.35}SiO_4$: 0.10Ce^{3+}, 0.10Li^+ crystallize in the $α'_L$-C_2S polymorph [37]. Further study is needed in order to explore the factors that stabilize the different polymorphs of belite upon doping for luminescent applications, as systematic studies have revealed the formation of all five belite polymorphs [38]. The formation of all five polymorphs of belite at room temperature upon doping with different cations results in the synthesis of materials in two very active research areas, i.e., belite-based cement products and photoluminescent ones. As these compounds are studied mostly in the form of polycrystalline materials, there is a need to have reliable powder diffraction diagrams in order to easily identify the most probable polymorph phases from the recorded diffraction pattern. Figure 9 presents all the simulated powder patterns of all five belite polymorphs which are discussed in the present work. For the calculations, the models obtained from the ICSD database which correspond to the entries listed in the Table 1 (or data from Table S2) were used, and they have been published in references [3,4]. Detailed structural models for all polymorphs are listed in Table S2. In Figure 9, the (hkl) indices for the main relaxions are also given. The indices of the hexagonal α-C_2S polymorph are given in gray color in the top pattern of Figure 9. In each pattern, the corresponding indices of the hexagonal system expressed with their values in the reference system of each polymorph are also given in gray color. The values for these indices are obtained by applying the relation $(h'k'l') = (hkl)P$, where P is the transformation matrix of the basis vectors for the related systems: $(a',b',c') = (a,b,c)P$. [39] The matrices used for all the transformations are listed in Table S3. No characteristic trend is observed concerning the presence of these reflections in the patterns of different polymorphs. In the patterns of all the polymorphs, additional peaks with low intensity are observed due to change in symmetry. The most characteristic observation concerns the peaks at around 32.5°, where splitting of intense peaks is observed in conformity with the experimental observations [38].

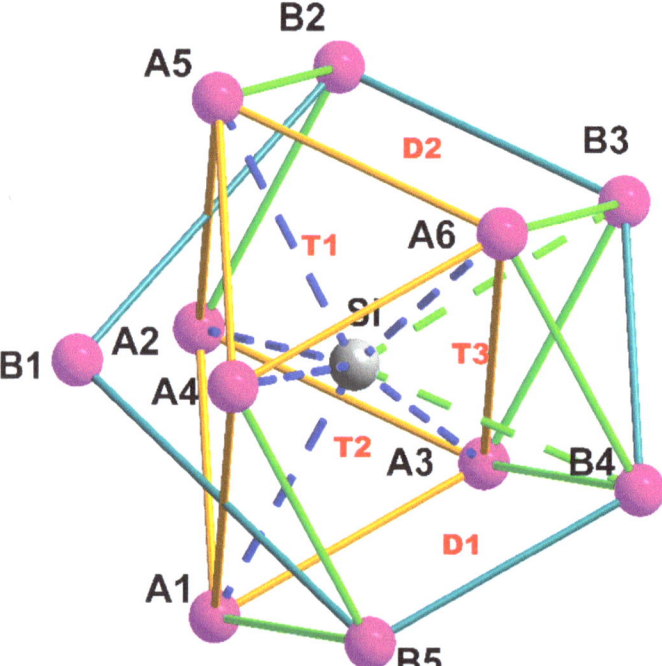

Figure 8. Arrangement of Ca cations around the Si^{+4} cations in all belite polymorphs. A1, A2 and A3 indicate the Ca^{+2} cations on the bottom trigonal face of the prism; A4, A5 and A6 indicate those at the top face of the prism; B1, ..., B5 indicate those at the equator plane. T1, T2 and T3 and D1 and D2 indicate the sites of oxygen anions in different belite polymorphs.

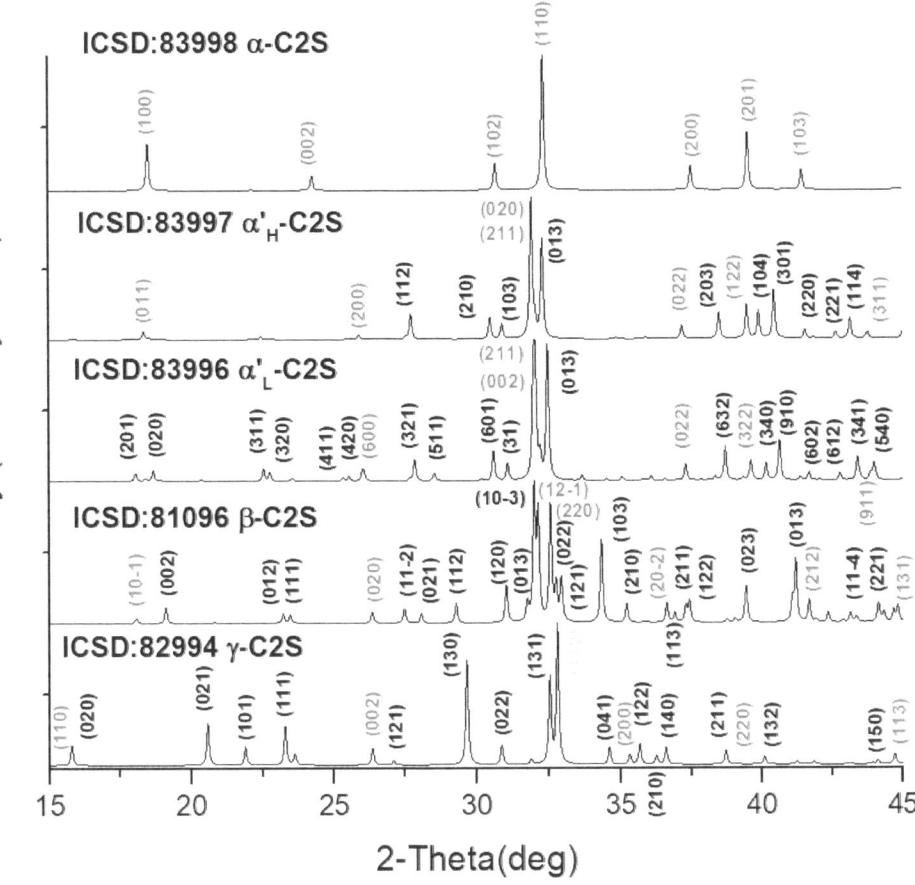

Figure 9. Simulated powder diffraction patterns for all the belite polymorphs (Cukα radiation) based on the ICSD data of the corresponding code numbers (Table 1 and Table S2).

4. Conclusions

In the present study, the structure–superstructure relations of the five polymorphs of belite were derived based on unit cell axes relations. The common polyhedron of all polymorphs is a trigonal prism of calcium cations that hosts the silicon one, which is surrounded by five calcium cations at an equator plane for the α-C2S trigonal/hexagonal polymorphs, thus forming a five-capped trigonal prism. For the rest of the polymorphs, the trigonal prisms are tri-capped ones, as three nearest neighbor calcium cations exist at the equator planes. In the α-C2S to γ-C2S a significant increase is observed, normal in planes of the "walls" of trigonal prisms in addition to the usually mentioned volume increase. Based on the common characteristics of the arrangement of cations for all of the polymorphs, the transformation mechanism could be considered as displacive and, on the sites occupied by oxygen anions, could be considered of diffusion type. The systematic study of silicon dicalcium silicates could be not only for the cement industry but for the field of photoluminescent compounds.

Supplementary Materials: The following supporting information can be downloaded at: https://www.mdpi.com/article/10.3390/cryst12121692/s1, Figure S1: Presentation of the trigonal prism and the atoms at the equator plane for the α-C$_2$S trigonal polymorph; Figure S2: Presentation of the trigonal prism and the atoms at the equator plane for the α-C$_2$S hexagonal polymorph; Figure S3: Presentation of the trigonal prism and the atoms at the equator plane for the α'H-C$_2$S polymorph; Figure S4: Presentation of the trigonal prism and the atoms at the equator plane for the α'L-C$_2$S polymorph; Figure S5: Presentation of the trigonal prism and the atoms at the equator plane for the β-C$_2$S trigonal polymorph; Figure S6: Presentation of the plane for the γ-C$_2$S polymorph; Table S1: Geometric parameters of trigonal prisms; Table S2: trigonal prism and the atoms at the equator Crystallographic parameters for all studied structures.; Table S3: Transformation matrices of the unit cell axes which relate the cell axes of each belite polymorph with the hexagonal axes of the α-C2S polymorph.

Author Contributions: Conceptualization, V.P. and M.C.; methodology, V.P., D.K. and M.C.; software, V.P and M.C.; validation, V.P., M.C., D.K., M.K. and M.B.-K.; writing—original draft preparation, V.P.; writing—review and editing, V.P., M.C., D.K., M.K. and M.B.-K.; supervision, V.P., M.B.-K. and M.K.; project administration, V.P. and M.K. All authors have read and agreed to the published version of the manuscript.

Funding: This research was supported by NCSR "Demokritos" Industrial Research Fellowship Program (E.E.-12149) funded by the Stavros Niarchos Foundation and TITAN Cement S.A.

Data Availability Statement: Data available in a publicly accessible repository that does not issue DOIs. Publicly available datasets were analyzed in this study and the terms and conditions for accessing and using these data are given in the link: https://www.ccdc.cam.ac.uk/access-structures-terms/).

Conflicts of Interest: The authors declare no conflict of interest.

References

1. Lawrence, C.D. The production of low-energy cements. In *Lea's Chemistry of Cement and Concrete*, 4th ed.; Butterworth-Heinemann: Oxford, UK, 2003.
2. Jost, K.H.; Ziemer, B.; Seydel, R. Redetermination of the structure of β-Dicalcium Silicate. *Acta Cryst.* **1977**, *B33*, 1696–1700. [CrossRef]
3. Mumme, W.G.; Hill, R.J.; Bushnell-Wye, G.; Segnit, E.R. Rietveld crystal structure refinements, crystal chemistry and calculated powder diffraction data for the polymorphs of dicalcium silicate and related phases. *Neues Jahrb. Fuer Mineral.–Abh.* **1995**, *169*, 35–68.
4. Mumme, W.G.; Cranswick, L.M.D.; Chakoumakos, B.C. Rietveld crystal structure refinements from high temperature neutron powder diffraction data for the polymorphs of dicalcium silicate. *Neues Jahrb. Fuer Mineral.–Abh.* **1996**, *170*, 171–188.
5. Mondal, P.; Jeffery, J.W. The Crystal Structure of Tricalcium Aluminate Ca$_3$Al$_2$O$_6$*. *Acta Cryst.* **1975**, *B31*, 689–697. [CrossRef]
6. Nishi, F.; Takeuchi, Y. The Al$_6$O$_{18}$ Rings of tetrahedra in the structure of Ca$_{8.5}$NaAl$_6$O$_{18}$. *Acta Cryst.* **1975**, *B31*, 1169–1173. [CrossRef]
7. Colville, A.A.; Geller, S. The crystal structure of brownmillerite, Ca$_2$FeAlO$_5$. *Acta Cryst.* **1971**, *B27*, 2311–2315. [CrossRef]
8. Colville, A.A.; Geller, S. Crystal structures of Ca$_2$Fe$_{1.43}$Al$_{0.57}$O$_5$ and Ca$_2$Fe$_{1.28}$Al$_{0.72}$O$_5$. *Acta Cryst.* **1972**, *28*, 3196–3200. [CrossRef]
9. Dunstetter, F.; de Noirfontaine, M.-N.; Courtial, M. Polymorphism of tricalcium silicate, the major compound of Portland cement clinker: 1. Structural data: Review and unified analysis. *Cem. Concr. Res.* **2006**, *36*, 39–53. [CrossRef]
10. Yamnova, N.A.; Zubkova, N.V.; Eremin, N.N.; Zadov, A.E.; Gazeev, V.M. Crystal Structure of larnite β-Ca$_2$SiO$_4$ and specific features of polymorphic transitions in dicalcium orthosilicate. *Struct. Inorg. Compd.* **2011**, *56*, 210–220. [CrossRef]
11. Courtial, M. Polumorphism of tricalcium silicate in Portland cement: A fast visual identification of structure and superstructure. *Powder Diffr.* **2012**, *18*, 7–15. [CrossRef]
12. de Noirfontaine, M.-N.; Dunstetter, F.; Courtial, M.; Gasecki, G.; Signes-Frehel, M. Tricalcium silicate Ca$_3$SiO$_5$, the major component of anhydrous Portland cement: On the conservation of distances and directions and their relationship to the structural elements. *Zertscrift Fur Krist.* **2003**, *218*, 8–25. [CrossRef]
13. de Noirfontaine, M.-N.; Dunstetter, F.; Courtial, M.; Gasecki, G.; Signes-Frehel, M. Polymorphism of tricalcium silicate, the major compound of Portland cement clinker 2: Modelling alite for Rietveld analysis an industrial challenge. *Cem. Concr. Res.* **2006**, *36*, 54–64. [CrossRef]
14. de Noirfontaine, M.-N.; Dunstetter, F.; Courtial, M.; Gasecki, G.; Signes-Frehel, M. Tricalcium silicate Ca$_3$SiO$_5$ superstructure analysis: A route towards the structure of the M1 polymorph. *Zeitchrift Fur Krist.* **2012**, *227*, 102–112. [CrossRef]
15. Manzano, H.; Durgun, E.; Qomi, M.J.A.; Ulm, F.J.; Pellenq, J.M.; Grossman, J.C. Impact of Chemical Impurities on the crystalline cement clinker phases determined by atomistic simulations. *Cryst. Growth Des.* **2011**, *11*, 2964–2972. [CrossRef]
16. Xia, Z.; Liu, Q. Progress in discovery and structural design of color conversion phosphors for LED. *Prog. Mater. Sci.* **2016**, *84*, 59–117. [CrossRef]
17. Chan, C.J.; Kriven, W.M.; Young, J.F. Physical Stabilization of the β→γ Transformation in Dicalcium Silicate. *J. Am. Ceram. Soc.* **1992**, *75*, 1621–1627. [CrossRef]
18. Regourd, M.; Chromy, S.; Hjorth, L.; Mortureux, B.; Guinier, A. Polumorphisme des solutions solidides du sodium dans l'aluminate tricalcique. *J. Appl. Crystallogr.* **1973**, *6*, 355–364. [CrossRef]

19. Ghosh, S.N.; Rao, P.B.; Paul, A.K.; Raina, K. The chemistry of dicalcium silicate mineral. *J. Mater. Sci.* **1979**, *14*, 1554–1566. [CrossRef]
20. Cuesta, A.; Lossilla, E.R.; Aranda, M.A.G.; De la Torre, A.G. Reactive belite stabilization mechanisms by boron-bearing dopands. *Cem. Concr. Res.* **2012**, *42*, 598–606. [CrossRef]
21. Moore, P.B.; Araki, T. The crystal structute of bredigite and the genealogy of some alkaline earth orthosilicates. *Am. Mineral.* **1976**, *61*, 74–87.
22. Pauling, L. *The Nature of the Chemical Bond and the Structure of Molecules and Crystals: An Introduction to Modern Structural Chemistry*, 3rd ed.; New York Cornell University Press: Ithaka, NY, USA, 1960; p. 505.
23. O'Keeffe, M.; Hyde, B.G. An Alternative Approach to Non-Molecular Crystal Structures with emphasis on the arrangements of cations. *Struct. Bond.* **1985**, *61*, 77–144.
24. Vegas, A. Cations in Inorganic Solids. *Crystallogr. Rev.* **2000**, *7*, 189–283. [CrossRef]
25. Vegas, A.; Mattesini, M. Towards a generalized vision of oxides: Disclosing the role of cations and anions in determining unit-cell dimensions. *Acta Cryst.* **2010**, *B66*, 338–344. [CrossRef] [PubMed]
26. Barbier, J.; Hyde, B.G. The Structures of the Polymorphs of Dicalcium Silicate, Ca_2SiO_4. *Acta Cryst.* **1985**, *B41*, 383–390. [CrossRef]
27. Butt, Y.M.; Timashev, V.V.; Malozohn, L.I. *Proceedings of the 5th International Symposium on the Chemistry of Cement*; Cement Association of Japan: Tokyo, Japan, 1968.
28. Koumpouri, D.; Angelopoulos, G.N. Effect of boron waste and boric acid addition on the production of low energy belite cement. *Cem. Conc. Comp.* **2016**, *68*, 1–8. [CrossRef]
29. Koumpouri, D.; Karatasios, I.; Psycharis, V.; Giannakopoylos, I.G.; Katsiotis, M.S.; Kilikoglou, V. Effect of clinkering conditions on phase evolution and microstructure of Belite Calcium-Sulpho-Aluminate cement clinker. *Cem. Conc. Res.* **2021**, *147*, 106529. [CrossRef]
30. *DIAMOND—Crystal and Molecular Structure Visualization, Crystal Impact*; Purtz, H.; Brandenburg, K. (Eds.) GbR, Kreuzherrenstr: Bonn, Germany, 2009; Volume 102, p. 53227. Available online: https://www.crystalimpact.de/diamond (accessed on 2 October 2022).
31. Raman, R.S.K.; Gupta, R.K.; Sujir, M.N. Lattice constants of B8 structure in Cu2In-Ni2In alloys. *J. Sci. Res. Banaras Hindu Univ.* **1964**, *14*, 95–99.
32. Eckerlin, P.; Wölfel, E. Die Kristallstruktur von Ca_2Si und Ca_2Ge. *Z. Anorg. Allg. Chem.* **1955**, *280*, 321–331. [CrossRef]
33. Johson, V. Diffusionless Orthorhombic to Hexagonal Transitions in Ternary Silicides and Germanides. *Inorg. Chem.* **1975**, *14*, 1117–1120. [CrossRef]
34. Jeitscek, W. A High-Temperature X-ray Study of the Displacive Phase Transition in MnCoGe. *Acta Cryst.* **1975**, *B31*, 1187–1190. [CrossRef]
35. Kim, Y.J.; Nettleship, I.; Kriven, W.M. Phase Transformations in Dicalcium Silicate: II, TEM Studies of Crystallography, Microstructure, and Mechanisms. *J. Am. Cheram. Soc.* **1992**, *75*, 2407–2419. [CrossRef]
36. O'Keeffe, M.; Hyde, B.G. Some Structures Topologically Related to Cubic Perovskite ($E2_1$), ReO_3 ($D0_9$) and $Cu_3Au(L1_2)$. *Acta Cryst.* **1977**, *B33*, 3802–3813. [CrossRef]
37. Xia, Z.; Miao, S.; Chen, M.; Molokeev, M.S.; Liu, Q. Structure, Crystallographic Sites, and Tunable Luminescence Properties of Eu^{2+} and Ce^{3+}/Li^+-Activated $Ca_{1.65}Sr_{0.35}SiO_4$ Phosphors. *Inorg. Chem.* **2015**, *54*, 7684–7691. [CrossRef]
38. Mao, Z.; Lu, Z.; Chen, J.; Fahlman, B.D.; Wang, D. Tunable luminescent Eu^{2+}-doped dicalcium silicate polymorphs regulated by crystal engineering. *J. Mater. Chem. C* **2015**, *3*, 9454–9460. [CrossRef]
39. Arnold, H. Transformations of the coordinate system (unit-cell transformations). In *International Tables for Crystallography, Volume A, Space Group Symmetry*; Hahn, T., Ed.; Springer: Dordrecht, The Netherlands, 2005; Chapter 5; p. 77.

Article

Removal of Organic Materials from Mytilus Shells and Their Morphological and Chemical-Physical Characterisation

Alberto Ubaldini [1,*], Flavio Cicconi [2], Sara Calistri [1,3], Stefano Salvi [2], Chiara Telloli [1], Giuseppe Marghella [1], Alessandro Gessi [1], Stefania Bruni [1], Naomi Falsini [1] and Antonietta Rizzo [1]

1. ENEA, Italian National Agency for New Technologies, Energy and Sustainable Economic Development, C.R. Bologna, Via Martiri di Monte Sole 4, 40129 Bologna, Italy; sara.calistri2@unibo.it (S.C.); chiara.telloli@enea.it (C.T.); giuseppe.marghella@enea.it (G.M.); alessandro.gessi@enea.it (A.G.); stefania.bruni@enea.it (S.B.); naomi.falsini@enea.it (N.F.); antonietta.rizzo@enea.it (A.R.)
2. ENEA, Italian National Agency for New Technologies, Energy and Sustainable Economic Development, C.R. Brasimone, 40032 Camugnano, Italy; flavio.cicconi@enea.it (F.C.); stefano.salvi@enea.it (S.S.)
3. Department of Pharmacy and Biotechnology, Alma Mater Studiorum, University of Bologna, 40126 Bologna, Italy
* Correspondence: alberto.ubaldini@enea.it

Abstract: A simple and effective method to eliminate the organic component from mussel shells is presented. It is based on the use of hot hydrogen peroxide. Mollusc shells are composite materials made of a calcium carbonate matrix with different polymorphs and numerous biomacromolecules. The described method was used on mussel shells, but it is generalisable and allows the complete removal of these organic components, without altering the inorganic part. Specimens were kept in a H_2O_2 40% bath for few hours at 70 °C. The organic layers found on the faces of the shells were peeled away in this way, and biomacromolecules were degraded and removed. Their fragments are soluble in aqueous solution. This easily permits the chemical-physical characterisation and the study of the microstructure. The quality of calcite and aragonite microcrystals of biogenic origin is very high, superior to that of materials of geological or synthetic origin. This may suggest various industrial applications for them. Calcium carbonate is a useful precursor for cements and other building materials, and the one obtained in this way is of excellent quality and high purity.

Keywords: chemical treatment; hydrogen peroxide; polymorphs; Raman spectroscopy; microstructure

1. Introduction

Inorganic materials are used by living organisms in an innumerable number of cases and for the most diverse purposes, including uses such as protection, defence, attack, and structures for resistance to loads and mechanical stresses. The mineralised component of vertebrate bones consists mainly of calcium phosphates in the form of apatite and hydroxyapatite [1]. Many marine invertebrates especially, but not only, including molluscs [2], brachiopods [3], and corals [4], use carbonates to build the hard parts of their body, including shells and corallites. Diatoms create their shells, called frustules, using silica (SiO_2) [5]. Similar structures are observed in Radiolaria [6], with the significant exception of the *Acantharea* group, whose shell is made up of strontium sulphate ($SrSO_4$, celestine) [7]. Sponge spicules (the rigid, structural elements in their endoskeleton) are made by calcium carbonate or silica, depending on the species [8]. Crystals based on magnetite are found in the beaks and heads of migratory birds [9], and a high concentration of zinc and manganese are found in scorpions' stings [10] and in the mouthparts of some insects [11]. To achieve the intended purpose of their function, these inorganic materials are organised and structured at the microscopic and sometimes even sub–microscopic level with absolute precision, due to the driving force of evolution rules. Evolution, in fact, selects organisms over time that are more efficiently able to exploit the materials available in their environment and that can create the best structures for their survival. These examples show how inorganic materials are exploited by living organisms, and in turn they can become useful starting materials for human purposes and industrial activities. Among others, the materials produced by molluscs appear to be the most potentially useful, and they are also produced in large quantities and therefore are easy to use. Many molluscs, as they are soft-bodied animals, have evolved a complex strategy for maintaining and protecting their soft tissues, which relies on the elaboration of an external calcified rigid structure, the so-called shells. The fabrication of the shells by these animals is a very complex set of processes [12]. They require extremely specialised cells, the biomineralised materials can be often very far from the thermodynamical equilibrium, and their shape and morphology are complex and hierarchically organised.

Although the methods of shell biosynthesis among the many different classes of molluscs can be even very different, they have also some general and common features. In practically all cases, the shells are made of calcium carbonate (the gasteropod Chrysomallon squamiferum, a species of snail living on deep-sea hydrothermal vent, has a complex shell wherein most of the external layer is made by iron sulphides, especially pyrite (FeS_2) and greigite (Fe_3S_4) [13]).

Sometimes, these biomineralised inorganic materials have different physical and chemical properties compared to their counterparts of geological origin or that have been synthesised in the laboratory using traditional chemical methods.

For instance, aragonite, one of the calcium carbonate polymorphs, is metastable at room conditions, but it does not transform into the more stable calcite for kinetic reasons; however a phase transition occurs when it is heated. It is interesting to note that in some cases for samples of biological origin, it has been reported [14,15] that this transition temperature can be even 80–100 °C lower (i.e., in the range between 250–380 °C) than the same transition temperature for the mineral that is 480–500 °C [16], also because of a different water content.

The nacre is the iridescent inner shell layer of some molluscs [17]. It is composed of 95 wt% aragonite and 5 wt% organic materials. Because of the presence of this small quantity of organic molecules and polymers and its microstructure, the nacre has strength and toughness 20–30 times higher that of normal aragonite [18]. Indeed, calcium carbonate of biological origin can also often be found in an amorphous state, something that is very rare in geological or synthetic samples [19]. This may be due to small differences between the same geological or synthetic materials and those of biological origin, such as the presence of specific secondary elements, different contents of crystallisation water, or different degrees of crystallinity or amorphous components, or it possibly can be due to the micro/nanostructure. It should also be kept in mind that inorganic components are often part of very complex composite materials, where organic components also play an important role. In the case of shells, there are proteins, carotenoids, polysaccharides, and other complex organic components that are used by the living animal for growth and to precisely hold together the present microcrystals.

Shells, moreover, are important not only from a biological point of view and for their inherent beauty and charm, but also due to many other characteristics. They are useful bioindicators to monitor the state of natural environments; recently, shells produced as waste from the food industry have been indicated as a valuable material for the production of building materials, possibly even replacing traditional cements. In fact, these consume a large amount of limestone, but the carbonates derived from the shells of molluscs have properties very comparable to it and a much lower environmental impact in terms of carbon dioxide released.

This means, incidentally, that it is of great importance to know in depth all the mechanical, (micro) structural, and thermodynamic properties of their organic and inorganic components. Deepening the knowledge of intrinsic features and morphological aspects of the inorganic part of shells would be important for these characterisations.

However, although it is the prevalent one in terms of mass and volume, the organic components are always present, and they would play a role in any measurement and in the control of properties. It is therefore interesting to identify chemical or physical methods to separate the two components. While it is relatively easy to eliminate the inorganic part, for example by means of an acid bath, it is more complicated to do the opposite. This can be achieved by, for example, immersing the shells in concentred sodium hypochlorite or ammonium thioglycolate solutions for sufficiently long times [20]. Diluted sodium hypochlorite is useful for removing deposits of algae or bacteria or possibly encrustations; when it is more concentrated, it can attack organic parts and layers of cellular origin, and a partial whitening is achieved. However, the colour of the most external layer on a shell, called periostracum, which is very rich in organic molecules, can be often preserved in this way, indicating that the removal of organic components is partial. Sodium hypochlorite is also used for achieving a fast chitin extraction from the shells of crab, crayfish, and shrimp [21], but, in this case, the extraction of the organic component is more important than the preservation of the starting sample.

Another possibility to whiten shells is to use hydrogen peroxide. It has been reported to be very efficient for this purpose [22–25]. Furthermore, hydrogen peroxide solution has the great advantage of being much less harmful to the environment than the agents mentioned previously, being the by-products essentially only oxygen and water vapour. Whitening was achieved by attacks with peroxide solutions at different concentrations for long periods of time of 24 h or more.

In this work, the action of removing organic components was achieved using solutions at rather high concentrations (40% m/v) at higher temperatures. The result was achieved within a few hours, which in itself is interesting for possible applications, because it makes the process attractive from an industrial point of view. It is interesting to observe that by using these conditions, it is also possible to remove the organic molecules that are present inside the shells and not just those present on the surfaces, which is more difficult not always achieved using milder conditions.

At the end of the process, only the inorganic part remains.

This work also aims to present a chemical-physical characterisation of the samples before and after treatment and the evolution as a function of time.

Some Mytilus shells have been studied, as this genus is among those with the greatest nutritional importance and one of the greatest geographic distributions, being cosmopolitan. Therefore, it can be considered emblematic for the study of the inorganic components. The properties depend on the structure; the different parts of the shells can have different ratios between calcite and aragonite, and their crystals can have different morphologies. Investigations on some specimens were carried out using characterisation techniques, such as Raman spectroscopy with an optical Raman microscope, X-ray diffraction, scanning electron microscopy (SEM-EDX), and ICP mass spectrometry, which have also been used to determine the presence and nature of trace secondary elements, which can be useful markers.

2. Experimental Section

2.1. Materials and Methods

Mytilus is a cosmopolitan genus of marine bivalve mollusc in the family *Mytilidae*. They have a shell about 5–10 centimetres long, depending on the species, with an elongated oval shape consisting of a right and left half of the shell (valves), which are held together with an elastic lock strap (ligament).

The specimens studied belong to the species *M. Chilensis* (Hupè, 1859) [26], a species widespread on the coasts of South America, being a subject of intensive aquaculture production.

In order to separate the organic part from the inorganic component, the fresh shells were immersed in a 40% (m/v) aqueous solution of hydrogen peroxide (Sigma Aldrich, St. Louis, MO, USA) for times ranging from 5 min to 5 h, in order to study the kinetics of the process. The treatments were performed at different temperatures, but the results described here were carried out at 70 °C. At the end of this process, the colour of the shells changes, and there is an average weight variation of 3–5%. This variation was determined measuring the weight of at least five specimens, at fixed intervals, in order to minimise random effects that could occur on a single sample. Before weight measurement, the samples were carefully dried in an oven.

Highly concentrated hydrogen peroxide is not so stable and tends to decompose over time. For this reason, fresh batches were used as often as possible and were kept in freezer at −18 °C. Under these conditions, it is considered stable, at least for weeks or more.

2.2. Instrumental Analysis

Raman spectra of the compounds and mixtures were acquired, at room temperature, by a BWTEK i-Raman plus spectrometer (B&W Tek, Plainsboro, NJ, USA) equipped with a 785 nm laser in the range of 100–3500 cm^{-1} with a spectral resolution of 2 cm^{-1}. The measurement parameters, such as acquisition time, number of repetitions, and laser power, were selected for each sample in order to maximise the signal-to-noise ratio. The standard acquisition, however, was 20 repetitions of 10 s each. This instrument has a maximal power of 350 mW, but in most cases, only 10% of it was used. For each spectrum, a reference acquisition was previously carried out with the same parameters to subtract the instrumental background.

X-ray powder diffraction (XRPD) investigations were performed to determine the crystalline phases, using a Philips X'Pert PRO 3040/60 diffractometer (Philips, Amsterdam, The Netherlands) operating at 40 kV, 40 mA, with Bragg–Brentano geometry, equipped with a Cu Kα source (1.54178 Å), Ni filtered, and with a curved graphite monochromator. PANalytical High Score software (version 4.1) was used for data elaboration. The XRD acquisitions were performed using these parametes: start position: 10.0125° [2θ]; end position: 99.9875° [2θ]; step size: 0.0250°; scan step time: 6.0000 s, scan type: continuous.

The characterisation, morphology, and composition of the samples were performed by scanning electron microscopy (SEM-FEI Inspect-S, FEI Company Hillsboro, OR, USA), coupled with energy dispersive X-ray spectroscopy (EDX, Oxford Xplore, Oxford Instruments plc, Abingdon, UK). Observations were carried out at different magnifications using both secondary electrons and backscattered electrons detectors at 10 mm working distance, with energy ranging from 10 to 20 KV. The elemental analysis was carried out in the most significant areas of the samples. Data were processed by the software Oxford AZtec One (AZTecLive 6.1 platform).

A triple quadrupole inductively coupled plasma mass spectrometer (ICP-MS-QQQ, 8800 model, Agilent Technologies, Santa Clara, CA, USA) equipped with two quadrupoles, one (Q1) before and one (Q2) after the octupole reaction system (ORS3), installed in a dedicated Clean Room ISO Class 6 (ISO 14644-1 Clean room [27]) with controlled pressure, temperature, and humidity was used for trace analysis of the secondary elements. Quality control standards (QCs) were added in the batch of analysis in order to control the accuracy of the analytical method. The collision cell in Helium Mode (MS/MS) configuration ensures the removal of any spectroscopic interferences caused by atomic or molecular ions that have the same mass-to-charge as analytes of interest.

The shells samples were finely grounded with a laboratory blender, and 0.50 g of grounded shells were mixed in a solution with HNO_3 and H_2O_2. TraceSELECT® grade 69% HNO_3 was used. High-purity de-ionised water (resistivity 18.2 MΩ cm^{-1}) was obtained from a Milli-Q Advantage A10 water purification system (Millipore, Bedford, MA, USA) for the dilutions. The samples were digested with a microwave digestion system, Speedwave Four model (Berghof, Germany), equipped with temperature and pressure control in PTFE vessels using 4 mL of nitric acid, 3 mL of hydrogen peroxide, and 13 mL of high purity de-ionised water. Complete dissolution was achieved keeping the pressure constant at 30 bar, while the temperature was linearly increased in 5 min from room condition to 200 °C and then maintained for 10 min. After the digestion, the acid mixture was left to cool at room temperature, and after that, it was quantitatively transferred into plastic vials (Falcon®

50 mL Polypropylene Conical Tubes, BD Biosciences, Cowley, UK), previously cleaned with nitric solution, and made up to a final volume of 50 mL with high purity de-ionised water.

The multi-element standard solution IV-ICPMS-71 A (10 µg/mL) and the Rare Earth Elements standard solution CCS-1 (100 µg/mL) supplied by Inorganic Ventures (Christiansburg, VA, USA) were used for calibration. The calibration curve was obtained with seven concentration points in the ppb range: 0, 1, 5, 10, 50, 100, and 200 ppb.

Limit of detection (LOD) and limit of quantification (LOQ) were estimated, respectively, as three and ten times the standard deviation (σ) of 10 consecutive measurements of the reagent blanks according to EURACHEM recommendation.

3. Results and Discussion

A bivalve shell is composed of two hinged parts called valves; they are, when the animal is alive, joined by a ligament, a kind of fibrous connective tissue.

A mollusc shell is formed, repaired, and maintained by a part of the anatomy called the mantle. The shells are made up of three layers [28]: the top layer of organic material (periostracum); the middle thick layer of lime (ostracum); and the innermost, silver-white, shiny mother-of-pearl layer (hypostracum). During the animal's life, the shells grow from one side, called the beak, while the raised area around it is known as the umbo. This region is therefore the oldest and the thickest of the valve, while the opposite end is younger and thinner.

Despite the fact that each species of bivalve has its own peculiarities and the compositions of both the organic and inorganic parts can vary from one case to another, many characteristics of the shells of the genus Mytilus could be seen as representative of the entire class of bivalves. In particular, the response to the very oxidising conditions used in this work seems to be general, even shells of molluscs belonging to other classes of this phylum behave in similar way suggesting a much wider field of application of the method.

Figure 1 shows the evolution (in time t) of a shell kept in the peroxide bath for a few hours, starting from its initial state: t = 0 (Figure 1A,B). During the process, many small bubbles, of oxygen and water vapour, form on the surfaces of shells.

The temperature has a role in the rate of transformation, as the higher it is, the faster the process. However, to avoid any other possible secondary process, the reaction was carried out at 70 °C. It is worth noting that concentrated solutions of hydrogen peroxide are weakly acid [29]. Thus, it may be possible that too much aggressive conditions lead to partial dissolution of the inorganic components too. Very thin sheets, mostly from the inner face, i.e., from the nacreous layer, can detach, very likely because of the mechanical action of bubbles Valves kept only in hot distilled water do not change in appearance, colour, or weight. Depending on the specimen, some dark lines are still evident in on the valves, even if in some cases white colour is more uniform.

Figure 1K shows an untreated clam shell, while Figure 1L shows the same specimen treated with the same procedure for 5 h. Whitening is also evident in this case, indicating that the method is completely generalisable to other samples, except for a possible different duration of treatment.

At the end of the process, morphology of the shells does not change, while their colour undergoes to a systematic evolution and a small weight loss is observed. After few hours of treatment, shells are white in area of beak and bluish with an almost metallic lustre on the opposite side. Figure 2 shows the trend of the weight variation, averaged over five different samples, and expressed as $\Delta P = (P_0 - P_t)/P_0 \times 100$, where P_0 is the initial weight and P_t is the weight at time t. Before each measurement, the shells were carefully dried in an oven to eliminate absorbed water.

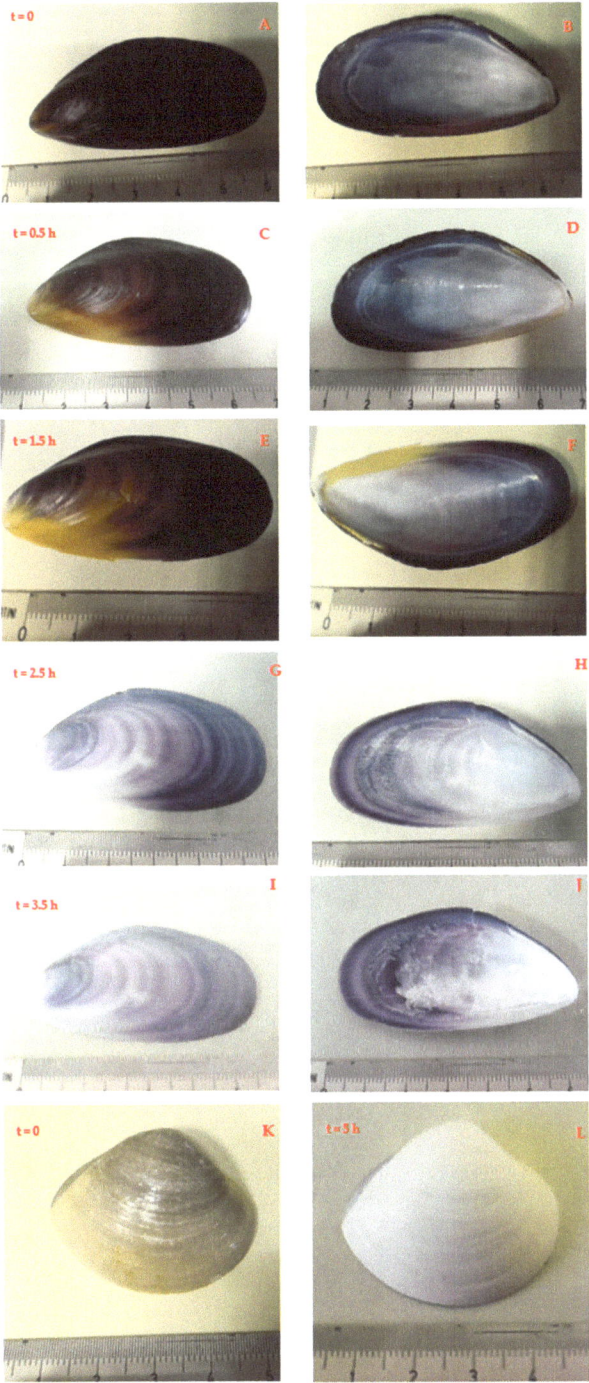

Figure 1. Evolution in time (t) of the appearance of Mytilus valves kept in hydrogen peroxide for different times ((**A**,**C**,**E**,**G**,**I**) external face; (**B**,**D**,**F**,**H**,**J**) inner face), and a clam valve before the treatment (**K**) and after 5 h (**L**).

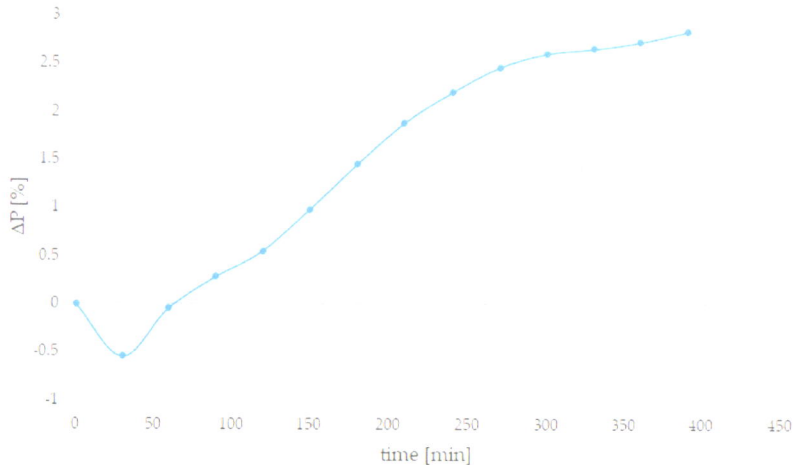

Figure 2. Weight variation ΔP (%) for shells kept in hydrogen peroxide bath as a function of time expressed in minutes (min).

It is interesting to note that the weight increased in the early stages of the process (i.e., ΔP is negative), suggesting that initially an absorption of water and, possibly, peroxide inside and on the surface of the shells occurred. After about an hour, the weight decreased monotonically. After about 2 h of treatment, the periostracum and the layer on the inner face began to peel away from the shells in small shreds, revealing the underlying structure and growth lines. This detachment did not occur uniformly across the entire surface of the valve, but preferentially from the edges and from the beak area, i.e., where the layer rich in organic compounds is thinner. Eventually, periostracum completely detached in the form of a thin yellow/brown veil. Eventually the valves were whiter, even if some dark lines were still present. These features could depend on the specific specimen. It should be kept in mind that the composition of valves depends on many factors, starting on the environment where the animal lived. Secondary elements can be absorbed from the marine waters. Parameters, such as temperature, salinity, and water hardness, can play a role in the formation of the shells. Both the living body of a bivalve and its shell are bioindicators of quality of the environment, the presence of any pollution, and the presence of various substances [30].

SEM-EDX analysis showed that the shells were composed of practically pure calcium carbonate, without the evident presence of other elements, at least with concentrations above the sensitivity of this type of measurement, i.e., about 1% or less. Figure 3 shows a typical EDX spectrum of a mussel specimen, and only the peaks due to calcium, carbon, and oxygen were evident, highlighting its remarkable chemical purity.

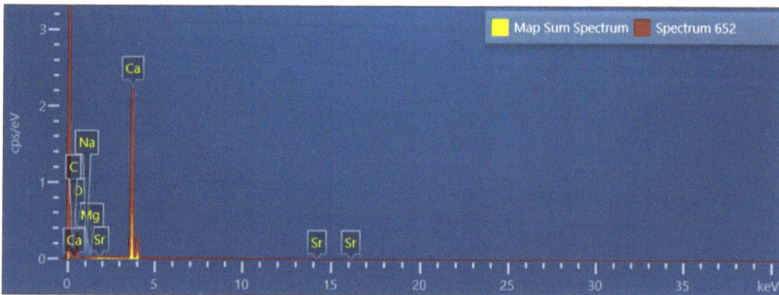

Figure 3. EDX spectrum of a Mytilus specimen.

Some specimens were also analysed with ICP-MS-QQQ. Table 1 shows the results obtained. RSD (relative standard deviation) for all measurements were below 5%.

Table 1. Mean secondary element concentration in the shells, expressed in ppm.

Element	Concentration (ppm)
Sr	83.392
Mg	75.761
K	2.494
Al	1.479
Fe	1.030
Cr	0.125
Mn	0.333
Co	0.996
Ni	0.873
Cu	0.545
Zn	0.423
Ba	0.186
Pb	0.172
U	1.86×10^{-3}

Despite the fact that the concentrations of a given element could be different, depending on the environmental conditions in which the animal lived, all these elements, with the only exception of strontium and magnesium, and in particular the heavy metals, were present in rather low concentrations, and this is very important for a possible use of these shells for industrial applications. Heavy metals incorporated into the shell can provide a record of environmental contamination or pollution. Strontium and magnesium, due to their characteristics and chemical similarity, can replace calcium in calcium carbonate, so it should not be surprising that these elements have a higher concentration than the others. It is worth remembering that natural mineral dolomite is a solid solution of calcium and magnesium carbonates, with an ideal ratio between them equal to 1:1 [31].

Chemically, calcium carbonate can be found in different polymorphs: calcite, which is most stable at room temperature and pressure, which crystallises in the hexagonal system; aragonite, which is the densest form; and the hexagonal vaterite, which is less common both in samples of geological origin and in samples of biological nature. Figure 4 shows their crystalline structures.

In a sample of biological origin, all these polymorphs can be found, even if calcite and aragonite are by far the most common, whereas vaterite is rare. Oyster shells are made almost completely by calcite, but in those of other species, both calcite and aragonite are present in different ratios in different points.

These polymorphs have different properties, including mechanical ones, and animals preferentially grow microcrystals of one or the other polymorph based on the functional needs of their shell. For example, where mechanical stresses are concentrated, aragonite is more often found, because it is harder, while in other parts of the shells, a prevalence of calcite is found. These microcrystals are normally organised into complex and ordered structures to better serve the purposes of the shells. In the case of Mytilus species, shells grow simultaneously in two directions: the shell grows longer with layers of calcite being added on at the growing edge, and it also thickens, adding layers of aragonite internally [12].

Micro-Raman spectroscopy is a powerful tool for carrying out point-by-point compositional analyses and identifying the present phases. Raman spectra of some points of a valve of an untreated specimen, sampled on both the external and internal surfaces and on the section as well, are shown in Figures 5A and 4B,C, respectively. The points, chosen randomly in such a way as to fully present the characteristics of the shell, represent areas where Raman spectra have been acquired which are then reported in the following figures. The external face appears bluish-dark grey with not very noticeable growth lines; on the contrary, the inner face is mother-of-pearl almost everywhere because of the presence of the so called nacre, even if on the extremity it is darker, and a rather dark area where the posterior adductor muscle of the animal was attached is easily recognisable (zone number 10); the section, not constant in thickness in all its parts, is instead whitish-yellowish.

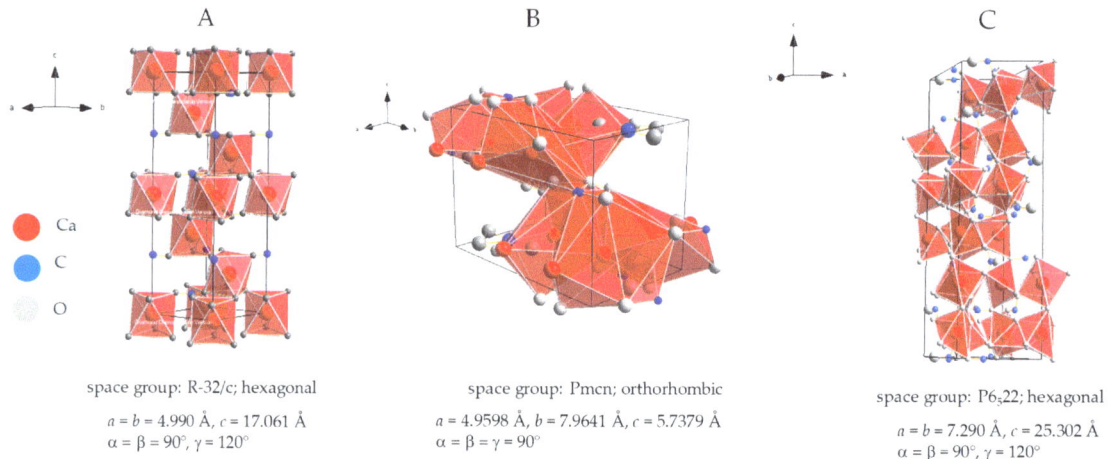

Figure 4. Crystal structure of three calcium carbonate polymorphs: calcite (**A**), aragonite (**B**), and vaterite (**C**). Polyhedra represent the oxygen coordination around calcium atoms, that is, octahedral in the case of calcite and more complex for the other two polymorphs: calcium is surrounded by eight oxygen atoms in total, distributed in two non-equivalent positions in vaterite and by nine oxygen atoms in total, distributed in five non-equivalent positions in aragonite. Cell parameters and space group were taken from [32] for calcite, from [33] for aragonite, and from [34] for vaterite. Diamond version 3.2k software was used to draw the crystal structures.

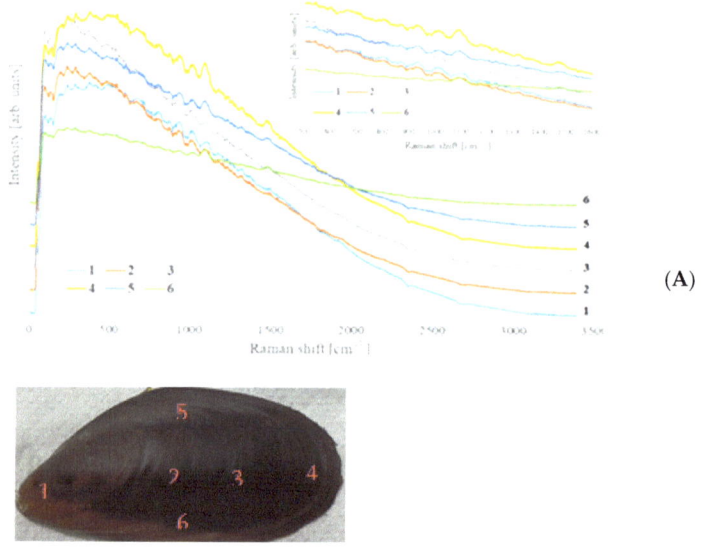

Figure 5. *Cont.*

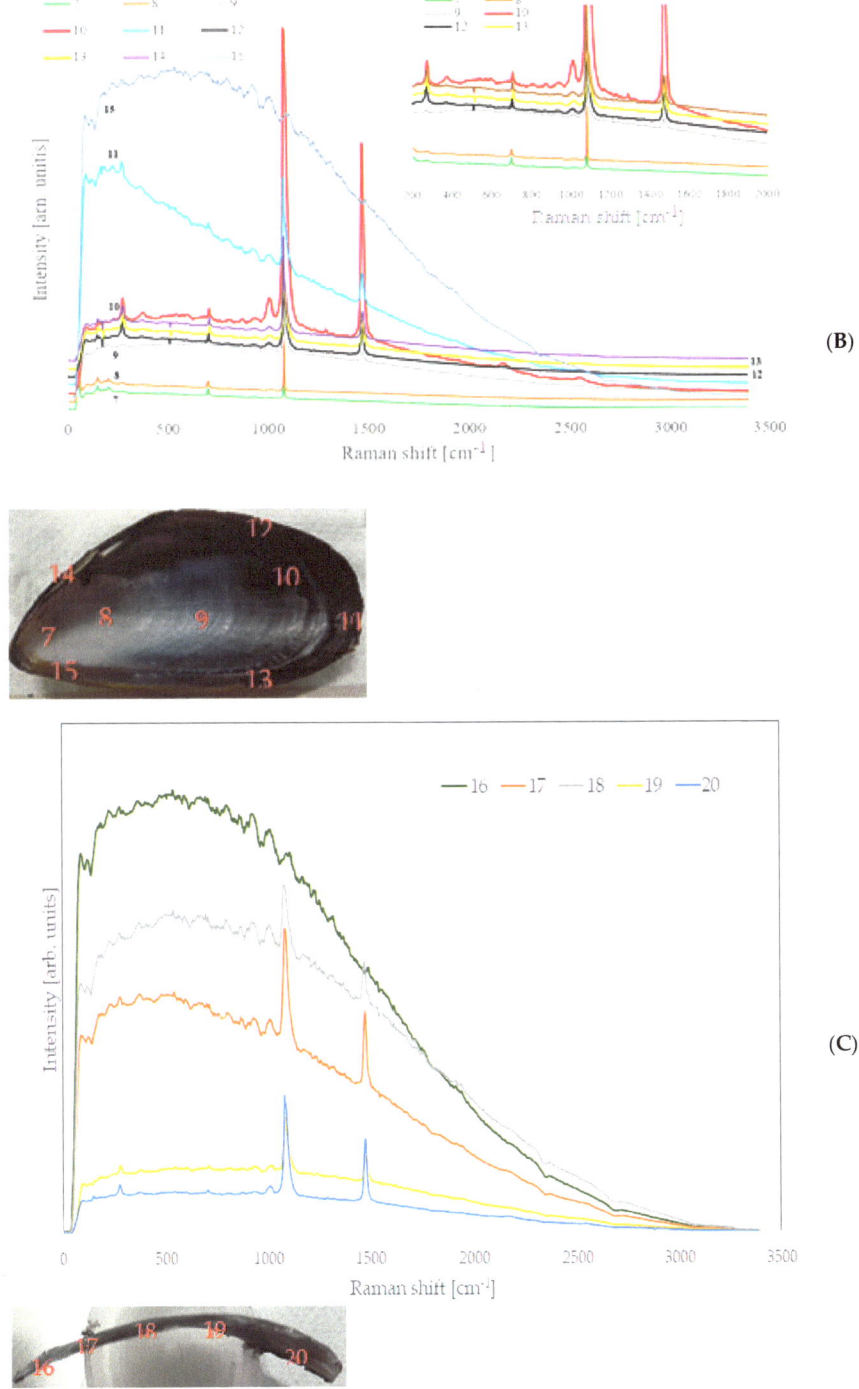

Figure 5. Raman spectra of different points on the external face (**A**), internal face (**B**), and section (**C**) of an untreated shell, as it is possible to see in the insets.

Due to the different nature and chemical-physical characteristics of the two faces, the acquisitions of the spectra were optimised in terms of laser power, duration of a single acquisition, and number of repetitions, specifically for each of them.

Raman spectra of the external face are characterised by predominant fluorescence, and different bands were present, as it can be seen in the inset showing the spectra in the characteristic zone between 500 and 1600 cm^{-1}. It is difficult to attribute these bands to specific chemical components, and even the inorganic part cannot be recognised immediately in any point because the characteristic peaks of calcite and aragonite fall in areas of the spectrum where many signals of different origin are present. In general, the same signals, regardless with their relative intensity and the fluorescence intensity, were present in all the spectra. This means that the same components were present, even if possibly in relative quantities that change as a function of the position. No evident or just quite weak signals could be observed in the high wavenumber part of the spectrum, that is, the characteristic region of hydrogen–heteroelements bonds (-OH, -NH$_2$, -NH -C$_{sp}^3$H, -C$_{sp}^2$H, -C$_{sp}$H); however, signals were observable around the 1500–1670 cm^{-1} region, which is normally associated with the vibrations of the double bond C=C and C=O, particularly the amide group C=O-N-, which normally characterise proteins [35]. Signals between about 1100 and 1000 cm^{-1} can be generally attributed to weak asymmetric stretching of C–O–C [36]. Some additional signals, originating from CH groups, were observed in two spectral regions (1350–1050 cm^{-1} and 880–680 cm^{-1}) [35].

The internal face (Figure 5B) was less uniform, and there were differences among the areas taken into consideration. The fluorescence was still present, but except for the edges where it was on the contrary very strong (spectrum of zone 15, which has been multiplied by 0.2 to be shown together with the others); this spectral feature was weaker than on the external face (Figure 4A). In this case, it is generally possible to recognise the inorganic mineral present, that is, aragonite or calcite, depending on the position.

As can be expected, the mother-of-pearl area (nacre) was mainly made up of aragonite; however, moving from the edge towards the centre, i.e., from zone 7 towards zone 9, the fluorescence became more intense, and bands that cannot be attributed to calcium carbonate appeared. In particular, in spectrum 9, there was a peak around 1500 cm^{-1}, which can be associated with organic functional groups, and in addition to other signals with lower wavenumbers also detectable. Bands in this position can be assigned to C=O or to C=C bond vibrations, in agreement with the presence of biomolecules. This trend may have been due to the structure of the nacre and the way in which the animal synthesises it. Nacre is composed of thin hexagonal platelets of aragonite arranged in a continuous parallel manner. Depending on the species, the shape of these tablets changes, but these layers are separated by sheets of organic matrix. Because the way the mollusc grows, near position 7, the nacre was thicker and more continuous so that the platelets formed a compact layer, whereas around position 9, it was thinner, and the new secreted platelets were partially separated. Therefore, the organic matter can be stimulated by the laser. This is particularly evident in position 10, and in subsequent ones, where the shell was macroscopically thinner and the carbonate components more separated. In position 10, traces of organic molecules may also have been due to the adductor muscle attached in that position. The band at about 1500 cm^{-1} was still present and became stronger and stronger.

All calcium carbonate polymorphs had a sharp peak at about 1080 cm^{-1} (1085 cm^{-1} for calcite, 1080 cm^{-1} for aragonite, 1090 cm^{-1} for vaterite that also has a secondary weaker peak at 1075 cm^{-1}), but in spectra of these positions, they were hidden by a much larger band that also had an evident shoulder at about 1095 cm^{-1}. A second very intense band was evident at about 1020 cm^{-1}. In general, it seems that the thinner the shell, the more intense these bands, also because the thickest parts of the shell are those formed first, while the thin ones were secreted by the animal at the end of its growth and can contain many biopolymers. In points close to other extremities, calcite also begins to be present.

Raman spectra were acquired in the shell section (Figure 5C). They were rather similar to each other. In the selected areas, fluorescence was observable, meaning that organic molecules were present, but generally it was less strong than on the faces, suggesting that the organic components were less abundant inside the shell, and decreased moving from the extremities toward the centre. As one of the main features of these spectra, the band around 1480 cm^{-1} was present; its relative intensity, however, decreased following the same trend of the fluorescence, being stronger on the extremities.

Inorganic components were more recognisable than on the inner face, and it was possible to discriminate areas where calcite was more abundant (external face and thin end), where aragonite was predominant (internal face thick end), or they coexisted (some central areas). After the complete treatment, the organic part was almost completely gone, as it is possible to observe in Figure 6, which shows Raman spectra collected in several points of the external and inner faces. Raman spectra collected on samples treated for a short period of time show the bands due the organic components. The longer the treatment, the weaker they become.

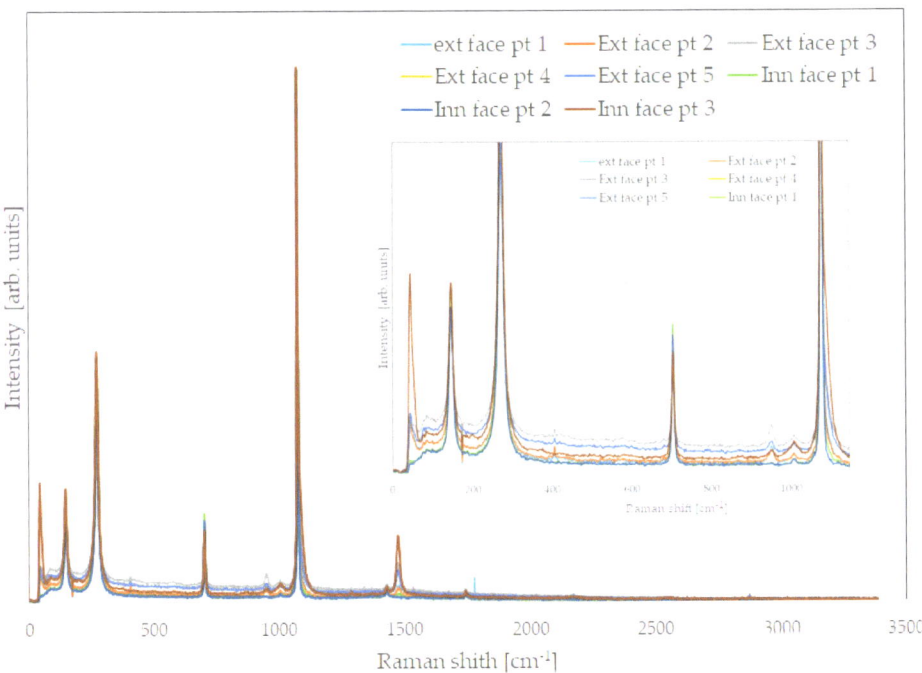

Figure 6. Raman spectra of some randomly chosen points on the external and internal face of a shell.

Despite traces of organic components in some places being still present, in general, only peaks of calcium carbonate were recognisable. Table 2 shows the theoretical peak positions of Raman spectra of calcite and aragonite (from [37] and [38], respectively).

In practically all the points where the spectra were acquired, they did not exhibit fluorescence signals, and this finding is in perfect agreement with the complete removal of organic components.

It is very important to note that all the carbonate Raman bands, which were possible to see in the treated specimens, were extremely narrow: for both calcite and aragonite, the most intense peak corresponded to the symmetric stretching of the CO_3 group, at approximately 1080 cm^{-1}, and they had a very small FWHM, much narrower than 5 cm^{-1}. Commercial or synthetic calcium carbonates have values much higher than this. This result indicates the high crystalline grade and confirms the high purity of the carbonate of biogenic origin, in good agreement with the ICP-MS results described previously. Impurities and crystalline defects can indeed induce a broadening and a certain degree of asymmetry in

Raman bands. Large size distribution of crystallites can also have a similar effect. The slow growth of crystals in the shells, mediated by proteins and biopolymers, minimise these effects, and the individual crystals also have a very constant and uniform size.

Table 2. Calcite and aragonite Raman peaks.

	Raman Shift (cm^{-1})	
	Calcite [33]	Aragonite [34]
1	155	145
2	282	155
3	713	182
4	1087	192
5	1437	208
6	1749	705
7		716
8		1085
9		1463
10		1576

In the shell section, the spectra were very similar to those observed on the two faces, meaning that the small quantity of organic components, which were found in the untreated specimens, were completely removed. This is an interesting observation, because it suggests that weight loss is not only due to the detachment of periostracum and hypostracum, but also to a real chemical process, and that, thanks to the action of peroxide, water-soluble molecules are released from the shells. It can be imagined, however, that the mechanism of this release is a rather complex path.

The complex proteins secreted by the external epithelial tissue of molluscs, i.e., by the mantle, form the so-called conchiolin. These proteins form a matrix that constitutes the environment in which calcium carbonate crystals nucleate and grow.

There are some methods for extracting the organic component from the shells as these biopolymers are intrinsically attractive, have excellent chemical-physical and mechanical properties, and can have some interesting applications: for example in the biomedical field or even as antibacterials. However, these methods require the mechanical destruction of the shells and the removal of the inorganic part by acid treatments, which are obviously not the aim of the present work, which is instead to keep the shells as little modified as possible.

It is not easy to indicate what the chemical mechanism underlying the process described here is, and further investigations would be required for highlighting this point. Nevertheless, hydrogen peroxide can perturb hydrogen bonds among the macromolecules, causing them to move away from each other [39,40]. Furthermore, it is well known that proteins themselves can degrade under the effect of high concentrations of hydrogen peroxide by action of free radicals, eventually leading to their hydrolysis [41,42].

Polysaccharides behave in a similar manner in respect to H_2O_2 [43]. Therefore, it can be imagined that hydrogen peroxide can cut biopolymers into smaller units that are more soluble in aqueous environments. In contrast, peroxide has limited or no effect on the inorganic part of the shells. It can therefore be suggested that the main action of the peroxide is to cut away the bonds between the organic layers and the crystalline matrix and possibly to break the large macromolecules into smaller and soluble residues. In part, this also justifies the detachment of the very small sheets of nacre, because the biopolymer chains that are located among them and which help to keep them bound are broken.

Once the organic part has been removed, it is possible to evaluate the ratio between the different polymorphs present overall. For doing this and to evaluate the possible presence of amorphous phases, a sample, after being treated by peroxide, was carefully ground in an agate mortar. The diffraction pattern (XRD) of the obtained powder was acquired (Figure 7); according to these results, it turned out that the valve was made mainly by calcite, about 75%, and the rest was aragonite.

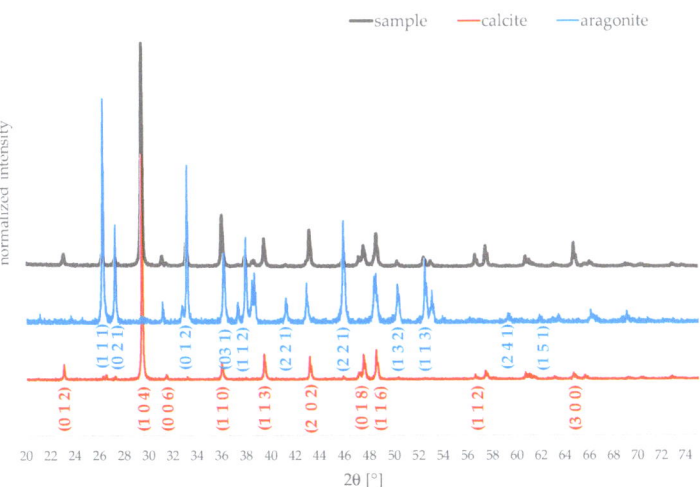

Figure 7. XRD analysis of a ground Mytilus shell.

Again, the quality of the material was very high and only calcite, and aragonite were present, without spurious phases, with just some very weak peaks due to Na_2O, that can form during the treatment. All peaks were extremely narrow, indicating the high degree of crystallinity and suggesting that valves are promising raw materials for many possible applications.

The quality of the calcium carbonate resulting from these shells was very high, in terms of purity and crystallinity, especially when compared with commercial materials. This may suggest useful applications as a raw material for industry. It is an absolutely non-toxic product and without any risk for humans and the environment. It should in fact be kept in mind that calcium carbonate is used as a starting material for cements and other construction compounds, being used as an additive in numerous sectors, for example in tires. It is used as a filler in the production of paints, varnishes, paper, rubbers, and glues. Also, it is used in the production of glass and crystals. All these sectors could have many advantages using a better raw starting material. Even the organic layers that detach from the shells could have applications, being rich in interesting biomolecules. They could, for example, have biotechnological applications or be used to produce biocompatible and environmentally friendly polymers and hydrogels.

This treatment also allows us to better analyse the microstructure of the valves, as it can also be easily seen in SEM images shown in Figure 8A–H.

The structure of the shells of bivalves and molluscs in general has been thoroughly investigated, and there are numerous scientific articles and reviews on the subject [12,44–46].

The external face (Figure 8A,B) but also for a certain degree the internal one of an untreated sample do not show characteristic details, because they are covered by very homogeneous layers. The section, however, is much more organised and the different microstructures are clearly evident. But, also for it, the treatment with hydrogen peroxide has a positive effect, and the characteristics are well highlighted. Sometimes, the microstructure was highlighted by more vigorous attacks than the milder treatment described in this work. Therefore, it may be imagined that it is less altered for observations.

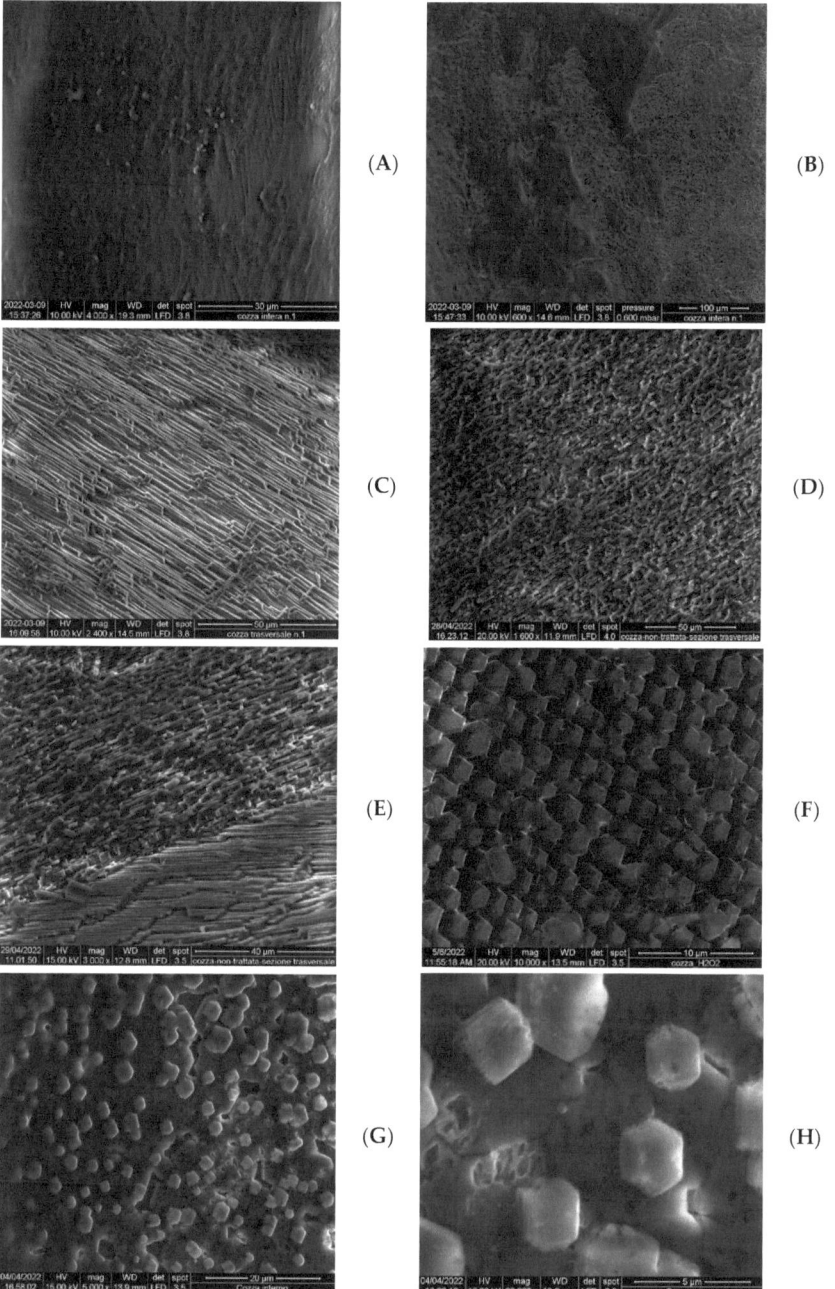

Figure 8. SEM micrographs of different areas of a Mytilus shell before and after the treatment with hydrogen peroxide. (**A,B**): external face of an untreated specimen; (**C,D**): images of the section of a shell specimen; (**E**): boundary between prismatic part and nacreous layer; (**F–H**): nacreous layer showing the hexagonal crystals.

The inner part of a shell is formed by prismatic crystal, as can be seen in Figure 8C,D, which shows the section of a treated shell. A very clear separation from the nacreous layer exists (Figure 8E).

Nacre, in itself, is composed of hexagonal platelets, arranged in a parallel manner, as can be easily seen in Figure 8F–H. These platelets have a very uniform size, just a few microns large. Depending on the position in the valve, they are more or less compact. This is due to the moment in which they were formed. In the most recent part of growing shell, these crystallites are separated and isolated. Furthermore, it can be observed that they grow preferentially at the joints between the grain boundaries of the underlying layer, which is much more compact. The size of these isolated crystallites is also roughly constant, and this strongly suggests that they begin forming at the same time. Indeed, even if the growth of the shell occurs in successive layers, each layer is formed when the one below is not completely compact.

4. Conclusions

This work presents a discussion about the use of concentrated hydrogen peroxide solution for removing and separating the organic components from the inorganic part in mollusc shells. Shells are an intriguing example of a composite material. In life, the animal produces calcium carbonate microcrystals in the form of various polymorphs, such calcite and aragonite, thanks to the action of specialised proteins and biopolymers, based on its functional needs.

The process of removal of the inorganic part and recover of organic molecules, which can successively have various applications, is relatively easy to perform using acid baths. The opposite process is more difficult.

A bath of a few hours in concentrated hot hydrogen peroxide completely removes the organic component from the valves without altering the inorganic structure.

The method discussed here is very effective, has a low cost, and is environmentally friendly because no pollutant and expansive reagents are required. Raman, XRD, and morphological analyses showed the excellent crystallographic quality of the calcium carbonate crystals that make up the shells, and ICP-MS instead showed their very high chemical purity.

This makes the shells extremely interesting starting raw materials for applications in the building field, especially with a view of limiting greenhouse gas emissions and CO_2, as well as being more environmentally sustainable.

A possible future development could be represented by the optimisation of the hydrogen peroxide concentration, the process temperature, and the time necessary to obtain the removal of the organic components. Furthermore, it could be systematically applied to other mollusc species.

Author Contributions: Conceptualisation, A.U.; methodology, A.U., F.C., N.F., G.M., A.G., S.S. and S.B.; validation, S.B. and A.R.; investigation, A.U., F.C., N.F., G.M., A.G., S.S. and S.B.; writing—original draft preparation, A.U.; writing—review and editing, S.C., C.T., N.F. and A.U. All authors have read and agreed to the published version of the manuscript.

Funding: This research received no external funding.

Data Availability Statement: The original contributions presented in this study are included in the article; further inquiries can be directed to the corresponding author.

Conflicts of Interest: The authors declare no conflicts of interest.

References

1. Omelon, S.; Georgiou, J.; Henneman, Z.J.; Wise, L.M.; Sukhu, B.; Hunt, T.; Wynnyckyj, C.; Holmyard, D.; Bielecki, R.; Grynpas, M.D. Control of vertebrate skeletal mineralization by polyphosphates. *PLoS ONE* **2009**, *4*, e5634. [CrossRef] [PubMed]
2. Huang, J.; Zhang, R. The Mineralization of Molluscan Shells: Some Unsolved Problems and Special Considerations. *Front. Mar. Sci.* **2022**, *9*, 874534. [CrossRef]
3. Simonet Roda, M.; Griesshaber, E.; Ziegler, A.; Rupp, U.; Yin, X.; Henkel, D.; Häussermann, V.; Laudien, J.; Brand, U.; Eisenhauer, A.; et al. Calcite fibre formation in modern brachiopod shells. *Sci. Rep.* **2019**, *9*, 598. [CrossRef] [PubMed]
4. Zaquin, T.; Pinkas, I.; Di Bisceglie, A.P.; Mucaria, A.; Milita, S.; Fermani, S.; Goffredo, S.; Mass, T.; Falini, G. Exploring Coral Calcification by Calcium Carbonate Overgrowth Experiments. *Crystal. Growth Des.* **2022**, *22*, 5045–5053. [CrossRef]

5. Sardo, A.; Orefice, I.; Balzano, S.; Barra, L.; Romano, G. Mini-Review: Potential of Diatom-Derived Silica for Biomedical Applications. *Appl. Sci.* **2021**, *11*, 4533. [CrossRef]
6. Lazarus, D.B.; Kotrc, B.; Wulf, G.; Schmidt, D.N. Radiolarians decreased silicification as an evolutionary response to reduced Cenozoic ocean silica availability. *Proc. Natl. Acad. Sci. USA* **2009**, *106*, 9333–9338. [CrossRef]
7. Somu, D.R.; Cracchiolo, T.; Longo, E.; Greving, I.; Merk, V. On stars and spikes: Resolving the skeletal morphology of planktonic Acantharia using synchrotron X-ray nanotomography and deep learning image segmentation. *Acta Biomater.* **2023**, *159*, 74–82. [CrossRef]
8. Łukowiak, M. Utilizing sponge spicules in taxonomic, ecological and environmental reconstructions: A review. *PeerJ* **2020**, *8*, e10601. [CrossRef] [PubMed] [PubMed Central]
9. Wiltschko, R.; Wiltschko, W. The magnetite-based receptors in the beak of birds and their role in avian navigation. *J. Comp. Physiol. A Neuroethol. Sens. Neural. Behav. Physiol.* **2013**, *199*, 89–98. [CrossRef] [PubMed] [PubMed Central]
10. Schofield, R.M.S.; Bailey, J.; Coon, J.J.; Devaraj, A.; Garrett, R.W.; Goggans, M.S.; Hebner, M.G.; Lee, B.S.; Lee, D.; Lovern, N.; et al. The homogenous alternative to biomineralization: Zn- and Mn-rich materials enable sharp organismal tools that reduce force requirements. *Sci. Rep.* **2021**, *11*, 17481. [CrossRef]
11. Hillerton, J.E.; Vincent, J.F.V. The Specific Location of Zinc in Insect Mandibles. *J. Exp. Biol.* **1982**, *10*, 333–336. [CrossRef]
12. Louis, V.; Besseau, L.; Lartaud, F. Step in Time: Biomineralisation of Bivalve's Shell. *Front. Mar. Sci.* **2022**, *9*, 906085. [CrossRef]
13. Okada, S.; Chen, C.; Watsuji, T.; Nishizawa, M.; Suzuki, Y.; Sano, Y.; Bissessur, D.; Deguchi, S.; Takai, K. The making of natural iron sulfide nanoparticles in a hot vent snail. *Proc. Natl. Acad. Sci. USA* **2019**, *116*, 20376–20381. [CrossRef] [PubMed]
14. Oertle, A.; Szabó, K. Thermal Influences on Shells: An Archaeological Experiment from the Tropical Indo-pacific. *J. Archaeol. Method Theory* **2023**, *30*, 536–564. [CrossRef]
15. Yoshioka, S.; Kitano, Y. Transformation of aragonite to calcite through heating. *Geochem. J.* **1985**, *19*, 245–249. [CrossRef]
16. Boettcher, A.L.; Wyllie, P.J. The Calcite-Aragonite Transition Measured in the System $CaO-CO_2-H_2O$. *J. Geol.* **1968**, *76*, 214–330. [CrossRef]
17. Nudelman, F.; Gotliv, B.A.; Addadi, L.; Steve, W. Mollusk shell formation: Mapping the distribution of organic matrix components underlying a single aragonitic tablet in nacre. *J. Struct. Biol.* **2006**, *153*, 176–187. [CrossRef] [PubMed]
18. Jackson, A.P.; Vincent JF, V.; Turner, R.M. The mechanical design of nacre. *Proc. R. Soc. B Biol. Sci.* **1988**, *234*, 415–440. [CrossRef]
19. Radhaa, A.V.; Forbesa, T.Z.; Killianb, C.E.; Gilbert PU, P.A.; Navrotsk, A. Transformation and crystallization energetics of synthetic and biogenic amorphous calcium carbonate. *Proc. Natl. Acad. Sci. USA* **2010**, *107*, 16438–16443. [CrossRef] [PubMed]
20. Sawada, N.; Toyohara, H.; Nakano, T. A New Cleaning Method for Accurate Examination of Freshwater Gastropod Shell Specimens Covered with Iron-rich Deposits. *Species Divers.* **2021**, *26*, 217–224. [CrossRef]
21. Kaya, M.; Baran, T.; Karaarslan, M. A new method for fast chitin extraction from shells of crab, crayfish and shrimp. *Nat. Prod. Res.* **2015**, *29*, 1477–1480. [CrossRef] [PubMed]
22. Falster, G.; Delean, S.; Tyler, J. Hydrogen peroxide treatment of natural lake sediment prior to carbon and oxygen stable isotope analysis of calcium carbonate. *Geochem. Geophys. Geosyst.* **2018**, *19*, 3583–3595. [CrossRef]
23. Smith, A.M.; Key Jr, M.M.; Henderson, Z.E.; Davis, V.C.; Winter, D.J. Pretreatment for removal of organic material is not necessary for X-ray-diffraction determination of mineralogy in temperate skeletal carbonate. *J. Sediment. Res.* **2016**, *86*, 1425–1433. [CrossRef]
24. Hou, Y.; Shavandi, A.; Carne, A.; Bekhit, A.A.; Ng, T.B.; Cheung, R.C.F.; Bekhit, A.E.D.A. Marine shells: Potential opportunities for extraction of functional and health-promoting materials. *Crit. Rev. Environ. Sci. Technol.* **2016**, *46*, 1047–1116. [CrossRef]
25. Lertvachirapaiboon, C.; Parnklang, T.; Pienpinijtham, P.; Wongravee, K.; Thammacharoen, C.; Ekgasit, S. Selective colors reflection from stratified aragonite calcium carbonate plates of mollusk shells. *J. Struct. Biol.* **2015**, *191*, 184–189. [CrossRef]
26. Available online: https://www.molluscabase.org/aphia.php?p=taxdetails&id=397041 (accessed on 10 May 2024).
27. Available online: https://www.iso.org/standard/53394.html (accessed on 10 May 2024).
28. Giribet, G. Bivalvia. In *Phylogeny and Evolution of the Mollusca*; University of California Press: Berkeley, CA, USA, 2008. [CrossRef]
29. Harris, M.G.; Torres, J.; Tracewell, L. pH and H_2O_2 concentration of hydrogen peroxide disinfection Systems. *Optom. Vis. Sci.* **1988**, *65*, 527–535. [CrossRef]
30. Strehse, J.S.; Maser, E. Marine bivalves as bioindicators for environmental pollutants with focus on dumped munitions in the sea: A review. *Mar. Environ. Res.* **2020**, *158*, 105006. [CrossRef] [PubMed]
31. Al-Awadi, M.; Clark, W.J.; Moore, W.R.; Herron, M.; Zhang, T.; Zhao, W.; Hurley, N.; Kho, D.; Montaron, B.; Sadooni, F. Dolomite: Perspectives on a perplexing mineral. *Oilfield Rev.* **2009**, *21*, 32–45.
32. Antao, S.; Hassan, I.; Mulder, M.; Lee, P.L.; Toby, B.H. In situ study of the R3c → R3m orientational disorder in calcite. *Phys. Chem. Miner.* **2009**, *36*, 159–169. [CrossRef]
33. Ulian, G.; Valdrè, G. The effect of long-range interactions on the infrared and Raman spectra of aragonite ($CaCO_3$, Pmcn) up to 25 GPa. *Sci. Rep.* **2023**, *13*, 2725. [CrossRef]
34. Christy, A.G. A Review of the Structures of Vaterite: The Impossible, the Possible, and the Likely. *Cryst. Growth Des.* **2017**, *17*, 3567–3578. [CrossRef]
35. Kuhar, N.; Sil, S.; Umapathy, S. Potential of Raman spectroscopic techniques to study proteins. *Spectrochim. Acta Part A Mol. Biomol. Spectrosc.* **2021**, *258*, 119712. [CrossRef] [PubMed]
36. Ettah, I.; Lorna Ashton, L. Engaging with Raman Spectroscopy to Investigate Antibody Aggregation. *Antibodies* **2018**, *7*, 24. [CrossRef]

37. King, H.E.; Geisler, T. Tracing Mineral Reactions Using Confocal Raman Spectroscopy. *Minerals* **2018**, *8*, 158. [CrossRef]
38. Tomić, Z.P.; Petre Makreski, P.; Gajic, B. Identification and spectra–structure determination of soil minerals: Raman study supported by IR spectroscopy and X-ray powder diffraction. *J. Raman Spectrosc.* **2010**, *41*, 582–586. [CrossRef]
39. Parida, C.; Chowdhuri, S. Effects of Hydrogen Peroxide on the Hydrogen Bonding Structure and Dynamics of Water and Its Influence on the Aqueous Solvation of the Insulin Monomer. *J. Phys. Chem. B* **2023**, *127*, 10814–10823. [CrossRef]
40. Le, H.-T.; Chaffotte, A.F.; Demey-Thomas, E.; Vinh, J.; Friguet, B.; Mary, J. Impact of Hydrogen Peroxide on the Activity, Structure, and Conformational Stability of the Oxidized Protein Repair Enzyme Methionine Sulfoxide Reductase A. *J. Mol. Biol.* **2009**, *393*, 58–66. [CrossRef]
41. Fligiel, S.E.; Lee, E.C.; McCoy, J.P.; Johnson, K.J.; Varani, J. Protein degradation following treatment with hydrogen peroxide. *Am. J. Pathol.* **1984**, *115*, 418–425.
42. Song, I.-K.; Lee, J.-J.; Cho, J.-H.; Jeong, J.; Shin, D.-H.; Lee, K.-J. Degradation of Redox-Sensitive Proteins including Peroxiredoxins and DJ-1 is Promoted by Oxidation-induced Conformational Changes and Ubiquitination. *Sci. Rep.* **2016**, *6*, 34432. [CrossRef] [PubMed]
43. Chen, X.; Sun-Waterhouse, D.; Yao, W.; Li, X.; Zhao, M.; You, L. Free radical-mediated degradation of polysaccharides: Mechanism of free radical formation and degradation, influence factors and product properties. *Food Chem.* **2021**, *365*, 130524. [CrossRef] [PubMed]
44. Checa, A.G. Physical and Biological Determinants of the Fabrication of Molluscan Shell Microstructures. *Front. Mar. Sci.* **2018**, *5*, 353. [CrossRef]
45. Kempf, H.L.; Gold, D.A.; Carlson, S.J. Investigating the Relationship between Growth Rate, Shell Morphology, and Trace Element Composition of the Pacific Littleneck Clam (*Leukoma staminea*): Implications for Paleoclimate Reconstructions. *Minerals* **2023**, *13*, 814. [CrossRef]
46. Duarte, C.M.; Rodriguez-Navarro, A.B.; Delgado-Huertas, A.; Krause-Jensen, D. Dense *Mytilus* Beds Along Freshwater-Influenced Greenland Shores: Resistance to Corrosive Waters Under High Food Supply. *Estuaries Coasts* **2020**, *43*, 387–395. [CrossRef]

Disclaimer/Publisher's Note: The statements, opinions and data contained in all publications are solely those of the individual author(s) and contributor(s) and not of MDPI and/or the editor(s). MDPI and/or the editor(s) disclaim responsibility for any injury to people or property resulting from any ideas, methods, instructions or products referred to in the content.

Article

Research on Microstructure, Synthesis Mechanisms, and Residual Stress Evolution of Polycrystalline Diamond Compacts

Peishen Ni [†], Yongxuan Chen [†], Wenxin Yang [†], Zijian Hu and Xin Deng *

School of Electromechanical Engineering, Guangdong University of Technology, Guangzhou 510006, China; 1111901024@mail2.gdut.edu.cn (P.N.); 2112101482@mail2.gdut.edu.cn (Y.C.); 1112201021@mail2.gdut.edu.cn (W.Y.); 1112101029@mail2.gdut.edu.cn (Z.H.)
* Correspondence: dengxin@gdut.edu.cn; Tel.: +86-020-3932-2925
[†] These authors contributed equally to this work.

Abstract: The microstructure and residual stress of polycrystalline diamond compact (PDC) play crucial roles in the performance of PDCs. Currently, in-depth research is still to be desired on the evolution mechanisms of microstructure and residual stress during high pressure high temperature (HPHT) synthesis process of PDCs. This study systematically investigated the influencing mechanisms of polycrystalline diamond (PCD) layer material design, especially the Co content of the PCD layer, on microstructure and residual stress evolution in PDCs via Raman spectroscopy and finite element micromechanical simulation. The research shows that when the original Co content of the PCD layer is higher than 15 wt.%, the extra Co in the PCD layer will migrate backwards towards the carbide substrate and form Co-enrichment regions at the PCD–carbide substrate interface. As the original Co content of the PCD layer increases from 13 to 20 wt.%, the residual compressive stress of diamond phase at the upper surface center of the PCD layer gradually decreases and transforms into tensile stress. When the original Co content of the PCD layer is as high as 30 wt.%, the residual stress transforms back into significant compressive stress again. The microstructure-based micromechanical simulation at the PCD–carbide substrate interface shows that the Co-enrichment region is the key for the transformation of the residual stress of the diamond phase from tensile stress into significant compressive stress.

Keywords: polycrystalline diamond compact; HPHT process; residual stress; finite element analysis; micromechanical simulation

Citation: Ni, P.; Chen, Y.; Yang, W.; Hu, Z.; Deng, X. Research on Microstructure, Synthesis Mechanisms, and Residual Stress Evolution of Polycrystalline Diamond Compacts. *Crystals* **2023**, *13*, 1286. https://doi.org/10.3390/cryst13081286

Academic Editors: Sanja Burazer and Lidija Androš Dubraja

Received: 29 July 2023
Revised: 15 August 2023
Accepted: 17 August 2023
Published: 20 August 2023

Copyright: © 2023 by the authors. Licensee MDPI, Basel, Switzerland. This article is an open access article distributed under the terms and conditions of the Creative Commons Attribution (CC BY) license (https://creativecommons.org/licenses/by/4.0/).

1. Introduction

Polycrystalline diamond compact (PDC) is a kind of superhard composite composed of a polycrystalline diamond (PCD) layer and carbide substrate [1–4]. During high pressure and high temperature (HPHT) synthesis processes, a significant diamond–diamond (D-D) bond forms between diamond particles by a dissolution and re-precipitation process, leading to the formation of the PCD layer, which is firmly bonded with the carbide substrate [1,5,6]. PDCs have been widely used in oil and gas drilling, ore mining, and the machining of non-ferrous metals and hard materials due to their high wear resistance and hardness, good impact resistance, and excellent brazing performance [2,7–9].

The thermal stability has always been a critical property for the application of PDCs. Due to the significant difference of the coefficient of thermal expansion between the PCD layer and the carbide substrate, residual stress will inevitably come into being during the cooling stage of the HPHT process, which can potentially result in cracks and complete failure of PDCs in various applications [10,11]. The magnitude and distribution of the residual stress of PDCs are closely related to the diamond grain size and thermal treatment of the PCD layer [12–14], the geometry design of the PCD-cemented carbide substrate [10], and the feedstock diamond powder morphology [15]. The existence of residual stress will have a significant impact on the thermal stability of PDCs. During the application of PDCs, as the number of impact and thermal cycles increase, the combination of residual stress and external stress may lead to micro-cracks, edge collapse, and even spalling of the PCD layer, resulting in the abnormal failure of PDCs [16–20]. Therefore, it is of great significance to study the mechanism of residual stress evolution in PDCs in order to better understand and reduce the negative effect.

Non-destructive testing methods such as laser Raman spectroscopy [21], X-ray diffraction [22] and Neutron diffraction [23] are generally used to measure the residual stress of PDCs. In order to evaluate the residual stress more accurately, especially to analyze the internal stress distribution of PDCs, finite element modeling (FEM) can be used to simulate the stress distribution inside PDCs [10].

McNamara et al. [12] investigated the effect of diamond grain size and the oil quenching process on the residual stress of PDCs. They found that a finer diamond grain size resulted in higher residual stress and oil quenching can lead to tensile residual stress in the PCD layer. Lin et al. [18] conducted a comprehensive FEM investigation on the effect of carbide substrate thickness on the residual stress of the PCD layer. They found that with the increase in carbide substrate thickness, the radial compressive stress of the PCD layer increased, which was favorite for the performance of PDCs. Krawitz et al. [23] evaluated the residual stress of PDCs using neutron diffraction. In view of the great penetration depth, the residual stress for both PCD layer and carbide substrate can be measured. It was found that the in-plane stress of the PCD layer increased with the thickness ratio of carbide substrate/PCD layer and there existed a significant stress gradient in the thickness direction for both the PCD layer and the carbide substrate. Debkumar [24] found that the leaching of Co from a PCD layer resulted in the decrease in residual compressive stress of the PCD layer. Yue et al. [14] found that vacuum annealing can effectively reduce the residual stress of a PCD layer. Chen et al. [25] added graphene to feedstock diamond powder and found that the residual stress of a PCD layer was significantly reduced. Peishen et al. [15] showed that feedstock diamond powder morphology had a significant effect on the residual stress of PDCs, and the higher specific surface area of diamond powder led to higher compressive stress of a PCD layer.

Over the current residual stress investigations on PDCs, the effect of the Co content of PCD layers on the residual stress of PDCs has rarely been reported yet. Moreover, in spite of the fact that FEM has been widely employed for residual stress investigation on PDCs, the current FEM method based on the macro-structure of PDCs simplifies the PCD layer and cemented carbide substrate into homogeneous material units. The influence of the characteristic microstructure of PDCs on the residual stress evolution, especially the residual stress of a diamond phase, has not been fully considered yet.

In this study, the feedstock diamond powder (average particle size 11 μm) was mixed with different contents of Co powder (average particle size 1 μm) to prepare the PDCs via the HPHT process. The effect of Co content of a PCD layer on the evolution of the microstructure and residual stress of a diamond phase was systematically investigated via both Raman spectrum and FEM. In particular, micromechanical FEM was made based on the microstructure of PCD-cemented carbide substrate interface to reveal the unique micromechanism of residual stress evolution from a new perspective.

2. Experimental Materials and Methods

2.1. Experimental Materials

The morphology, particle size distribution and Raman spectrum test results of the feedstock diamond powder employed in this study (Zhecheng Huifeng Diamond Technology Co., Ltd., Shangqiu, China) are shown in Figure 1. The median particle size D50 of the diamond powder is 11μm. The feedstock diamond powder used in this study is the same batch of the regular diamond powder used in [15]. The average particle size of Co powder in this study (Shanghai ST-Nano Material Technology Co., Ltd., Shanghai, China) is 1μm, and the purity is 99.9%. The dimension of the WC-13 wt.% Co cemented carbide substrate (Zhuzhou Cemented Carbide Group Co., Ltd., Zhuzhou, China) is 16.5 mm diameter and 12 mm height, and the microstructure of the carbide substrate is shown in Figure 2. A flat PCD layer–carbide substrate interface was adopted in this study to remove any extra residual stress possibly caused by interface structure design.

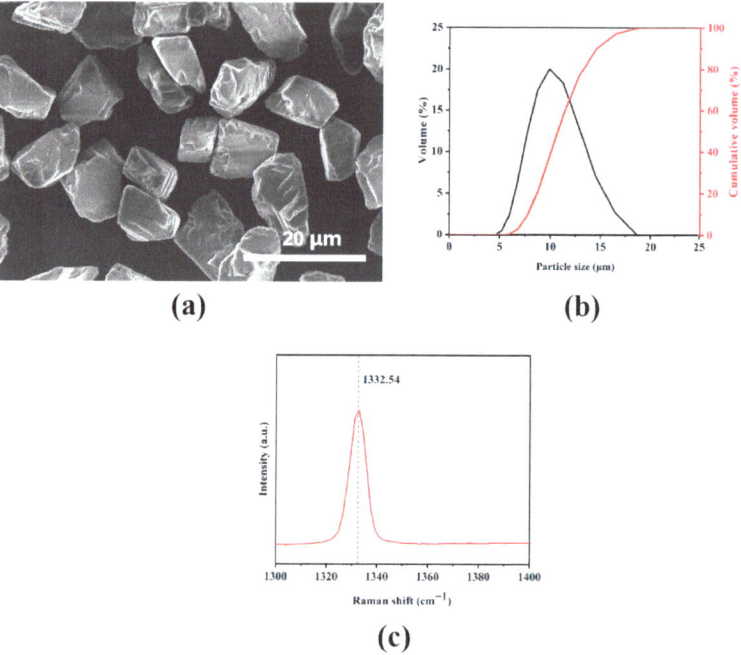

Figure 1. Characterization of feedstock diamond powder, (a) diamond powder morphology, (b) particle size distribution, and (c) Raman spectrum.

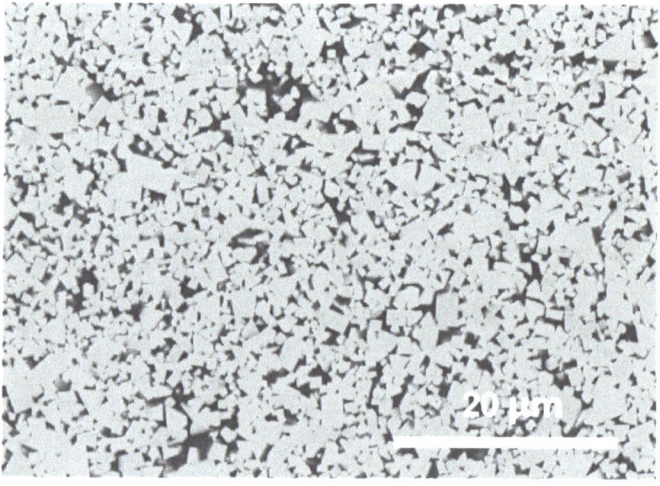

Figure 2. Microstructure of WC-13 wt.% Co carbide substrate.

2.2. Experimental Methods

2.2.1. Preparation of Initial Powder of PDCs

As shown in Table 1, the composition design of PCD layer in this study includes four different Co contents ranging from 13 to 30 wt.%. During the diamond–Co powder milling process, anhydrous ethanol was used as the dispersing medium. A horizontal ball mill (LHK-1.5-I, Haoqiang Machinery Factory, Yixing, China) was used for powder milling. A 4 mm diameter YG6 (WC-6 wt.% Co) milling ball was used and the ball/powder ratio was 1:1. The milling speed was 120 rpm and the milling time was 4 h. Then, the mixed diamond–Co powders were dried in a rotary evaporation instrument (RE-52C, Shanghai Leighton Industrial Co., Ltd., Shanghai, China), and purified in vacuum (5×10^{-3} Pa) at 500 °C for 1.5 h in a tube furnace (GSL-1500X-OTF, Hefei Kejing Material Technology Co., Ltd., Hefei, China).

Table 1. Composition design of PCD layer for PDCs synthesized in this study.

PDC Designation	Diamond Powder Content, wt.%	Co powder Content, wt.%
PCD-13	87	13
PCD-15	85	15
PCD-20	80	20
PCD-30	70	30

2.2.2. HPHT Process

In this study, cubic press (CS-6X29500KN, Guilin Guiye Machinery Co., Ltd., Guilin, China) was used for the HPHT synthesis process. The above purified diamond–Co powder mixture and cemented carbide substrate were encapsulated in niobium and molybdenum cups, and then assembled into a pyrophyllite composite block, which was then put into a cubic press for HPHT process. The sintering temperature was 1580 °C and pressure was 5.5 GPa. The encapsulation and HPHT processes used in this study are similar to those adopted in [15,23]. The selection of sintering temperature and pressure is to ensure the melting of Co and the thermal stability of diamond during the HPHT process. The HPHT process can have an apparent effect on the residual stress of PDCs, while in this study, the research focus is the Co content of PCD layer, so only a set of commonly used HPHT process has been employed according to the general synthesis routine in PDC manufacturing industry. After HPHT process, the PDCs were removed from the pyrophyllite composite block, the metal cups on the surface of the PDCs were removed by sand blasting, and then the PDCs were finally cleaned and dried for further testing.

2.2.3. Residual Stress Evaluation

In this study, the residual stress of PDCs was determined via the position shift of the characteristic diamond peak from its stress-free position in Raman spectrum per the following equation [12,13]:

$$\sigma = \frac{v_0 - v}{\alpha} \tag{1}$$

where σ is the residual stress at the test point (positive value means tensile residual stress and negative value means compressive residual stress), v_0 is the wave number of characteristic diamond peak without any residual stress, v is the measured wave number of characteristic diamond peak at the testing point, and α is the stress deviation coefficient, which is taken as 1.92 cm^{-1}/GPa according to [26]. In this study, v_0 was taken as 1332.54 cm^{-1}, corresponding to the wave number of stress free diamond powder in Figure 1c.

2.2.4. Property Characterization of Diamond Powder and Microstructure Analysis of PDCs

In this study, the particle size of the feedstock diamond powder was measured by laser particle size analyzer (Malvern, Mastersizer 3000). For HPHT-processed PDCs, metallographic samples were made by first EDM cutting the PDCs into half and then mounting the sectioned sample with bakelite. The mounted samples were then ground with 50 micron diamond slurry. The ground samples were then, respectively, polished with the 20, 10, 6, 3, and 1 micron diamond pastes in sequence. Field emission scanning electron microscopy (Hitach, SU8220) was used to analyze the morphology of the diamond powder and the microstructure of PDCs. Raman spectra of both diamond powder and PDCs were made by HORIBA Jobin Yvon (LabRAM HR Evolution) confocal Raman spectrometer. The metallographic samples of PDCs were prepared by a single-sided lapping machine (SS-15H, Lemat Walters (Shenyang) Precision Machinery Co., Ltd., Shenyang, China) and polishing equipment (Buehler, MetaServ 250).

2.3. FEM Simulation of Residual Stress

In this study, the residual stress was investigated based on the real-time simulation of the temperature field evolution of PDCs and the shrinkage behavior of PCD layer and cemented carbide substrate during the cooling stage. It is generally believed that the uneven distribution of temperature and the difference of physical parameters such as thermal expansion coefficient and elastic modulus between PCD layer and carbide substrate result in the residual stress of PDCs. According to Fourier heat transfer law and energy conservation law, the thermal conduction equation of the object can be described as [27]

$$\frac{\partial}{\partial x}(k_x \frac{\partial T}{\partial x}) + \frac{\partial}{\partial y}(k_y \frac{\partial T}{\partial y}) + \frac{\partial}{\partial z}(k_z \frac{\partial T}{\partial z}) + \rho Q = \rho C_p \frac{\partial T}{\partial t} \quad (2)$$

where ρ is the density, C_p the specific heat, k is the thermal conductivity, T is temperature, t is time, x–z are 3 directions, and Q is the thermal intensity inside the object.

The thermal strain caused by differences in material properties can be expressed as [27]

$$\varepsilon_{th} = \alpha_T \cdot \Delta T(x, y, z) \quad (3)$$

where is ε_{th} thermal strain, α_T is coefficient of thermal expansion, and $\Delta T(x, y, z)$ is thermal gradient.

Based on the thermal strain information per Equation (3), the residual stress of PDCs was calculated by generalized Hooke's law.

As shown in Figure 3, an axisymmetrical PDC model was established in this study to simulate the residual stress evaluation during the cooling stage. The simulation details will be given in the experimental results and discussion section.

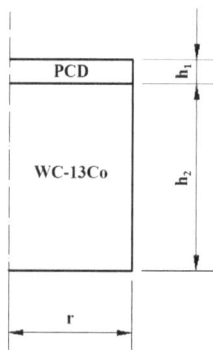

Figure 3. Schematic of the axisymmetric structure of PDCs for FEM simulation.

The physical properties of diamond, cobalt and carbide substrate used in the FEM simulation in this study are shown in Table 2.

Table 2. Physical properties of the constituents of PDCs for FEM simulation [10,12,28].

Property	Diamond	Cobalt	Carbide Substrate
Elastic modulus (GPa), E	900	200	579
Density (kg/m^3), ρ	3500	8500	15000
Poisson's ratio, υ	0.07	0.33	0.22
Thermal conductivity (W/(m·K)), k	2000	69.2	100
Specific Heat (J/(kg·K)), C_p	471	440	230
Coefficient of thermal expansion (10^{-6}·K^{-1}), α	3.2	14.4	5.2

Since the PCD layer itself is a diamond–cobalt composite, the physical properties of the PCD layer for FEM simulation were calculated by the equivalent principle formula [29]. The equivalent physical properties of PCD layers with different cobalt contents and the dimensions of PDCs (Figure 3) are shown in Table 3. It should be mentioned that for experimental testing, only four kinds of PDCs have been synthesized, i.e., PCD-13, 15, 20, and 30, while for FEM simulation, many more PDCs with different Co contents (0–50 wt.%) have been utilized in order to make a more systematic investigation of the effect of Co content of PCD layers.

Table 3. Physical properties of PCD layer and the dimension of PDCs (Figure 3) for FEM simulation.

Designation of PDCs (Number Corresponds to wt.% of Co in PCD Layer)	Dimension of PDCs in Figure 3			ρ (kg/m^3)	E (GPa)	υ	α (10^{-6}·K^{-1})	C (J/(kg·K))	k (W/(m·K))
	r (mm)	h_1 (mm)	h_2 (mm)						
PCD-0	8.25	1.645	12	3500.000	900.000	0.070	3.200	471.000	2000.000
PCD-5	8.25	1.597	12	3606.061	877.762	0.076	3.362	470.342	1921.703
PCD-10	8.25	1.548	12	3718.750	854.677	0.082	3.536	469.644	1841.776
PCD-15	8.25	1.500	12	3838.710	830.693	0.089	3.724	468.900	1760.166
PCD-20	8.25	1.452	12	3966.667	805.757	0.096	3.928	468.107	1676.821
PCD-25	8.25	1.403	12	4103.448	779.808	0.103	4.150	467.259	1591.684
PCD-30	8.25	1.355	12	4250.000	752.781	0.110	4.393	466.350	1504.697
PCD-35	8.25	1.306	12	4407.407	724.606	0.119	4.659	465.374	1415.799
PCD-40	8.25	1.258	12	4576.923	695.204	0.127	4.952	464.323	1324.926
PCD-45	8.25	1.210	12	4760.000	664.488	0.137	5.276	463.188	1232.012
PCD-50	8.25	1.161	12	4958.333	632.363	0.147	5.637	461.958	1136.986

3. Experimental Results and Discussion

3.1. Microstructure of the PDCs

The microstructure of the PCD layer of the PDCs is shown in Figure 4, in which the gray region is a diamond phase, and the bright region and the concave pore-like region are actually a Co phase, which was almost totally removed during the polishing process and looks like pores. The pore-like Co phase was quite commonly observed in polished PDC samples [30] due to the extremely significant hardness difference between the diamond and Co phases. It can be observed that after the HPHT process, the original diamond particles have formed an interconnected diamond skeleton structure in all four PDCs, and the Co phase is distributed in between the diamond skeleton structure.

The formation of a diamond skeleton in Figure 4 is mainly attributed to the diamond dissolution and re-precipitation process during the HPHT sintering process [1,5,6]. During the HPHT process, when the sintering temperature is over the melting point of Co, liquid Co will flow between diamond particles. Due to the thermal stability difference between finer and coarser diamond particles or between the sharp corner and flat surface (or concave surface, especially the diamond-diamond contact region) of diamond particles, the finer diamond particles and the sharp corner of diamond particles will dissolve into liquid Co first and then re-precipitate onto the coarser diamond particles or a flat or concave surface of diamond particles, resulting in the diamond grain growth as well as the flattening and connecting of diamond particles. Eventually, the interconnection of diamond particles (D-D bond) will happen between diamond particles to form the diamond skeleton in Figure 6 via Ostwald ripening [31]. Such a dissolution and re-precipitation process of diamond particles will happen only in liquid Co, Ni, or Fe due to their significant catalytic effect for carbon–diamond transformation processes.

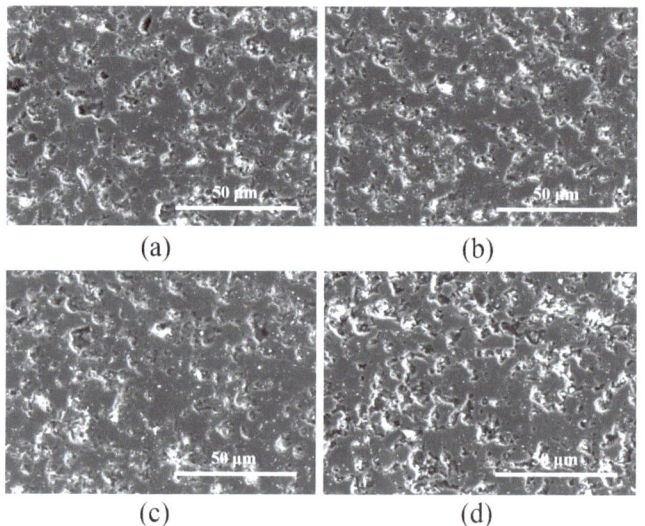

Figure 4. Microstructure of PCD layers for PDCs, (**a**) PCD-13, (**b**) PCD-15, (**c**) PCD-20, and (**d**) PCD-30.

The microstructure of the PCD–carbide substrate interface is shown in Figure 5. It can be observed that for PCD-13 and -15, there are clear interfaces between the PCD layer and carbide substrate, while for PCD-20 and -30, significant Co-enrichment regions exist at the PCD–carbide substrate interface. The formation of Co enrichment regions can also be explained by the dissolution and re-precipitation process of diamond particles during the HPHT process.

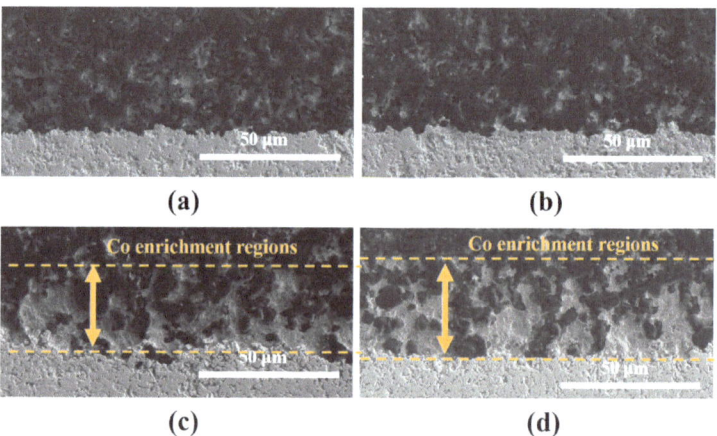

Figure 5. Microstructure of PCD-carbide interface of PDCs, (**a**) PCD-13, (**b**) PCD-15, (**c**) PCD-20 and (**d**) PCD-30.

As shown in Figure 6, during the HPHT process, liquid Co will penetrate into the PCD layer from the carbide substrate under high pressure. Both the liquid Co penetrating into the PCD layer and the Co in the initial PCD layer will interact with the diamond powder to complete the above mentioned dissolution and re-precipitation process of diamond particles, resulting in a D-D bond and the growth of diamond particles into a skeleton structure [1,4,32]. With the growth of diamond particles becoming more and more significant and the gaps between the diamond skeleton structure becoming smaller, the extra Co between diamond particles will be expelled back to the carbide substrate, resulting in the reverse migration of Co. As compared with PCD-13 and 15, more Co in PCD-20 and 30 were involved in the dissolution-re-precipitation process of diamond particles and more significant reverse migration of Co will happen. Due to the fact that the carbide substrate cannot absorb the extra Co involved in reverse migration in a short sintering time, the Co-enrichment regions will form at the PCD–carbide substrate interface as shown in Figure 5. Compared with PCD-13 and 15, PCD-20 and 30 have noticeable Co-enrichment regions at the PCD–carbide substrate interface. The Co-enrichment region may enhance the bonding between the PCD layer and the carbide substrate. In addition, due to the significant CTE difference between diamond and Co, a Co-enrichment region can possibly alter the residual stress of a PCD layer, while up until now, related research has rarely been reported.

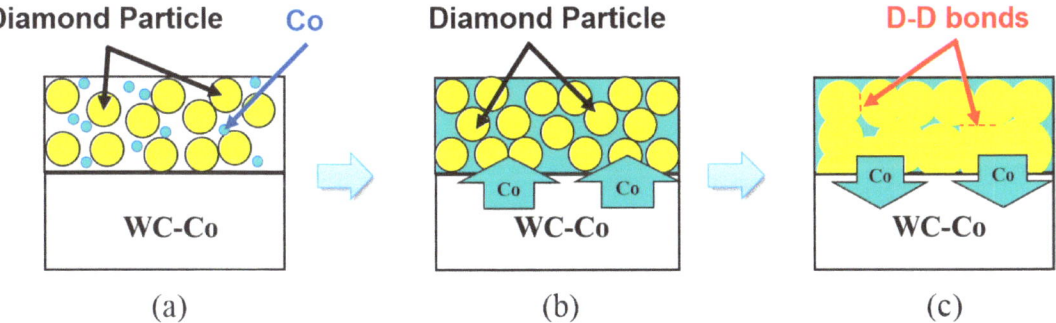

Figure 6. Schematic of PDC sintering process, (**a**) initial stage, (**b**) infiltration of Co from carbide substrate, and (**c**) reverse migration of Co.

3.2. Raman Spectrum Analysis

As shown in Figure 7, Raman spectroscopy was used to evaluate the characteristic diamond peaks at the center of the upper surface of the PCD layer of all the PDCs. The diamond peak around 1332.54 cm^{-1} has been observed and no graphite peaks appear for all the PDCs. The residual stress results calculated per Equation 1, based on the accurate diamond peak position in Figure 7, are shown in Figure 8. The residual stress at the center of the upper surface of the PCD layer of PCD-13 is compressive stress, and with the increase in Co content in the PCD layer, the compressive stress gradually decreases and turns into tensile stress when the Co content in the PCD layer is as high as 20 wt.% (PCD-20). A further increase in Co content results in significant compressive stress (PCD-30). Comprehensive FEM simulation has to be made in order to investigate the main reason for the abrupt transition of residual stress from tensile (PCD-20) to significant compressive (PCD-30).

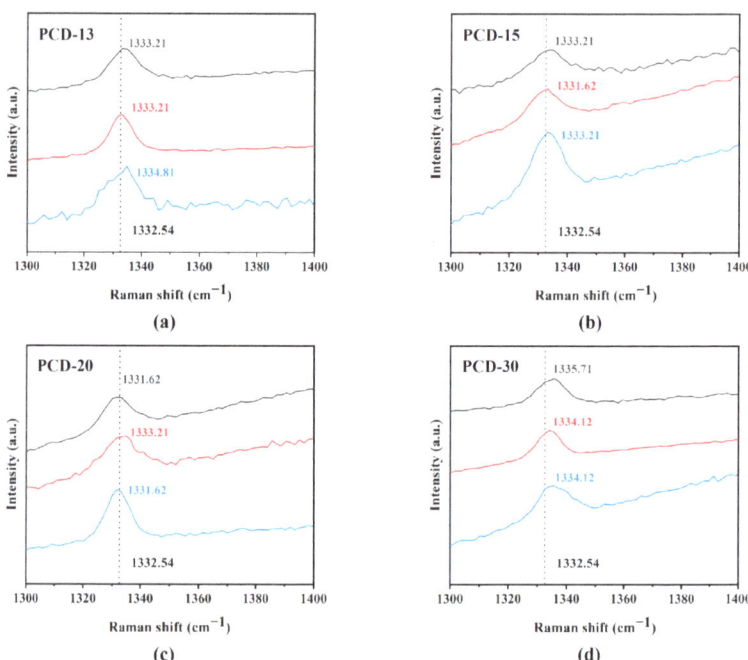

Figure 7. Raman spectra of the center of upper surface of PCD layer for PDCs, (**a**) PCD-13, (**b**) PCD-15, (**c**) PCD-20, and (**d**) PCD-30.

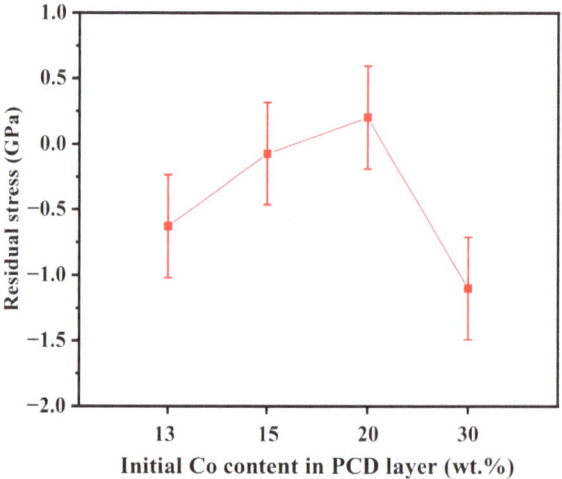

Figure 8. Residual stress of diamond phase at the center of upper surface of PCD layer for PDCs as the function of Co content of PCD layer.

3.3. FEM Simulation of Residual Stress for PDCs

It is generally believed that the residual stress of PDCs is mainly caused by the difference of the thermal expansion coefficient and Young's modulus between the PCD layer and carbide substrate [10,12]. Based on the two-dimensional planar structure model commonly used in residual stress simulation, this study carried out FEM simulation of the residual stress. According to the actual synthesis process, the initial temperature for

FEM was set to 1580 °C, the ambient temperature was 25 °C, and the length of the analysis step was set to 480. The transient temperature-displacement coupling method was used for simulation calculation. The factors of the axisymmetric interface wedge condition were set to XSYMM, structured grid partitioning was used for meshing process, and the mesh cell type was set to CAX4T, which shows good mesh quality and high mesh sensitivity. In addition, the surface heat transfer coefficient of the upper and lower surfaces and sides of the PDCs was set to 0.042.

Figure 9 shows the mesh elements and residual stress (mises, S11 and S22) nephograms for PCD-15. Figure 10 shows the residual stress on the top surface of the PCD layer as the function of radial distance from the top center of the PCD layer (Figure 10a,b) as well as the residual stress at the center of the top surface of the PCD layer as the function of the Co content of the PCD layer (Figure 10c,d) for all the PDCs (PDC-0-50). The FEM results in this study shows the similar trends to the research results of CHEN et al. [10]. Figure 9 shows that the maximum stress is concentrated at the PCD–carbide substrate interface, and the stress is smaller at the positions away from the interface. Figure 10a,b shows that the horizontal radial residual stress (S11) is much more significant compared with the vertical axial stress in the PCD layer thickness direction (S22). Figure 10c,d shows that with the increase in Co content, both the radial stress S11 and axial stress S22 monotonically increases and changes from compressive stress to tensile stress, which is inconsistent with the experimental residual stress profile shown in Figure 8, where the residual stress transforms back into compressive stress from tensile stress when Co content is more than 20%.

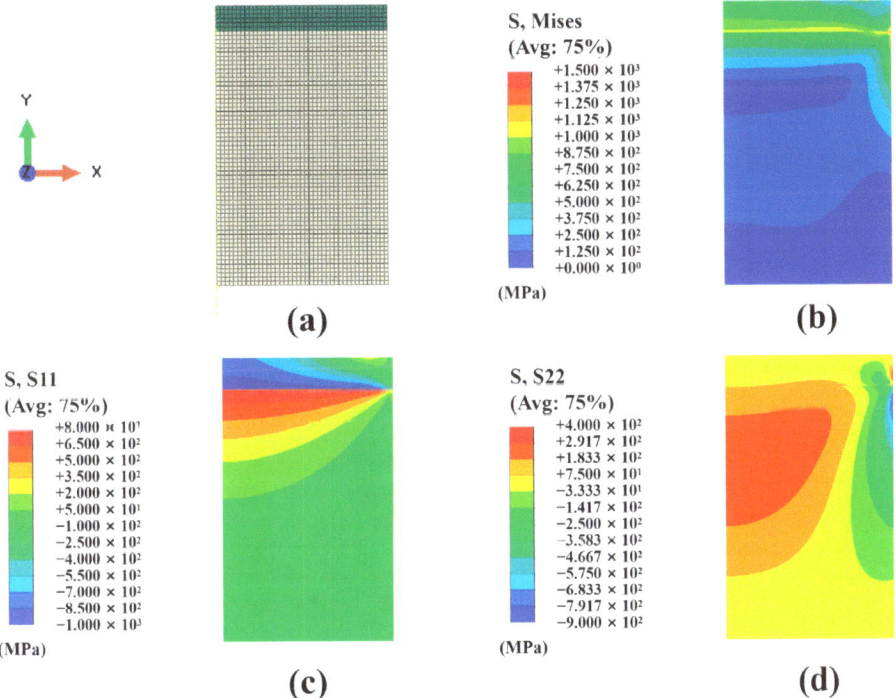

Figure 9. FEM for PCD-15, (a) mesh elements, and distribution of residual stresses including (b) Mises, (c) S11, and (d) S22.

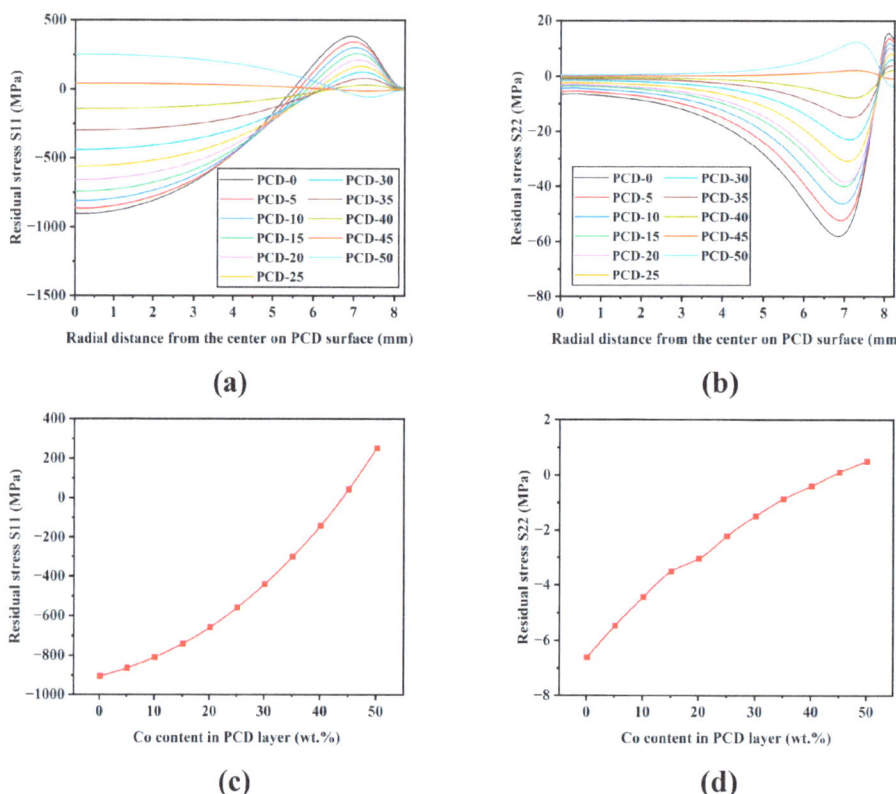

Figure 10. FEM residual stress (**a**) S11 and (**b**) S22 of the upper surface of PCD layers as the function of radial distance from the center of PCD surface, and FEM residual stress (**c**) S11 and (**d**) S22 of the center of upper surface of PCD layer as the function of Co content of PCD layer.

From both experimental residual stress measurement and FEM simulation results, it can be seen that when the cobalt content in the PCD layer is low (0–15%), the residual compressive stress will form in the PCD layer after the HPHT process mainly due to the more significant shrinkage of the carbide substrate as compared with that of the PCD layer. With a further increase in Co content, the thermal expansion coefficient of the PCD layer continues to increase, and the Young's modulus continues to decrease, so the thermal and elastic property difference between the PCD layer and carbide substrate decreases, leading to the decrease in residual compressive stress of the PCD layer and the transformation of compressive stress into tensile stress. However, when the Co content is greater than 20 wt.%, Figure 8 shows that the experimental residual stress of the PCD layer transforms abruptly from tensile stress back into compressive stress again, while the FEM simulation in Figure 10 shows that the residual compressive stress monotonically decreases with the increase in Co content of the PCD layer and gradually develops into tensile stress when the Co content of the PCD layer is high enough, but there is no transformation of tensile stress back into compressive stress. From the inconsistency trends between the experimental residual stress and the FEM simulation results, it can be inferred that the difference of physical properties between the PCD layer and carbide substrate is not the only factor affecting the residual stress evolution behavior of PDCs. In addition, it should be noticed that the traditional FEM method (Figures 9 and 10) takes the whole PCD layer (diamond + Co) as a unit and cannot reveal the residual stress of the diamond phase, which can be another reason for the inconsistency trends between the experimental residual stress and the FEM simulation results. In order to investigate the actual residual stress of the diamond phase, a micromechanical simulation based on the microstructure of PDCs was employed in this study.

3.4. Micromechanical Simulation Based on Microstructure of PCD–Carbide Substrate Interface

As shown in Figure 5, when the Co content is more than 20 wt.%, there is a noticeable reverse migration of Co from the PCD layer to carbide substrate, resulting in a Co enrichment region at the PCD–carbide interface. The Co enrichment region may contribute to the transformation of residual stress of the diamond phase from tensile into compressive stress for PDC-30. In order to reveal the micromechanism of residual stress evolution, micromechanical simulation was carried out based on the actual microstructure of the PCD–carbide substrate interface.

As shown in Figure 11, the specific micromechanical simulation process was based on the PCD–carbide substrate interface microstructure of PDCs. The microstructure was first binarized by MATLAB software, and then vectorized by RasterVect software. Finally, the vectorized file was imported into ABAQUS finite element simulation software to establish a simulation model.

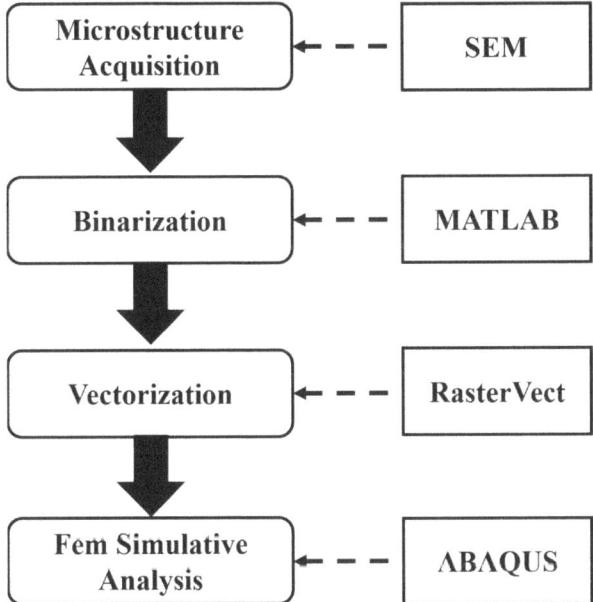

Figure 11. Micromechanical simulation procedure based on microstructure of PCD–carbide interface.

Figure 12 is the SEM micrograph of the PCD–carbide substrate interface of PCD-15, PCD-20 and PCD-30. According to the procedure in Figure 11, the SEM micrographs of PCD-15, PCD-20 and PCD-30 were binarized and vectorized, and the finite element simulation models were shown in Figure 13, where the unique role of the Co-enrichment region was emphasized. The static general method was used for simulation calculation. Free meshing was adopted to address the complicated microstructure feature, and the mesh cell type was set to CPS4R. Mesh analysis shows that both the error rate and warning rate of all the meshes are 0%.

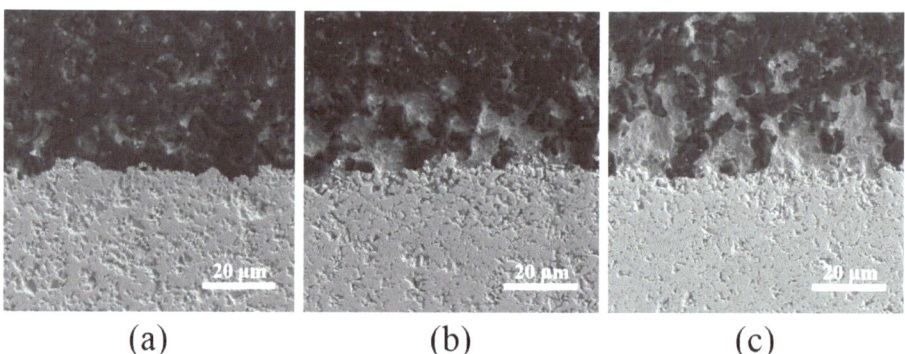

Figure 12. Microstructure of PCD–carbide substrate interface, (**a**) PCD-15, (**b**) PCD-20, and (**c**) PCD-30.

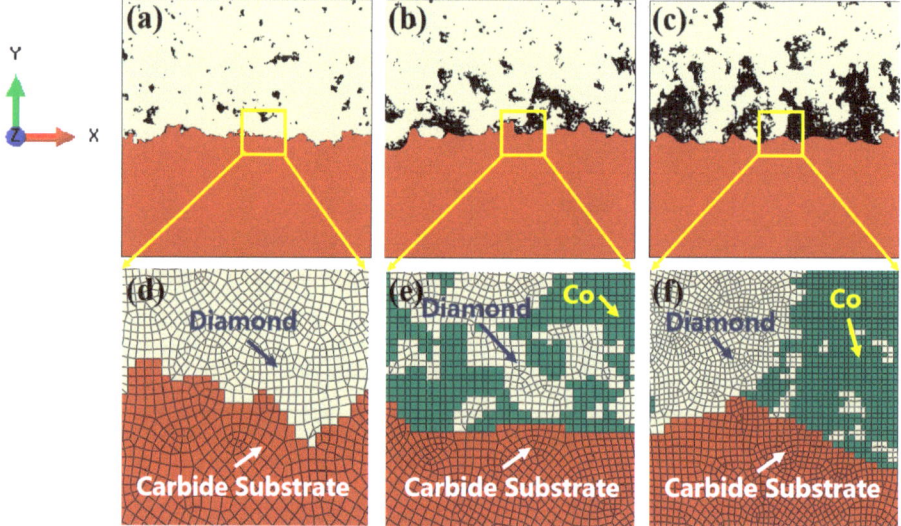

Figure 13. Simulation models and meshing of PDCs based on the actual microstructure in Figure 12, (**a**) PCD-15, (**b**) PCD-20, and (**c**) PCD-30, and meshing of (**d**) PCD-15, (**e**) PCD-20, and (**f**) PCD-30.

Figure 14 is the residual stress nephogram in X (S11) and Y-axis (S22) based on the actual microstructure in Figure 12. The positive means tensile stress, while the negative means compressive stress. It can be observed that the Co-enrichment region is subjected to clearly tensile stress. Compared with PCD-15, there is more evident compressive stress of the diamond phase in PCD-20 and 30.

Figure 15 shows the statistically average stress of the diamond phase based on the residual stress results in Figure 14. It can be observed that the average stress of the diamond phase is compressive and the compressive stress increases with the Co content in the PCD layer, which is completely opposite to the trend in Figure 10c,d. It can be inferred from Figure 15 that when the actual microstructure of the PCD–cemented carbide interface, especially the Co enrichment region, is included in FEM simulation, the evolution trend of residual stress is significantly different from the traditional FEM (without considering the effect of the Co enrichment region) in Figures 9 and 10. The results show that the Co-enrichment region at the PCD–cemented carbide interface introduces significant compressive residual stress to the diamond phase.

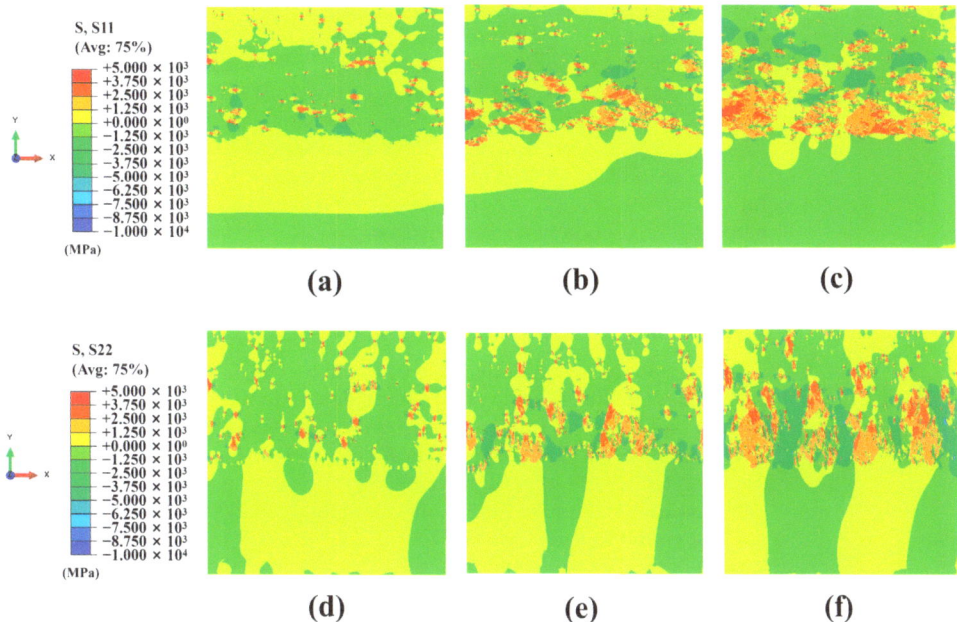

Figure 14. S11 (**a**–**c**) and S22 (**d**–**f**) distributions at PCD–carbide interface based on the actual microstructure of PDCs, (**a**,**d**) PCD-15, (**b**,**e**) PCD-20, and (**c**,**f**) PCD-30.

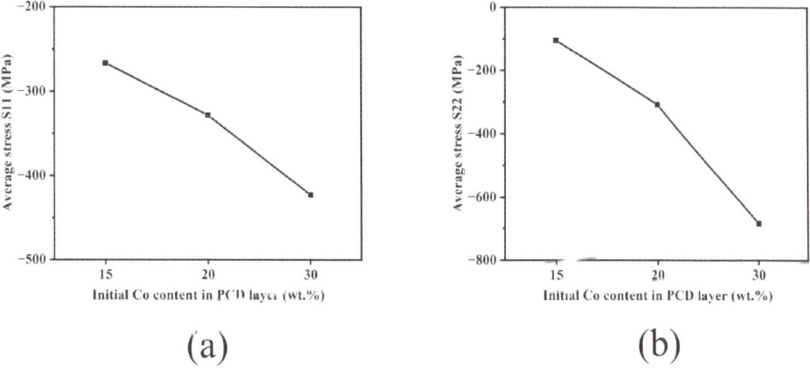

Figure 15. The average (**a**) S11 and (**b**) S22 of diamond phase in Figure 14 as the function of Co content of PCD layer.

Per a comprehensive comparison of the traditional axisymmetric simulation model (Figure 9) and the micromechanical simulation model based on interfacial microstructure (Figure 14), it can be found that the residual stress of the diamond phase in PDCs is the combination result of both the macro-scale stress caused by physical property differences between the PCD layer and the carbide substrate and the micro-scale stress caused by physical property differences between the diamond phase and the Co-enrichment region. When the Co content of the PCD layer is lower than 20 wt.%, the residual stress of diamond is mainly affected by the physical property difference between the PCD layer and carbide substrate, demonstrating clearly a compression state. When Co content of the PCD layer is increased to 20 wt.%, the thermal expansion coefficient of the PCD layer gradually increases from a level significantly below the carbide substrate to a level slightly higher than the carbide substrate, and the compressive residual stress of diamond becomes slightly tensile stress. When the Co content of the PCD layer is more than 20 wt.%, the Co enrichment region at the PCD–carbide substrate interface exerts a significant compressive stress on the diamond phase (Figure 15), resulting in the transformation of tensile stress back into compressive stress of the diamond phase shown in Figure 8.

In spite of the fact that micro-mechanical simulation is only based on a very small region due to the limitation of the calculation resource, Figures 14 and 15 demonstrate the critical role of the Co-enrichment region on the residual stress evolution of the diamond phase, especially for PDCs with high Co content of PCD layers. Moreover, both Figures 8 and 15 show that the Co content of the PCD layer is critical to the residual stress of the diamond phase, which has rarely been reported before.

4. Conclusions

Based on the above study, it can be found that the Co content of PCD layers, as well as the Co-enrichment region at PCD–carbide interfaces are critical to the residual stress evolution of diamond phases following the HPHT process. This study shows that, by optimizing the PCD layer material design and tailoring the microstructure of the PCD–carbide interface, it can be highly possible to precisely manipulate the residual stress of PDCs in order to improve the lifetime and stability of PDCs in various applications. The conclusions in this study include:

(1) For HPHT-processed PDCs, the Co content of the PCD layer is found to have a significant effect on the microstructure evolution of PDCs. When the Co content of the PCD layer is lower than 20 wt.%, the microstructure of both the PCD layer and carbide substrate are homogeneous without any noticeable phase segregation. When the Co content of the PCD layer is more than 20 wt.%, there is noticeable reverse migration of Co from the PCD layer to the carbide substrate, resulting in the prevalent Co-enrichment regions at the PCD–carbide substrate interface.

(2) The Co content of the PCD layer is critical to the residual stress evolution of the diamond phase in PDCs. When the Co content of the PCD layer is lower than 20 wt.%, the residual stress of the diamond phase at the center of the upper surface of the PCD layer is compressive. With the increase in Co content, the residual compressive stress gradually decreases and even develops into tensile stress. When the Co content is more than 20 wt.%, the tensile residual stress transforms back into a significant compressive stress.

(3) The wide spread Co-enrichment region at the PCD–carbide substrate of PDC with 30 wt.% Co of the PCD layer is the main reason for the transformation of residual stress of the diamond phase from tensile back into compressive.

Author Contributions: Conceptualization, P.N. and X.D.; data curation, W.Y.; formal analysis, Z.H.; funding acquisition, X.D.; investigation, P.N., Y.C. and W.Y.; methodology, P.N. and Y.C.; project administration, X.D.; resources, X.D.; supervision, X.D.; writing—original draft, P.N.; writing—review and editing, Y.C. and X.D. All authors have read and agreed to the published version of the manuscript.

Funding: This research was funded by Foshan Science and Technology Innovation Team project (FS0AA-KJ919-4402-0023), and Jihua Laboratory Project (Grant No. X190061UZ190).

Data Availability Statement: The data presented in this study are available on request from the corresponding author.

Acknowledgments: The authors appreciate the financial support from Foshan Science and Technology Innovation Team project (Grant No.: FS0AA-KJ919-4402-0023), and Jihua Laboratory Project (Grant No.: X190061UZ190).

Conflicts of Interest: The authors declare no conflict of interest.

References

1. Wentorf, R.H.; Devries, R.C.; Bundy, F.P. Sintered superhard materials. *Science* **1980**, *208*, 873. [CrossRef] [PubMed]
2. Li, G.; Rahim, M.Z.; Pan, W.; Wen, C.; Ding, S. The manufacturing and the application of polycrystalline diamond tools—A comprehensive review. *J. Manuf. Process.* **2020**, *56*, 400. [CrossRef]
3. Miyazaki, K.; Ohno, T.; Karasawa, H.; Takakura, S.; Eko, A. Performance evaluation of polycrystalline diamond compact percussion bits through laboratory drilling tests. *Int. J. Rock Mech. Min. Sci.* **2016**, *87*, 1. [CrossRef]
4. Bochechka, O.O. Production of Polycrystalline Materials by Sintering of Nanodispersed Diamond Nanopowders at High Pressure. Review. *J. Superhard Mater.* **2018**, *40*, 325. [CrossRef]
5. Miess, D.; Rai, G. Fracture toughness and thermal resistance of polycrystalline diamond compacts. *Mater. Sci. Eng. A* **1996**, *209*, 270. [CrossRef]
6. Clark, I.; Bex, P. The use of PCD for petroleum and mining drilling. *Ind. Diam. Rev.* **1999**, *59*, 43.
7. Abbas, R.K.; Musa, K.M. Using Raman shift and FT-IR spectra as quality indices of oil bit PDC cutters. *Petroleum* **2019**, *5*, 329. [CrossRef]
8. Bellin, F.; Dourfaye, A.; King, W.; Thigpen, M. The current state of PDC bit technology. *World Oil* **2010**, *231*, 53.
9. Kim, D.; Beal, A.; Kwon, P. Effect of Tool Wear on Hole Quality in Drilling of Carbon Fiber Reinforced Plastic-Titanium Alloy Stacks Using Tungsten Carbide and Polycrystalline Diamond Tools. *J. Manuf. Sci. Eng.-Trans. Asme* **2016**, *138*, 031006. [CrossRef]
10. Chen, F.; Xu, G.; Ma, C.; Xu, G. Thermal residual stress of polycrystalline diamond compacts. *Trans. Nonferrous Met. Soc. China* **2010**, *20*, 227. [CrossRef]
11. Paggett, J.; Drake, E.; Krawitz, A.; Winholtz, R.; Griffin, N. Residual stress and stress gradients in polycrystalline diamond compacts. *Int. J. Refract. Met. Hard Mater.* **2002**, *20*, 187. [CrossRef]
12. McNamara, D.; Alveen, P.; Damm, S.; Carolan, D.; Rice, J.H.; Murphy, N.; Ivanković, A. A Raman spectroscopy investigation into the influence of thermal treatments on the residual stress of polycrystalline diamond. *Int. J. Refract. Met. Hard Mater.* **2015**, *52*, 114. [CrossRef]
13. Erasmus, R.M.; Comins, J.D.; Mofokeng, V.; Martin, Z. Application of Raman spectroscopy to determine stress in polycrystalline diamond tools as a function of tool geometry and temperature. *Diam. Relat. Mater.* **2011**, *20*, 907. [CrossRef]
14. Yue, T.; Yue, W.; Li, J.; Wang, C. Effect of vacuum annealing temperature on tribological behaviors of sintered polycrystalline diamond compact. *Int. J. Refract. Met. Hard Mater.* **2017**, *64*, 66. [CrossRef]
15. Ni, P.; Zhao, Z.; Yang, W.; Deng, X.; Wu, S.; Qu, Z.; Jin, F. Effect of feedstock diamond powder property on microstructure and mechanical properties of polycrystalline diamond compacts. *Int. J. Refract. Met. Hard Mater.* **2023**, *111*, 106102. [CrossRef]
16. Gu, J.; Huang, K.J.D.; Materials, R. Role of cobalt of polycrystalline diamond compact (PDC) in drilling process. *Diam. Relat. Mater.* **2016**, *66*, 98. [CrossRef]
17. Kong, C.; Liang, Z.; Zhang, D.J.M. Failure analysis and optimum structure design of PDC cutter. *Mechanics* **2017**, *23*, 567. [CrossRef]
18. Lin, T.; Hood, M.; Cooper, G.A.; Smith, R.H. Residual stresses in polycrystalline diamond compacts. *J. Am. Ceram. Soc.* **1994**, *77*, 1562. [CrossRef]
19. Vhareta, M.; Erasmus, R.M.; Comins, J.D. Investigation of fatigue-type processes in polycrystalline diamond tools using Raman spectroscopy. *Diam. Relat. Mater.* **2014**, *45*, 34. [CrossRef]
20. Yadav, V.; Jain, V.K.; Dixit, P.M. Thermal stresses due to electrical discharge machining. *Int. J. Mach. Tools Manuf.* **2002**, *42*, 877. [CrossRef]
21. Catledge, S.A.; Vohra, Y.K.; Ladi, R.; Rai, G. Micro-Raman stress investigations and X-ray diffraction analysis of polycrystalline diamond (PCD) tools. *Diam. Relat. Mater.* **1996**, *5*, 1159. [CrossRef]
22. Rats, D.; Bimbault, L.; Vandenbulcke, L.; Herbin, R.; Badawi, K.F. Crystalline quality and residual stresses in diamond layers by Raman and x-ray diffraction analyses. *J. Appl. Phys.* **1995**, *78*, 4994. [CrossRef]
23. Krawitz, A.D.; Winholtz, R.A.; Drake, E.F.; Griffin, N. Residual stresses in polycrystalline diamond compacts. *Int. J. Refract. Met. Hard Mater.* **1999**, *17*, 117. [CrossRef]
24. Mukhopadhyay, D. Identifying the causes of residual stress in polycrystalline diamond compact (PDC) cutters by X-ray diffraction technique. *Results Mater.* **2021**, *11*, 100216. [CrossRef]
25. Chen, Z.; Ma, D.; Wang, S.; Dai, W.; Li, S.; Zhu, Y.; Liu, B. Effects of graphene addition on mechanical properties of polycrystalline diamond compact. *Ceram. Int.* **2020**, *46*, 11255. [CrossRef]
26. Prawer, S.; Nemanich, R.J. Raman spectroscopy of diamond and doped diamond. *Philos. Trans. R. Soc. Lond. Ser. A-Math. Phys. Eng. Sci.* **2004**, *362*, 2537. [CrossRef]
27. Zeng, P. *Fundamentals of Finite Element Analysis*; Higher Education Press: Beijing, China, 2009; p. 290.
28. Kanyanta, V.; Ozbayraktar, S.; Maweja, K. Effect of manufacturing parameters on polycrystalline diamond compact cutting tool stress-state. *Int. J. Refract. Met. Hard Mater.* **2014**, *45*, 147. [CrossRef]
29. Zhu, J.; Zhou, H.; Qin, B.; Zhao, Z. Design, fabrication and properties of TiB_2/TiN/WC gradient ceramic tool materials. *Ceram. Int.* **2020**, *46*, 6497. [CrossRef]
30. Shin, T.; Oh, J.; Oh, K.H.; Lee, D.N. The mechanism of abnormal grain growth in polycrystalline diamond during high pressure-high temperature sintering. *Diam. Relat. Mater.* **2004**, *13*, 488. [CrossRef]

31. Vengrenovitch, R.D. On the ostwald ripening theory. *Acta Metall.* **1982**, *30*, 1079. [CrossRef]
32. Kanyanta, V. *Microstructure-Property Correlations for Hard, Superhard, and Ultrahard Materials*; Springer: London, UK, 2016; p. 35.

Disclaimer/Publisher's Note: The statements, opinions and data contained in all publications are solely those of the individual author(s) and contributor(s) and not of MDPI and/or the editor(s). MDPI and/or the editor(s) disclaim responsibility for any injury to people or property resulting from any ideas, methods, instructions or products referred to in the content.

Article

The Influence of Inserted Metal Ions on Acid Strength of OH Groups in Faujasite

Glorija Medak, Andreas Puškarić and Josip Bronić *

Division of Materials Chemistry, Ruđer Bošković Institute, Bijenička 54, 10000 Zagreb, Croatia
* Correspondence: josip.bronic@irb.hr; Tel.: +385-1456-0991

Abstract: The number and the strength of acid sites in catalysts have paramount importance on their efficiency. In zeolites chemistry, increased content of framework Al in zeolites gives a higher number of strong acid sites. Their strength can be a disadvantage in catalytic reactions (e.g., methanol to olefins conversion) due to undesired secondary reactions of coke formation. Here, the Faujasite type of zeolite with higher content of Al has been used for investigating the role of defects in structure and inserted (wet impregnation and thermal treatment) metal cations (Mg, Co, Ni, Zn) on the strength of OH acid sites. Desorption of deuterated acetonitrile, as a probe molecule, was used for OH groups acid strength measurements at different temperatures (150, 200, and 300 °C).

Keywords: porous materials; zeolite X; wet impregnation; Brønsted and Lewis acid site; solvothermal synthesis

1. Introduction

Modified zeolite Y (ZY) is the most important catalyst used in chemistry for fuel production via hydrocracking of crude oil (Fluid Catalytic Cracking—FCC). On the other hand, "green" synthesis of hydrocarbons from CO_2 using solid-state acid catalysts, can be made in several steps, including conversion to CH_4 or CH_3OH, then to higher alkanes.

ZY is a high-silica zeolite of the Faujasite (FAU) topology. If the content of Al in the framework increases, it becomes ZX, but there is no widely accepted exact value for the Si/Al ratio when zeolite Y becomes ZX (the Si/Al ratio range for association to ZX name varies from 1 to 5, due to different authors).

A large number of articles are dedicated to modifications of ZY with rare-earth cations, and to understanding its structural characteristics and their correlation to catalytic activity [1]. An increase in efficiency, product selectivity, and longevity are just some of the properties that we are constantly trying to improve in existing commercially available catalysts. One of the most popular methods for catalyst modification is increasing the active surface of the zeolite either by reducing crystal size to nanoscale [2] or creating mesopores in larger crystals through the usage of mesoporosity templates such as cetrimonium bromide (CTAB) [3,4] and/or etching solutions [5]. Many of these materials have been further functionalized by the introduction of cations inside the zeolite framework, but this type of treatment can also lead to the formation of metal oxide/hydroxide nanoparticles on the zeolite's internal or external surface that are also catalytically active [6–8].

Special attention is dedicated to the understanding of the location, number, type, and accessibility of acid sites in zeolite structures [9], which are needed to fine-tune the catalytic properties of zeolites. The acid strength of OH groups can be measured using various probe molecules such as carbon monoxide [10], pyridine, or deuterated acetonitrile at low-pressure conditions. Even though many papers have reported a detailed analysis of Brønsted (BAS) and Lewis acid sites (LAS) after various post-synthesis treatments and modifications [11,12], only a few of them mentioned the effect of position and the cation species in the framework on the ratio between acid sites (AS).

Recent papers in the research of zeolites put a lot of emphasis on understanding the acidity [13] of metal cations in the zeolite framework and how they affect the mechanisms of chemical reactions. The amount of metal cations in the framework directly influences the occupancies of specific binding sites [14,15]. Changes in the coordination number and connectivity of donor atoms directly influence the binding affinity of probe molecules to the acid sites in the vicinity of the metal center. To further explore this concept and find out how much it affects specific metal cations, we have chosen three low-silica FAU zeolites of different crystal morphologies (size as well), with different Si/Al ratios (1.5, 2.0, and 2.3) and inserted (insert = wet impregnation + consecutive thermal treatment) several cations into the structure under the same conditions, while keeping the molar ratio of loaded metal to aluminum (M/Al) around 0.3 (about 60% of acid sites of ZX). To see the difference in acid strength of OH groups near the positions of exchangeable cations, relatively high loading of Mg^{2+}, Co^{2+}, Ni^{2+}, and Zn^{2+} in ZX were investigated by desorption of D_3-acetonitrile. These cations were chosen due to their similar ionic radii (from 69 to 74.5 pm) [16] and the stability of their II^+ oxidation state with a low expectation of forming large amounts of metal oxide, unlike Fe^{2+} or Cu^{2+}. Additionally, all of the chosen metals have shown potential for various applications such as gas storage and catalytic reactions. It is of paramount importance to better understand chemical and structural factors that affect catalyst acidity and activity in reactions such as CO_2 methanation, isopropilation of phenolic compounds from biomass, and hydro-dechlorination of chlorinated compounds [17–20].

2. Materials and Methods

2.1. Sample Preparation and Cation Introduction

Here, zeolites were prepared using three different procedures. Nano-crystallites labeled ZX-n (Si/Al = 1.44) were prepared from the gel of oxide composition 8.0 Na_2O: 0.7 Al_2O_3: 10.0 SiO_2: 160.0 H_2O [21] while the preparation of crystals with higher Si/Al ratio, the composition of 4 Na_2O:Al_2O_3:10 SiO_2:158 H_2O was used [22]. The preparation of the sample with Si/Al ratio of 2.3 was made using the procedure described by Bosnar et al. The post-synthesis desilication was performed using 0.2 mol dm^{-3} basic solution (mixture of NaOH and TPAOH) [12]. The treatment was carried out at 60 °C for 60 min, in a slurry with a ratio of zeolite to etching mixture 1:33 and the sample was labeled ZX-6060. To get the crystals with Si/Al ratio of 2.0, after 24 h of synthesis, the gelatinous solution of CTAB in alkaline water was added to the reaction gel and stirred vigorously until homogenized. The ratio of CTAB to solvent was 1:1, while the ratio of Si:CTAB was 1:0.4. The rest of the synthesis was carried out as already written. This sample was labeled ZX-Ct. All types of synthesized samples were washed, dried at 60 °C, and calcined at 550 °C for 8 h.

All prepared samples were impregnated using nitrate salts of magnesium(II), cobalt(II), nickel(II), and zinc(II) dissolved in redistilled water. The ratio of zeolite to a salt solution was 1:10. Concentration of 1 mol dm^{-3} was chosen for magnesium and zinc nitrate, while for the cobalt(II) and nickel nitrate, it was diluted to 0.5 mol dm^{-3} due to the solution's lower pH. The exchange procedure was repeated twice, after which all samples were thoroughly washed with redistilled water, dried at 60 °C, and calcined at 550 °C for 6 h.

2.2. Methods of Sample Characterization (PXRD, SEM, FAAS, FTIR, UVVis-DRS)

Powder X-ray diffraction was used for the phase analysis of the synthesized samples after each modification step. Diffraction patterns were collected using copper K_α radiation on an Empyrean (Malvern Panalytical, Malvern, UK) diffractometer with the Bragg–Brentano optics at 2θ angles from 5° to 50°.

High-resolution field emission scanning electron microscope (SEM) images of the samples were made using a JSM-7000F (JEOL) microscope.

The elemental composition of the samples was measured using Flame Atomic Absorption Spectroscopy (FAAS) on Aanalyst 200 (Perkin Elmer, Waltham, MA, USA). All solutions for the analysis were prepared in the accordance with prescribed procedures.

Qualitative and quantitative analysis of acid sites was determined using Fourier-transformed infrared spectroscopy (FTIR) on a Frontier (Perkin-Elmer) instrument, in the transmission mode under the pressure of 5×10^{-5} mBar and resolution of 4 cm^{-1}. Self-supported pellets (d = 13 mm) for the analysis were prepared from around 10 mg of sample and activated at 400 °C for 3 h. To determine acid site strength, desorption of deuterated acetonitrile was measured at 25, 150, 200, and 300 °C.

The concentration of the specific acid site (Brønsted and Lewis) was calculated from the integral intensity of the corresponding bands in the IR spectra after adsorption of CD$_3$CN, using the formula

$$C\left(\mu mol\, g_{cat}^{-1}\right) = \frac{IA(X)}{\varepsilon(X)*\sigma} \qquad (1)$$

where $IA(X)$ is the integrated absorbance of the peak of the acid species X (Lewis or Brønsted), σ is the "density" of the wafer normalized to the value of 10 mg cm^{-2} (actually it is wafer thickness after making at predefined pressure of 2 T/cm^2), while ε is the molar extinction coefficient (2.05 ± 0.1 cm/µmol and 3.6 ± 0.2 cm/µmol, for Brønsted and Lewis acid sites, respectively) as described by Wichterlova et al. [23].

UV–Vis diffuse reflectance spectroscopy (DRS) measurements were made using the instrument model UV 3600 (Shimadzu, Kyoto, Japan), equipped with Integrating Sphere, using BaSO$_4$ as standard.

3. Results and Discussion

Comparing the XRD patterns of all three samples (ZX-n, ZX-Ct, and ZX-6060), one can observe lower intensity but wider peaks of nanocrystals, and there is no significant difference in the crystallinity of the ZX-6060 and ZX-Ct samples (Figure 1A). According to SEM photos, the size of nanocrystals is 60–100 nm and can be better seen at higher magnification (33,000×, insert in Figure 1B). The twin ZX crystals, made with the addition of CTAB during the synthesis, consist of several smaller fragments (size around 0.5 µm) and large voids between them (Figure 1C). Changes on the surface of the ZX-6060 sample after treatment with etching solution are the creation of voids, which are visually similar to those of ZX-Ct. The difference in surface defects compared to the parent sample can be seen in the insert of the SEM image in Figure 1D.

All PXRD patterns of samples after cation introduction (except for Mg in ZX-n) display a decrease in crystallinity compared to the starting material (Figure 2). Despite using a lower concentration of Co and Ni salts (0.5 M) during wet impregnation, the highest degradation of ZX structure (amorphization) was observed on systems with Co, due to lower pH caused by hydrolysis. So, the influence of cation type on ZX structure amorphization is in the following order Mg < Ni < Zn < Co. The decrease is more pronounced for the samples of the ZX-n series, which have larger outside crystal surfaces and lower crystallinity to begin with (lower intensity but wider peaks are a result of much smaller—nanosized—zeolite crystals). After sample calcination at 550 °C, an increase in the peak intensity was observed. This can be explained by partial reintegration of the silanol nest to framework inducing the "healing effect" on the structural defects, which were introduced by acid solution light amorphization of the external crystal surface during wet impregnation. In the case of insertion of Mg^{2+} to ZX nanocrystals, crystallinity is even better than that of starting material, indicating lower amorphization during wet impregnation and a more pronounced healing effect of (reintegrated) silanol nests. There are no additional peaks between 35° and 50°, which indicates that there is no detectable amount (at least by PXRD technique) of metal oxide formed on the zeolite internal or external crystal surface [24–27].

The effect of change in Si/Al ratio on OH groups in zeolite is best visible in FTIR spectra (Figure 3). A comparison of parent samples shows that all of them have a broadband at ca. 3400 cm^{-1} which belongs to silanol nests [28], and the intensity of that band increases with the decrease of the Si/Al ratio (also with a decrease of crystal size). It is surprising for sample ZX-6060, which was modified by desilication of "solid" micrometer-sized crystals. It is also an indication that most of silanol nests are close to the external crystal surfaces, including mesopores. After the insertion of metal cation into the structure, the intensity of the band decreases for all samples compared to the parent sample, which indicates more interactions between inserted cation and silanol nests.

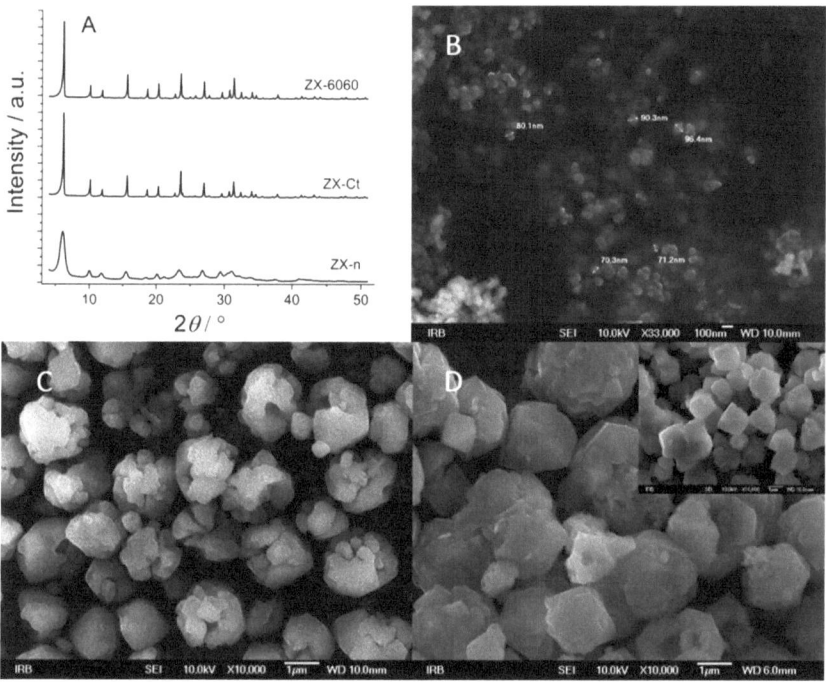

Figure 1. XRD patterns of parent samples before cation insertion (**A**). SEM images of nanocrystals, ZX-n. Insert—magnification of 33,000× (**B**). The sample prepared with the addition of CTAB during synthesis, ZX-Ct (**C**), and sample exposed to a mixture of NaOH + TPAOH solution and consecutive thermal treatment (insert—ZX crystals before etching), ZX-6060 (**D**).

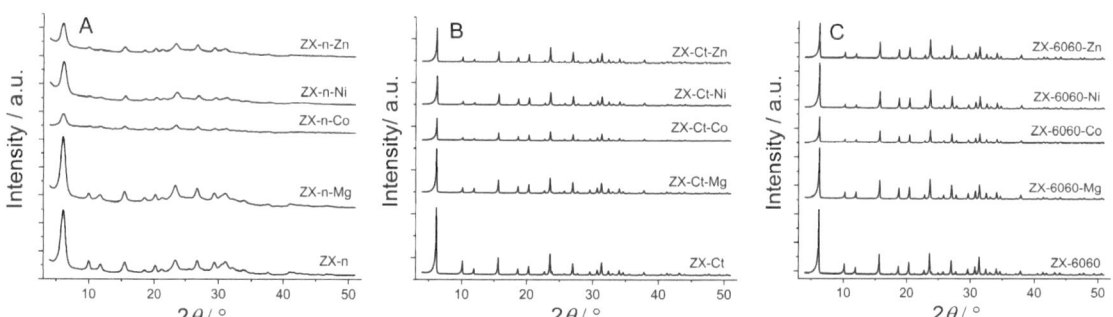

Figure 2. XRD patterns for the series of ZX-n (**A**), ZX-Ct (**B**), and ZX-6060 (**C**) samples before and after wet impregnation (Zn, Ni, Co, Mg) and consecutive thermal treatment. The scale for ZX-n samples is expanded for better visibility.

Besides the external surface of ZX crystals (determined by the size of crystals and macro-voids of twin crystals), the intensity of the band at 3745 cm^{-1} assigned to ν(SiOH), largely depends on the structural defects of the ZX framework introduced by amorphization during wet impregnation. The band intensity rises proportionally to the decrease in crystallinity, as can be confirmed from PXRD patterns in Figure 2.

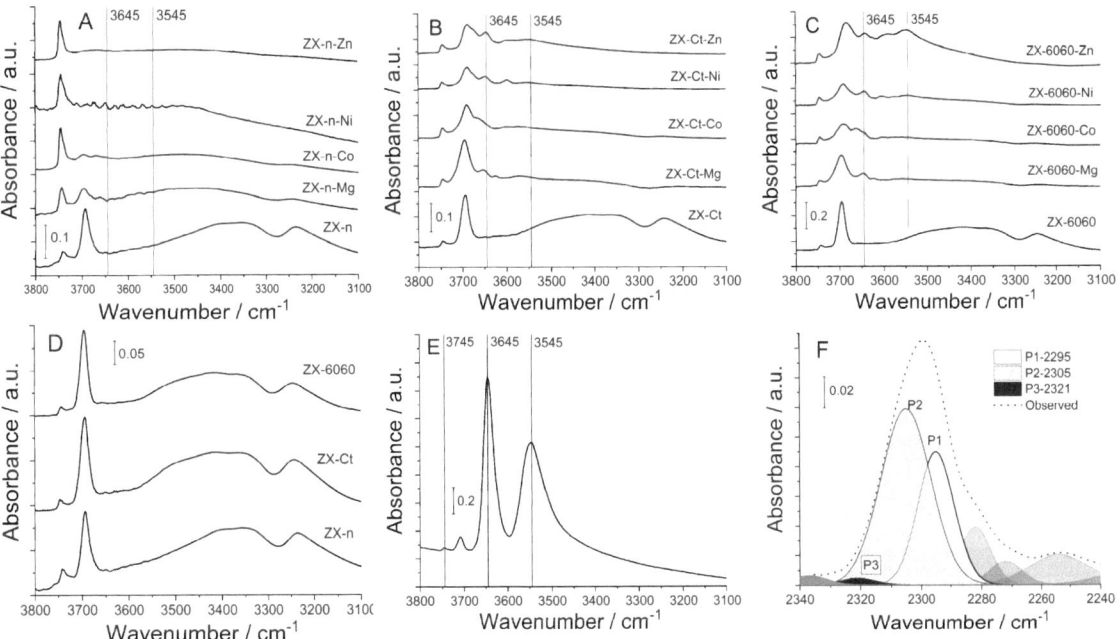

Figure 3. FTIR spectra of OH groups in the region from 3800 to 3100 cm^{-1} of prepared samples before and after insertion of Mg^{2+}, Co^{2+}, Ni^{2+}, and Zn^{2+} cations of samples: ZX-n (**A**), ZX-Ct (**B**), ZX-6060 (**C**), OH groups of used parent samples (**D**), OH groups of H-form of ZX-(**E**), an example of deconvolution of adsorbed acetonitrile FTIR spectra for ZX-Ct-Co sample at 150 °C (**F**).

Traces of metal M-OH stretching can also be observed in other samples. Both ZX-Ct-Co and ZX-6060-Co have a band at 3675 cm^{-1} assigned to Co-OH, ZX-Ct-Ni and ZX-6060-Ni have a Ni-OH shoulder at 3672 cm^{-1}, [29] while ZX-Ct-Zn and ZX-6060-Zn have a shoulder at 3672 cm^{-1}, as well as another two at 3656 and 3639 cm^{-1}, but one of highest intensities is at 3646 cm^{-1}, as observed here [30]. Comparing the rest of the acquired spectra to that of zeolite's H-form, one can observe that position of bands at 3645 cm^{-1} matches that of H positioned in the super cage (Figure 4 position II) and is not present in the samples before impregnation. Aside from that, samples ZX-Ct-Zn, ZX-6060-Zn, and ZX-6060-Ni also have the prominent band at 3546 cm^{-1} matching the protonated bridging oxygen atom in the sodalite cage (Figure 3 position I') [31]. Other samples from the ZX-Ct and ZX-6060 series have the mentioned band, but it is of much lower intensity. It is difficult to differentiate any peaks in that area from the background noise for the samples of the ZX-n series. The reason for the presence of protonated bridging atoms is H$_3$O$^+$ generated from hydrolysis of the used salts (pHs of Zn, Ni, and Co nitrate solutions were between 4 and 5, while Mg(NO$_3$)$_2$ solution has a pH value 5.7) [32,33]. This process is further enhanced by the interaction of the metal cation with the zeolite framework.

Detailed analysis of the atomic composition of the samples also indicates that there is a slight deficiency in the positive charge needed to fully compensate negative charge of the framework (Table 1), as well as slight dealumination caused by an acidic environment during cation insertion [34–36]. The presence of those acid sites is further confirmed by the adsorption of D$_3$-acetonitrile.

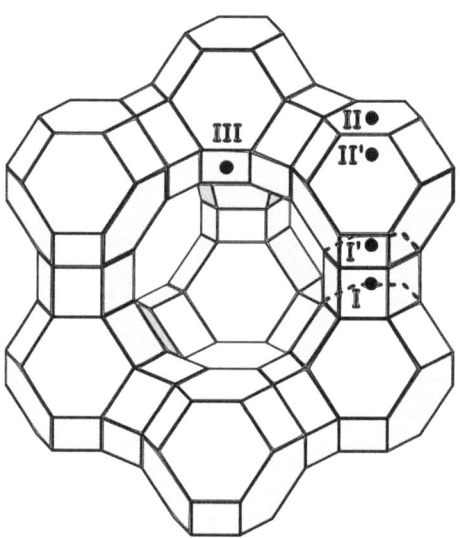

Figure 4. FAU structure with marked positions of exchangeable cations.

Table 1. Molar quantity of elements determined by flame atomic absorption spectroscopy (FAAS) and Si/Al ratio.

Sample	O	Na	Al	Si	Metal	Si/Al
ZX-n-Na	4.64	0.69	0.69	1.00		1.44
ZX-n-Mg	4.99	0.32	0.67	1.00	0.19	1.50
ZX-n-Co	4.15	0.26	0.68	1.00	0.20	1.47
ZX-n-Ni	4.77	0.29	0.67	1.00	0.21	1.50
ZX-n-Zn	4.48	0.18	0.70	1.00	0.23	1.48
ZX-Ct-Na	4.21	0.50	0.50	1.00		1.99
ZX-Ct-Mg	4.44	0.16	0.48	1.00	0.15	2.09
ZX-Ct-Co	4.30	0.18	0.48	1.00	0.15	2.09
ZX-Ct-Ni	4.04	0.15	0.45	1.00	0.14	2.21
ZX-Ct-Zn	4.15	0.15	0.48	1.00	0.14	2.10
ZX-6060-Na	3.94	0.44	0.44	1.00		2.27
ZX-6060-Mg	4.16	0.16	0.43	1.00	0.12	2.27
ZX-6060-Co	3.99	0.17	0.43	1.00	0.14	2.30
ZX-6060-Ni	4.03	0.15	0.44	1.00	0.14	2.28
ZX-6060-Zn	4.09	0.14	0.44	1.00	0.13	2.28

All samples after cation insertion have a band in the range of 2295 to 2298 cm^{-1} from D$_3$-acetonitrile adsorbed on type A or B of Brønsted acid sites as described by Pelmenschikov et al. [37]. At room temperature, the amount of adsorbed D$_3$-acetonitrile varies greatly from sample to sample due to surface adsorption. After desorption at 150 °C, it can be observed that samples ZX-Ct-Mg and ZX-Ct-Zn have the highest amount of D$_3$-acetonitrile still adsorbed on Brønsted acid sites (BAS) out of all samples (Figure 5), but the strength of those sites differs. ZX-Ct-Zn has by far the largest amount of probe molecules adsorbed on BAS, but only around 15% of them are strong enough to bind D$_3$-acetonitrile after desorption at 300 °C; while in the ZX-Ct-Mg sample, strong acid sites make almost 50% of all detected Brønsted acid sites. Even though samples with Co and Ni have a similar amounts of probe molecules adsorbed on BAS compared to Mg and Zn at room temperature, most of those sites are weak and most of D$_3$-acetonitrile was desorbed at 150 °C. The possible reason for that is the tendency of Mg^{2+} and Zn^{2+} to occupy site II' in the sodalite cage (Figure 4) [38], which leads to the increase in the acidity of the proton present near that binding site. A similar effect was already described in the presence of the Al^{3+} Lewis acid site (LAS) in the sodalite cage by Li et al. [39].

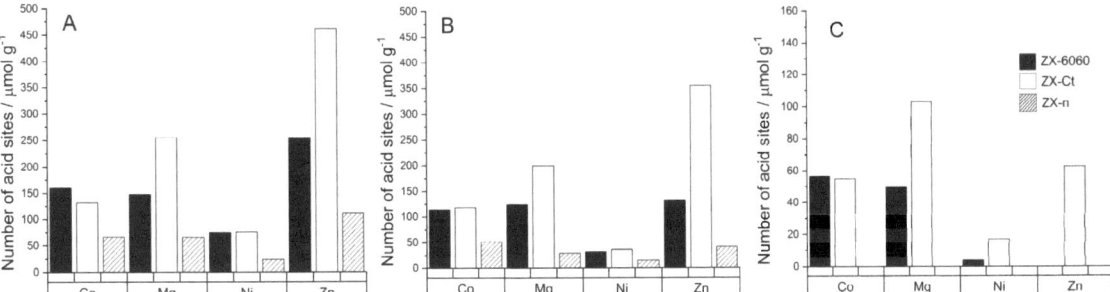

Figure 5. Number of Brønsted acid sites per gram of zeolite for samples after cation insertion, determined from the amount of desorbed D$_3$-acetonitrile molecules at BAS at 150 °C (**A**), BAS at 200 °C (**B**), BAS at 300 °C (**C**). The legend on graph (**C**) is valid for all graphs (**A**–**C**).

Unlike Brønsted acid sites, all samples have multiple bands of adsorbed probe molecule that match Lewis acid sites (Figure 5). Those sites can be roughly divided into three different regions: type 1 (from 2303 to 2317 cm^{-1}), type 2 (from 2321 to 2324 cm^{-1}), and type 3 (around 2327 cm^{-1}). Their ratios vary from sample to sample and they are prone to shifts and disappearance at higher temperatures due to cation migrations and probe molecule protonation.

Type 1 sites are caused by the presence of the metal cations in the zeolite framework and usually consist of one or two bands depending on the number of different sites in zeolite that accommodate the cation. The number of those LAS does not always increase with the increase in a number of metal atoms in the framework. For samples exchanged with Co^{2+}, the number of adsorbed probe molecules on Lewis acid sites at 150 °C slightly decreases with the increase of the Si/Al ratio but the amount of D$_3$-acetonitrile bound to stronger acid sites stays unchanged for the samples with a Si/Al ratio over 2 (Figure 6).

Comparing the spectra of samples to those of D$_3$-acetonitrile absorbed on H-FAU we were able to conclude that type 2 sites are related to the presence of H$_3$O$^+$ in the framework and are the result of protonation of the probe molecule, which is in accordance with Hadjiivanov [40]. Type 3 sites indicate the presence of extra framework aluminum (EF-Al) which is further confirmed by the weak band around 3600 cm^{-1} (Figure 3) [41–43]. The presence of the EF-Al in the framework can propagate the formation of Brønsted acid sites [44]. If we compare the number of D$_3$-acetonitrile molecules adsorbed on EF-Al Lewis acid sites between the samples of the ZX–Ct series (Figure 7) it can be seen that even though samples with Mg, Ni, and Zn have a fairly similar numbers of EF-Al, ZX-Ct-Ni has far fewer BAS than the other two. That further solidifies the assumption that the increase of acid site strength is caused by the presence of the exchanged cation in the sodalite cage rather than the EF-Al.

Changes in the Lewis acidity for these cations indicate that the number of those acid sites mainly depends on the number of metal cations in the binding sites I' and II'. Since the preferred binding site of Co^{2+} and Ni^{2+} is site I, it is coordinatively saturated (mostly by Ox, which means oxygen directly binds to the zeolite framework) and does not allow binding of D$_3$–acetonitrile to a metal center.

On the other hand, the preferred positions for Mg^{2+} and Zn^{2+} are II' and I'. Therefore, they occupy the position more accessible (within the alpha cage) for D$_3$-acetonitrile probe molecules adsorbed to the Lewis site, and the result is a higher number of determined acid sites (Figure 6).

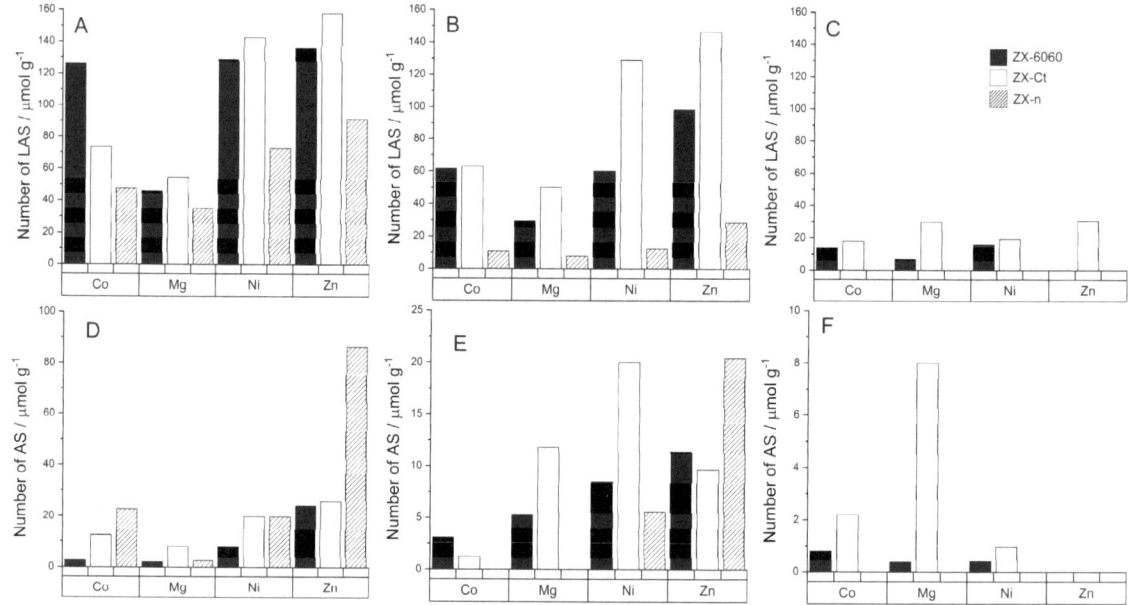

Figure 6. Number of specific acid sites per gram of zeolite determined from the amount of adsorbed D_3–acetonitrile molecules at different temperatures: LAS at 150 °C (**A**), LAS at 200 °C (**B**), LAS at 300 °C (**C**). Unspecified acid sites: AS at 150 °C (**D**), AS at 200 °C (**E**), AS at 300 °C (**F**). The legend on graph C is valid for all graphs (**A**–**F**).

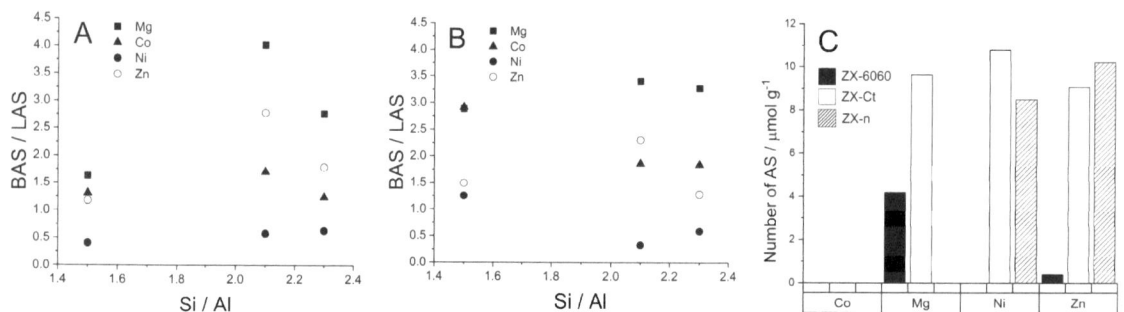

Figure 7. Dependence of ratio of Brønsted to Lewis acid sites (BAS/LAS) on the Si/Al ratio at 150 °C (**A**) and 200 °C (**B**). The number of extra framework aluminum (EF-Al) Lewis acid sites (**C**), obtained from deconvolution of the band at 2327 cm^{-1}. Si/Al ratio was also calculated from the number of adsorbed D_3-acetonitrile molecules.

If we look at the ratio between probe molecules bound to Brønsted and Lewis acid sites, it can be seen that all cations create a larger number of BAS than LAS, except Ni^{2+} which preferably creates more Lewis sites (Figure 7). Approximately linear growth of the BAS/LAS ratio with the Si/Al ratio at 150 °C shows usual behavior at higher temperature: faster elimination of D_3-acetonitrile from LAS than from BAS with increased temperature of desorption. Additionally, it shows that the most thermally stable systems are the ZX-Ct series at all temperatures, and systems ZX-n are the most unstable. The availability of BAS and LAS in ZX-n can be explained by a more open structure (also positions for exchangeable cations) due to the much larger outer crystal surface.

The DRS spectra of samples after wet impregnation with Co, Ni, Mg, and Zn are shown in Figure 8. It is evident that the strongest bands are associated with Co^{2+} species in all three ZX samples. Absorption spectra of ZX-Ct-Co and ZX-6060-Co have a band in the visible region at 520 nm assigned to the $^4T_1(P) \leftarrow {}^4A_2$, near 600 nm from $^4A_{2g} \leftarrow {}^4T_{1g}$ and a broadband from 1000 to 1400 nm from $^4T_{2g} \leftarrow {}^4T_{1g}$. All of these transitions indicate the presence of octahedrally coordinated $[Co(H_2O)_6]^{2+}$ in the binding site III′ in hydrated samples [45,46]. The shoulder at 620 nm can be attributed to the small amount of tetrahedrally coordinated Co^{3+} [47], but the exact amount is hard to determine due to overlapping with the band at 720 nm from octahedrally coordinated Co^{2+} at binding site I [48]. On the other hand, ZX-n-Co has typical tetrahedral triplet bands which did not disappear, even after being exposed to air moisture for a week [49]. This indicates the formation of very stable pseudo tetrahedral $Co(Ox)_3H_2O$ species within the framework (Ox stands for oxygen directly bound to the zeolite framework). This is in accordance with the work published by Egerton et al. [50], which mentions the tendency of cobalt(II) to have tetrahedral coordination at high loadings. All samples have weak bands from 250–400 nm. The band at 370 nm can be assigned to the presence of $Co(OH)_2$ while the shoulder at 400 nm in ZX–6060–Co and ZX–n–Co samples can be assigned to the mixed cobalt oxide, Co_3O_4 [51].

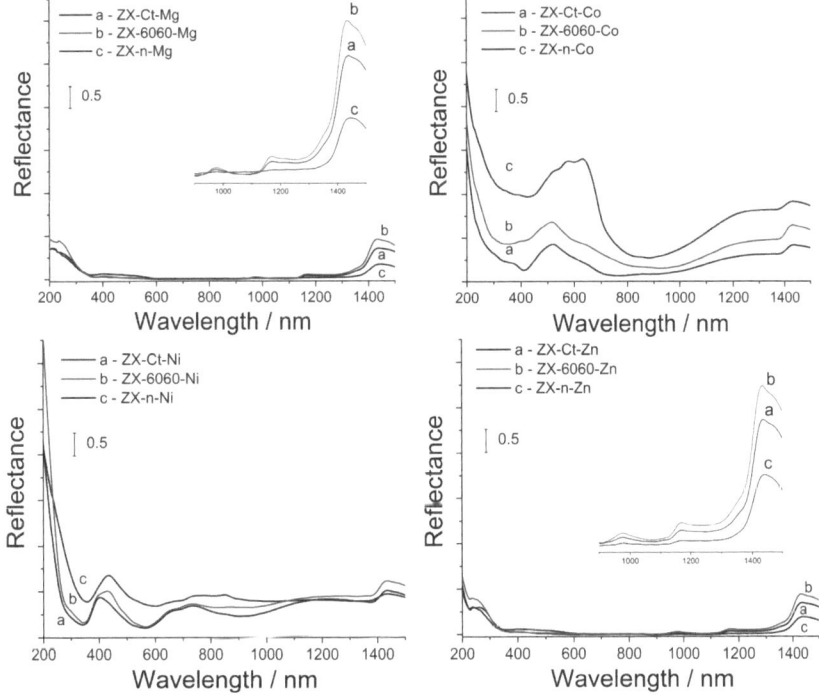

Figure 8. UV–Vis DRS spectra of samples after insertion of Co^{2+}, Ni^{2+}, Mg^{2+}, and Zn^{2+} cations and consecutive thermal treatment.

The situation is similar for the samples impregnated with nickel. Sample ZX-Ct-Ni has a band in the visible region at 400 nm assigned to the $^3T_{1g}(P) \leftarrow {}^3A_{2g}$ and two bands between 600 and 800 nm from $^1E_g \leftarrow {}^3A_{2g}$ and $^3T_{1g} \leftarrow {}^3A_{2g}$. All of these transitions indicate presence of octahedral coordinated $[Ni(Ox)_6]^{2+}$ [52–54] in the binding site I. In the ZX-6060-Ni and ZX-n-Ni samples, another band at 430 nm can be seen. This is from $Ni(OH)_2$, which also has an additional band at 384 nm [55,56] that overlaps with those of octahedral nickel(II). Weak bands around 300 nm indicate the presence of small amounts of NiO. Samples containing Mg and Zn have no visible bands in the region from 200 to 1500 nm, so we can only assume that those samples would follow the same trend in metal coordination. The weak bands at 970, 1180, and 1430 nm come from the zeolite network and are shown as inserts in Figure 8. This implies that samples with low Si/Al ratio, and larger accessible surface have a tendency to form a small amount of metal oxide particles, not detectable by PXRD [57], that have very weak Lewis acid sites. The formation of these particles also leads to a decrease in the overall number of acid sites.

4. Conclusions

Samples with different Si/Al ratios treated with M^{2+} salts under the same conditions were analyzed to determine the strength and number of acid sites. In spite of similar M^{2+} relatively high loadings, the total amount and ratio between Brønsted and Lewis acid sites varied greatly. The amount and strength of acid sites were primarily influenced by the location of the metal cation in the framework (positions II' and I' within the sodalite cage), the amount of structural defects generated by wet impregnation, and the amount of metal oxide formed during cation insertion.

The structural healing effect is observed for silanol nests after insertion (wet impregnation + thermal treatment) of any of the used Me cations (Mg, Zn, Co, Ni) in all three ZX samples. The most pronounced example (from PXRD and FTIR data) is the insertion of Mg^{2+} cations into nanosized ZX crystals (sample ZX-n-Mg).

Samples ZX-Ct-Mg and ZX-Ct-Zn had the highest amount of strong Brønsted acid sites, which is explained by an increase of the number of cations in the binding site II' that stabilize H^+ in the sodalite cage mostly with stable octahedral coordination. Insertion of Ni^{2+} in the structure preferably formed more Lewis than Brønsted acid sites, while other cations had the opposite effect. Furthermore, with the decrease in the Si/Al ratio, the proportion of BAS/LAS slightly increases, due to the formation of metal oxide particles that have weak Lewis acidity and whose formation leads to the mitigation of stronger acid sites.

Author Contributions: Conceptualization, experimental research, data collection, and analysis draft writing: G.M.; data analysis, visualization, and validation: A.P.; concept, resources, writing—review and editing, and supervision: J.B. All authors have read and agreed to the published version of the manuscript.

Funding: This research received no external funding.

Data Availability Statement: No additional data available.

Acknowledgments: This work was financially supported by Croatian Science Foundation through the project IP-2016-06-2214.

Conflicts of Interest: The authors declare no conflict of interest. The funders had no role in the design of the study; in the collection, analysis, or interpretation of data; in the writing of the manuscript; or in the decision to publish the results.

References

1. Haas, A.; Harding, D.; Nee, J. FCC catalysts containing the high-silica faujasites EMO and EMT for gas-oil cracking. *Micropor. Mesopor. Mater.* **1999**, *28*, 325–333. [CrossRef]
2. Rodionova, L.; Knyazeva, E.; Konnov, S.V.; Ivanova, I. Application of Nanosized Zeolites in Petroleum Chemistry: Synthesis and Catalytic Properties (Review). *Pet. Chem.* **2019**, *59*, 455–470. [CrossRef]
3. Guzmán-Castillo, M.; Armendáriz-Herrera, H.; Pérez-Romo, P.; Hernández-Beltrán, F.; Ibarra, S.; Valente, J.; Fripiat, J. Y zeolite depolymerization-recrystallization: Simultaneous formation of hierarchical porosity and Na dislodging. *Micropor. Mesopor. Mater.* **2011**, *143*, 375–382. [CrossRef]
4. Sachse, A.; Grau-Atienza, A.; Jardim, E.; Linares, N.; Thommes, M.; García-Martínez, J. Development of Intracrystalline Mesoporosity in Zeolites through Surfactant-Templating. *Cryst. Growth Des.* **2017**, *17*, 4289–4305. [CrossRef]
5. Verboekend, D.; Nuttens, N.; Locus, R.; Van Aelst, J.; Verolme, P.; Groen, J.; Pérez-Ramírez, J.; Sels, B. Synthesis, characterisation, and catalytic evaluation of hierarchical faujasite zeolites: Milestones, challenges, and future directions. *Chem. Soc. Rev.* **2016**, *45*, 3331–3352. [CrossRef]

6. Otto, T.; Zones, S.; Iglesia, E. Synthetic strategies for the encapsulation of nanoparticles of Ni, Co, and Fe oxides within crystalline microporous aluminosilicates. *Micropor. Mesopor. Mater.* **2018**, *270*, 10–23. [CrossRef]
7. Kostyniuk, A.; Bajec, D.; Likozar, B. Catalytic hydrogenation, hydrocracking and isomerization reactions of biomass tar model compound mixture over Ni-modified zeolite catalysts in packed bed reactor. *Renew. Energy* **2021**, *167*, 409–424. [CrossRef]
8. Choo, M.; Oi, L.; Ling, T.; Ng, E.; Lin, Y.; Centi, G.; Juan, J. Deoxygenation of triolein in green diesel in the H2-free condition: Effect of transition metal oxide supported on zeolite Y. *J. Anal. Appl. Pyrolysis* **2020**, *147*, 104797. [CrossRef]
9. Lakiss, L.; Kouvatas, C.; Gilson, J.; Aleksandrov, H.; Vayssilov, G.; Nesterenko, N.; Mintova, S.; Valtchev, V. Unlocking the Potential of Hidden Sites in Faujasite: New Insights in a Proton Transfer Mechanism. *Angew. Chem. Int. Ed.* **2021**, *133*, 26906–26913. [CrossRef]
10. Datka, J.; Gil, B.; Kawałek, M.; Staudte, B. Low temperature IR studies of CO sorbed in ZSM-5 zeolites. *J. Mol. Struct.* **1999**, *511–512*, 133–139. [CrossRef]
11. Sadowska, K.; Góra-Marek, K.; Datka, J. Accessibility of acid sites in hierarchical zeolites: Quantitative IR studies of pivalonitrile adsorption. *J. Phys. Chem. C* **2013**, *117*, 9237–9244. [CrossRef]
12. Thibault-Starzyk, F.; Travert, A.; Saussey, J.; Lavalley, J. Correlation between activity and acidity on zeolites: A high temperature infrared study of adsorbed acetonitrile. *Top. Catal.* **1998**, *6*, 111–118. [CrossRef]
13. Huang, Z.; Li, T.; Yang, B.; Chang, C. Role of surface frustrated Lewis pairs on reduced CeO2(110) in direct conversion of syngas. *Chin. J. Catal.* **2020**, *41*, 1906–1915. [CrossRef]
14. Olszowka, J.; Lemishka, M.; Mlekodaj, K.; Kubat, P.; Rutkowska-Żbik, D.; Dedecek, J.; Tabor, E. Determination of Zn Speciation, Siting, and Distribution in Ferrierite Using Luminescence and FTIR Spectroscopy. *J. Phys. Chem. C* **2021**, *125*, 9060–9073. [CrossRef]
15. Montanari, T.; Bevilacqua, M.; Resini, C.; Busca, G. UV-Vis and FT-IR Study of the Nature and Location of the Active Sites of Partially Exchanged Co-H Zeolites. *J. Phys. Chem. B* **2004**, *108*, 2120–2127. [CrossRef]
16. Shannon, R.D. Revised Effective Ionic Radii and Systematic Studies of Interatomic Distances in Halides and Chalcogenides. *Acta Cryst.* **1976**, *32*, 751–767. [CrossRef]
17. Bacariza, M.; Maleval, M.; Graça, I.; Lopes, J.; Henriques, C. Power-to-Methane over Ni/Zeolites: Influence of the Framework Type. *Micropor. Mesopor. Mater.* **2019**, *274*, 102–112. [CrossRef]
18. Afreen, G.; Patra, T.; Upadhyayula, S. Zn-Loaded HY Zeolite as Active Catalyst for Iso-Propylation of Biomass-Derived Phenolic Compounds: A Comparative Study on the Effect of Acidity and Porosity of Zeolites. *Mol. Catal.* **2017**, *441*, 122–133. [CrossRef]
19. Kim, H.; Choi, S.; Lim, W. Preparation and Structural Study of Fully Dehydrated, Highly Mg^{2+}-Exchanged Zeolite Y (FAU, Si/Al = 1.56) from Undried Methanol Solution. *J. Porous Mater.* **2014**, *21*, 659–665. [CrossRef]
20. Imre, B.; Konya, Z.; Hannus, I.; Halasz, J.; Nagy, J.; Kiricsi, I. Hydrodechlorination of Chlorinated Compounds on Different Zeolites. *Stud. Surf. Sci. Catal.* **2002**, *142*, 927–934.
21. Awala, H.; Gilson, J.; Retoux, R.; Boullay, P.; Goupil, J.; Valtchev, V.; Mintova, S. Template-free nanosized faujasite-type zeolites. *Nat. Mater.* **2015**, *14*, 447–451. [CrossRef] [PubMed]
22. Bosnar, S.; Bosnar, D.; Ren, N.; Rajić, N.; Grżeta, B.; Subotić, B. Positron lifetimes in pores of some low-silica zeolites: Influence of water content, crystal size and structural type. *J. Porous Mater.* **2013**, *20*, 1329–1336. [CrossRef]
23. Wichterlova, P.; Tvarużkova, B.; Sobalik, Z.; Sarv, Z. Determination and properties of acid sites in H-ferrierite A comparison of ferrierite and MFI structures. *Micropor. Mesopor. Mater.* **1998**, *24*, 223–233. [CrossRef]
24. Srivastava, V.; Gusain, D.; Sharma, Y. Synthesis, characterization and application of zinc oxide nanoparticles (n-ZnO). *Ceram. Int.* **2013**, *39*, 9803–9808. [CrossRef]
25. Bulavchenko, O.; Cherepanova, S.V.; Malakhov, V.V.; Dovlitova, L.; Ishchenko, A.V.; Tsybulya, S.V. In situ XRD study of nanocrystalline cobalt oxide reduction. *Kinet. Catal.* **2009**, *50*, 192–198. [CrossRef]
26. Dharmaraj, N.; Prabu, P.; Nagarajan, S.; Kim, C.; Park, J.; Kim, H. Synthesis of nickel oxide nanoparticles using nickel acetate and poly(vinyl acetate) precursor. *Mater. Sci. Eng. B Solid State Mater. Adv. Technol.* **2006**, *128*, 111–114. [CrossRef]
27. Aramendía, M.; Benítez, J.; Borau, V.; Jiménez, C.; Marinas, J.; Ruiz, J.; Urbano, F. Characterization of Various Magnesium Oxides by XRD and 1H MAS NMR Spectroscopy. *J. Solid State Chem.* **1999**, *144*, 25–29. [CrossRef]
28. Palčić, A.; Moldovan, S.; El Siblani, H.; Vicente, A.; Valtchev, V. Defect Sites in Zeolites: Origin and Healing. *Adv. Sci.* **2022**, *9*, 2104414. [CrossRef]
29. Tang, Y.; Liu, Y.; Yu, S.; Guo, W.; Mu, S.; Wang, H.; Zhao, Y.; Hou, L.; Fan, Y.; Gao, F. Template-free hydrothermal synthesis of nickel cobalt hydroxide nanoflowers with high performance for asymmetric supercapacitor. *Electrochim. Acta* **2015**, *161*, 279–289. [CrossRef]
30. Noei, H.; Qiu, H.; Wang, Y.; Löffler, E.; Wöll, C.; Muhler, M. The identification of hydroxyl groups on ZnO nanoparticles by infrared spectroscopy. *Phys. Chem. Chem. Phys.* **2008**, *10*, 7092–7097. [CrossRef]
31. Li, X.; Han, H.; Xu, W.; Hwang, S.J.; Lu, P.; Bhan, A.; Tsapatsis, M. Enhanced Reactivity of Accessible Protons in Sodalite Cages of Faujasite Zeolite. *Angew. Chem. Int. Ed.* **2022**, *61*, e202111180. [CrossRef]
32. Chizallet, C.; Costentin, G.; Che, M.; Delbecq, F.; Sautet, P. Infrared characterization of hydroxyl groups on MgO: A periodic and cluster density functional theory study. *J. Am. Chem. Soc.* **2007**, *129*, 6442–6452. [CrossRef] [PubMed]
33. Knözinger, E.; Jacob, K.; Singh, S.; Hofmann, P. Hydroxyl groups as IR active surface probes on MgO crystallites. *Surf. Sci.* **1993**, *290*, 388–402. [CrossRef]

34. Seo, S.; Lim, W.; Seff, K. Single-crystal structures of fully and partially dehydrated zeolite y (FAU, Si/Al = 1.56) Ni 2+ exchanged at a low pH, 4.9. *J. Phys. Chem. C* **2012**, *116*, 13985–13996. [CrossRef]
35. Seo, S.; Moon, D.; An, J.; Jeong, H.; Lim, W. Time-dependent Ni^{2+}-ion exchange in zeolites y (FAU, si/Al = 1.56) and their single-crystal structures. *J. Phys. Chem. C* **2016**, *120*, 28563–28574. [CrossRef]
36. Kim, C.; Jung, K.; Heo, N.; Seff, K. Crystal Structures of Vacuum-Dehydrated Ni^{2+}-Exchanged Zeolite Y (FAU, Si/Al = 1.69) Containing Three-Coordinate Ni^{2+}, $Ni_8O_4 \cdot xH_2O^{8+}$, $x \leq 4$, Clusters with Near Cubic Ni_4O_4 Cores, and H^+. *J. Phys. Chem. C* **2009**, *113*, 5164–5181. [CrossRef]
37. Pelmenschikov, A.; Van Santen, R.; Jänchen, J.; Meijer, E. CD3CN as a probe of Lewis and Bronsted acidity of zeolites. *J. Phys. Chem.* **1993**, *97*, 11071–11074. [CrossRef]
38. Frising, T.; Leflaive, P. Extraframework cation distributions in X and Y faujasite zeolites: A review. *Micropor. Mesopor. Mater.* **2008**, *114*, 27–63. [CrossRef]
39. Li, S.; Zheng, A.; Su, Y.; Zhang, H.; Chen, L.; Yang, J. Brønsted/Lewis Acid Synergy in Dealuminated HY Zeolite: A Combined Solid-State NMR and Theoretical Calculation Study. *J. Am. Chem. Soc.* **2007**, *129*, 11161–11171. [CrossRef]
40. Hadjiivanov, K. Chapter 2: Identification and Characterization of Surface Hydroxyl Groups by Infrared Spectroscopy. *Adv. Catal.* **2014**, *57*, 99–318. [CrossRef]
41. Batool, S.; Sushkevich, V.; van Bokhoven, J. Correlating Lewis acid activity to extra-framework aluminum species in zeolite Y introduced by Ion-exchange. *J. Catal.* **2022**, *408*, 24–35. [CrossRef]
42. Xu, B.; Bordiga, S.; Prins, R.; Van Bokhoven, J. Effect of framework Si/Al ratio and extra-framework aluminum on the catalytic activity of Y zeolite. *Appl. Catal. A General* **2007**, *333*, 245–253. [CrossRef]
43. Lutz, W.; Rüscher, C.; Heidemann, D. Determination of the framework and non-framework $[SiO_2]$ and $[AlO_2]$ species of steamed and leached faujasite type zeolites: Calibration of IR, NMR, and XRD data by chemical methods. *Micropor. Mesopor. Mater.* **2002**, *55*, 193–202. [CrossRef]
44. Bhering, D.; Ramírez-Solís, A.; Mota, C. A density functional theory based approach to extraframework aluminum species in zeolites. *J. Phys. Chem. B* **2003**, *107*, 4342–4347. [CrossRef]
45. Han, J.; Woo, S. UV/VIS diffuse reflectance spectroscopic (DRS) study of cobalt-containing Y zeolites dehydrated at elevated temperatures. *Korean J. Chem. Eng.* **1991**, *8*, 235–239. [CrossRef]
46. Sebastian, J.; Jinka, K.; Jasra, R. Effect of alkali and alkaline earth metal ions on the catalytic epoxidation of styrene with molecular oxygen using cobalt (II)-exchanged zeolite X. *J. Catal.* **2006**, *244*, 208–218. [CrossRef]
47. Smeets, P.; Woertink, J.; Sels, B.; Solomon, E.; Schoonheydt, R. Transition-Metal Ions in Zeolites: Coordination and Activation of Oxygen. *Inorg. Chem.* **2010**, *49*, 3573–3583. [CrossRef]
48. Verberckmoes, A.; Weckhuysen, B.; Schoonheydt, R. Spectroscopy and coordination chemistry of cobalt in molecular sieves 1. *Micropor. Mesopor. Mater.* **1998**, *22*, 165–178. [CrossRef]
49. Verberckmoes, A.; Weckhuysen, B.; Pelgrims, J.; Schoonheydt, R. Diffuse reflectance spectroscopy of dehydrated cobalt-exchanged faujasite-type zeolites: A new method for Co^{2+} siting. *J. Phys. Chem.* **1995**, *99*, 15222–15228. [CrossRef]
50. Egerton, T.; Hagan, A.; Stone, F.; Vickerman, J. Magnetic Studies of Zeolites. Part 1.—The magnetic properties of CoY and CoA. *J. Chem. Soc. Faraday Trans.* **1971**, *68*, 723–735. [CrossRef]
51. Alrehaily, L.; Joseph, J.; Biesinger, M.; Guzonas, D.; Wren, J. Gamma-radiolysis-assisted cobalt oxide nanoparticle formation. *Phys. Chem. Chem. Phys.* **2013**, *15*, 1014–1024. [CrossRef] [PubMed]
52. Bacariza, M.; Graça, I.; Westermann, A.; Ribeiro, M.; Lopes, J.; Henriques, C. CO_2 Hydrogenation over Ni-Based Zeolites: Effect of Catalysts Preparation and Pre-reduction Conditions on Methanation Performance. *Top. Catal.* **2016**, *59*, 314–325. [CrossRef]
53. Schoonheydt, R.; Roodhooft, D.; Leeman, H. Coordination of Ni_2^+ to lattice oxygens of the zeolites X and Y. *Zeolites* **1987**, *7*, 412–417. [CrossRef]
54. Lever, A. *Inorganic Electronic Spectroscopy*, 1st ed.; Elsevier Publishing Company: Amsterdam, The Netherlands, 1968.
55. Graça, I.; González, L.V.; Bacariza, M.; Fernandes, A.; Henriques, C.; Lopes, J.; Ribeiro, M. CO_2 hydrogenation into CH_4 on NiHNaUSY zeolites. *Appl. Catal. B Environ.* **2014**, *147*, 101–110. [CrossRef]
56. Qi, Y.; Qi, H.; Li, J.; Lu, C. Synthesis, microstructures and UV-vis absorption properties of β-Ni(OH)2 nanoplates and NiO nanostructures. *J. Cryst. Growth* **2008**, *310*, 4221–4225. [CrossRef]
57. Zhang, Z.; Xiao, Q.; Gu, J. Effective synthesis of zeolite-encapsulated Ni nanoparticles with excellent catalytic performance for hydrogenation of CO_2 to CH_4. *Dalt. Trans.* **2020**, *49*, 14771–14775. [CrossRef]

Disclaimer/Publisher's Note: The statements, opinions and data contained in all publications are solely those of the individual author(s) and contributor(s) and not of MDPI and/or the editor(s). MDPI and/or the editor(s) disclaim responsibility for any injury to people or property resulting from any ideas, methods, instructions or products referred to in the content.

Article

Fine Tuning of Hierarchical Zeolite Beta Acid Sites Strength

Ivana Landripet [1], Andreas Puškarić [1,2,*], Marko Robić [1] and Josip Bronić [1]

[1] Ruđer Bošković Institute, Bijenička 54, 10000 Zagreb, Croatia; ivana.landripet@irb.hr (I.L.); marko.robic@irb.hr (M.R.); josip.bronic@irb.hr (J.B.)
[2] National Institute of Chemistry, Hajdrihova 19, 1000 Ljubljana, Slovenia
* Correspondence: andreas.puskaric@irb.hr; Tel.: +385-1456-1111 or +385-1456-1310

Abstract: Two different synthesis methods to obtain hierarchical Beta zeolite are investigated: direct synthesis using cetyltrimethylammonium bromide (CTAB) as a mesoporous template and post-synthesis desilication by etching with NaOH and TPAOH. The main focus of this study is to show the possibility of fine tuning of the acid site (OH) strength (Brønsted and Lewis acid sites) through wet impregnation of these hierarchical Beta zeolites with divalent metal cations (Mg^{2+}, Co^{2+}, Ni^{2+}, Cu^{2+}, and Zn^{2+}), which are important for various applications. Fourier transform infrared spectroscopy (FTIR) and deuterated acetonitrile as the probe molecule were used as a powerful technique to analyze the quantity and number of Brønsted/Lewis acid sites in the modified zeolite Beta structure. Investigating the influence of different divalent metal cations with a comparable ionic radius on the acidity of the hierarchical Beta zeolites, the present research aims to shed light on the structure–activity relationship that determines their catalytic behavior, for the development of efficient and environmentally friendly catalysts for various industrial applications.

Keywords: hierarchical zeolite Beta; Brønsted acid site; Lewis acid site; FTIR spectroscopy

1. Introduction

In recent years, the design and synthesis of advanced materials with tailored properties have become imperative. The unique physicochemical properties of the zeolites, such as their controlled acidity, adsorption capacity, ion-exchange properties, and thermal stability, as well as uniform channels and cavities crystallographically ordered in size and position, determine their effectiveness in catalytic processes. As a solid acid catalyst, zeolite plays an irreplaceable role and is currently the most widely used solid acid catalyst in the petrochemical and fine chemical fields [1–3]. The H^+ from the OH group connected to the framework aluminum atom in zeolite can act as a proton donor, thus playing the role of a Brønsted acid site (BAS). The weaker acid sites dominantly come from non-framework aluminum and other balancing cations and act as Lewis acid sites (LAS) [4].

Despite the microporous structure, there are limitations that decrease the efficiency of zeolite as a catalyst. One of the greatest obstacles is intracrystalline diffusion of reactants and products during the process of catalysis. Such limitations can be avoided creating mesoporous voids in microcrystals, shortening the diffusion path. There are two approaches to increase the available number of active OH groups at the internal and external surface of crystals: a decrease in the particles size to nanosize and the creation of mesopores in micron-sized crystals by direct synthesis or by postsynthesis etching (desilication/dealumination) [5].

The primary focus lies in comprehending the precise positioning, quantity, categorization, and ease of access to acid sites within zeolite structures. This understanding is crucial for the meticulous adjustment of zeolites' catalytic attributes. For example, "green" syntheses of hydrocarbons from CO_2 using solid-state acid catalysts (zeolite) can be made in several steps, including conversion to CH_3OH, then to higher olefins (methanol to olefins-MTO). MTO conversion is a typical acid-catalyzed reaction. The active intermediates, e.g.,

polymethyl benzenes and the corresponding carbenium ions, can be formed at the BAS and induce the MTO conversion. On the other hand, coke compounds are formed faster at the BAS, causing catalyst deactivation [6,7].

For the characterization of zeolite and zeotype materials, among the different vibrational techniques, such as infrared (IR), Raman, inelastic neutron scattering (INS), and electron energy loss spectroscopy (EELS), to date, the most used is IR [8]. The acid strength and accessibility of OH groups can be measured using various probe molecules such as pyridine [9], carbon monoxide [10], and different nitriles [11] in low-pressure conditions and studied by transmission Fourier transform infrared spectroscopy (FTIR). Acetonitrile (CH_3CN) appears to be an attractive probe for zeolite acidity study since it can be attached to strong and weak AS and show the difference between them [12]. Since the acetonitrile C≡N spectral region is strongly influenced by the Fermi resonance between the symmetric C≡N stretching and the combination of symmetric CH_3 deformations and symmetric C–C stretching, deuterated acetonitrile (CD_3CN) is used instead.

In particular, zeolite Beta is a versatile and widely used zeolitic material, and its acid properties were studied by several authors [13–15]. In those studies, the strength and accessibility of different acid sites were investigated by IR spectroscopy, using weakly basic CO as a probe molecule. The authors found that CO readily interacts with different OH groups, shifting the positions of the vibrational bands. Based on the shift of the bands, it is possible to finely distinguish the strength of the acid sites, indicating their heterogeneity in the zeolite framework.

This paper explores two distinct synthesis methods for generating hierarchical Beta zeolite: one through the use of cetyltrimethylammonium bromide (CTAB) as a mesoporous template during synthesis [16], and the other is a post-synthetic treatment using a mixture of sodium hydroxide (NaOH) and tetrapropylammonium hydroxide (TPAOH) [17] solutions to etch the parent crystals.

The primary focus of this study is to show how a different way of zeolite Beta preparation can influence the strength of acid sites (Brønsted and Lewis AS) through wet impregnation of these hierarchical Beta zeolites with divalent metal cations (Mg^{2+}, Co^{2+}, Ni^{2+}, Cu^{2+}, and Zn^{2+}), which can be important in various catalytic/industrial applications. For example, zeolite beta catalyzes different types of reactions, such as (trans)alkylations, acylations, isomerizations, disproportionations, cracking, etc. [18–23], while metal exchanged/impregnated beta shows catalytic activity in NOx decomposition [24,25], acetone and methanol conversion to hydrocarbons [26,27] or oxidation of different organic compounds [28,29]. Also, the aim of this research is to shed a light on the structure–reactivity relationship governing the zeolite Beta acid strength, through design and development of efficient and environmentally friendly materials.

2. Materials and Methods

2.1. Zeolite Beta Synthesis and Cations Insertion

NaOH, NH_4NO_3, $Mg(CH_3COO)_2 \times 4H_2O$, $Co(CH_3COO)_2 \times 4H_2O$, $Ni(CH_3COO)_2 \times 4H_2O$, $Cu(CH_3COO)_2 \times H_2O$, and $Zn(CH_3COO)_2 \times 2H_2O$ were obtained from Kemika, Zagreb, Croatia. Tetraethylammonium hydroxide (TEAOH), tetrapropylammonium hydroxide (TPAOH), Al(i-OPr)$_3$, colloidal silica (Ludox HS-40) and cetyltrimethylammonium bromide (CTAB) were purchased from Sigma Aldrich, St. Louis, MO, USA.

Two different mesoporous parent zeolites, HB-1 and HB-C, were prepared using two different procedures from the similar Si/Al ratio in the gel composition.

Nano-crystallites of Beta zeolite were prepared by classical hydrothermal synthesis from the gel of oxide composition 2 $Na_2O \times 30$ $(TEA)_2O \times Al_2O_3 \times 165 SiO_2 \times 1980 H_2O$ adopted from [30]. The synthesis was made in Teflon-lined autoclaves as a one-pot synthesis in the following way: after the addition of an aqueous solution of NaOH, redistilled water, 35 wt% aqueous solution of tetraethylammonium hydroxide (TEAOH), and Al(i-OPr)$_3$, mixture was stirred using magnetic stirrer until complete hydrolysis (about 30 min). Then, there was the addition of colloidal silica (Ludox HS-40) and stirring for 24 h, followed by heating for 5 days at 100 °C and 24 h at 150 °C. The obtained product was washed, dried, and slowly calcined at 600 °C in the stream of air. The H-form of the samples, labeled as HB-0, were obtained by ion exchange with NH_4NO_3 (c = 0.8 mol dm^{-3}) followed by calcination. The post-synthesis desilication was performed using a mixture of NaOH and TPAOH (c = 0.2 mol dm^{-3}). The treatment was carried out at 65 °C for 30 min, with a ratio of zeolite to etching mixture of 1:30. Finally, parent material (labeled HB-1) was obtained by 2 cycles of ion exchange with NH_4NO_3 and calcination as described above.

Direct synthesis of hierarchically structured zeolite Beta was made using cetyltrimethylammonium bromide (CTAB) as a mesoporous template. The oxide form of the chemical composition of the starting mixture was

$$2\, Na_2O \times 30\, (TEA)_2O \times Al_2O_3 \times 165\, SiO_2 \times 1980\, H_2O \times 10\, CTAB$$

After stirring for 24 h at room temperature, reaction mixture was transferred into the Teflon-lined autoclave and heated for 5 days at 100 °C and 2 more days at 150 °C. The obtained product was washed, dried, and slowly calcined at 600 °C in the stream of air. The H-form of the samples was obtained by 2 cycles of ion exchange (0.8 mol dm^{-3} NH_4NO_3) followed by calcination. Obtained material was labeled as HB-C.

All prepared samples were impregnated using acetate salts of magnesium(II), cobalt(II), nickel(II), copper(II), and zinc(II) dissolved in redistilled water. The molar ratio of a zeolite to a salt solution was calculated to have M(II)/Al = 4. A concentration of 0.5 mol dm^{-3} was chosen. The cation exchange procedure was carried out at 60 °C with stirring and repeated three times, after which all samples were thoroughly washed with redistilled water, dried at 60 °C, and calcined at 350 °C with a ramp of 2 °C per minute for 180 min.

2.2. Methods of Sample Characterization

Powder X-ray diffraction (PXRD) was used for the phase analysis of the synthesized samples after each modification step. Diffraction patterns were collected using copper Kα radiation on an Empyrean (Malvern Panalytical, Malvern, UK) diffractometer with the Bragg–Brentano optics at 2θ angles from 5° to 50°, as well as from 0.5° to 5°.

The elemental composition of the samples was measured using Flame Atomic Absorption Spectroscopy (FAAS) on Aanalyst 200 (Perkin Elmer, Waltham, MA, USA). All solutions for the analysis were prepared in accordance with prescribed procedures.

UV–Vis diffuse reflectance spectroscopy (DRS) was applied to identify metal species. Measurements were made using the instrument model UV 3600 (Shimadzu, Kyoto, Japan), equipped with Integrating Sphere, and using $BaSO_4$ as standard.

The size and the morphology of the crystals were observed using a high-resolution field emission scanning electron microscope (FE-SEM) model JSM 7000F (JEOL, Tokyo, Japan).

Textural properties were determined by isothermal adsorption of nitrogen at 77 K, after a pre-treatment in a vacuum at 250 °C for 12 h (Autosorb iQ3, Anton Paar, Graz, Austria). The specific surface area was determined using the Brunauer–Emmett–Teller (BET) method. Micropore volumes and the micropore surface area were obtained using the t-plot method. Pore size distribution was calculated by non-linear DFT (NLDFT) method, using N_2 at 77 K on silica with cylindrical/spherical pores as a model.

Qualitative and quantitative analysis of acid sites was performed using Fourier transformed infrared spectroscopy (FTIR) on a Frontier (Perkin-Elmer) instrument, in the transmission mode under the pressure of 5×10^{-5} mbar and resolution of 4 cm^{-1}. Self-supported pellets (d = 13 mm) for the analysis were prepared from around 5 mg of sample and activated at 350 °C for 3 h. To determine acid site strength (AS), desorption of deuterated acetonitrile (CD_3CN) was measured at 100, 150, 200, and 300 °C. The concentration of the specific acid site (Brønsted and Lewis) was calculated from the integral intensity of the corresponding bands in the IR spectra after the adsorption of CD$_3$CN, using the formula

$$C\left(\mu mol\ g_{cat}^{-1}\right) = \frac{IA(X)}{\varepsilon(X) \times \sigma} \qquad (1)$$

where $IA(X)$ is the integrated absorbance of the peak of the acid species X (Lewis or Brønsted), σ is the "density" of the wafer (actually it is wafer thickness after making at predefined pressure of 1.25 T cm^{-2}), while ε is the molar extinction coefficient (2.05 ± 0.1 cm µmol^{-1} and 3.6 ± 0.2 cm µmol^{-1}, for Brønsted and Lewis acid sites, respectively) as described by Wichterlová et al. [31].

3. Results and discussion

3.1. Powder X-ray Diffraction (PXRD)

PXRD profile of all samples is consistent with the standards available in the database of the International Zeolite Association relative to Beta zeolite phases. XRD patterns of the two different parent samples (HB-1 and HB-C) modified with divalent metal cations are presented in Figure 1, showing no differences in the diffraction patterns of the zeolites, indicating that the metal incorporation did not alter the zeolitic structure.

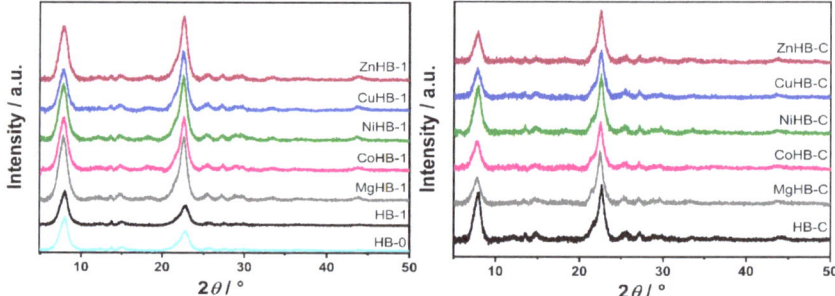

Figure 1. PXRD patterns of HB-0 (H-form of untreated parent zeolite) and HB-1 (desilicated zeolite) series (**left**) and HB-C (zeolite synthesized using CTAB) series (**right**).

However, increase in the intensity of the diffraction maxima in HB-1 series upon loading of metal cations can be explained by the "healing" effect of the impregnation procedure. Desilication (etching) with a NaOH/TPAOH solution creates defects (silanol nests) and possibly some amorphous debris. After the process of wet impregnation, subsequent washing, and slow calcination, part of the structural defects is "healed", and the intensity of healing depends on the type of metal cation used. At the same time, if present, the amorphous content is removed. Therefore, samples become more structured, that is, more crystalline as evidenced by the increase in the intensity of XRD signals.

Furthermore, no additional peaks between 35° and 50° were observed indicating that samples do not contain any crystalline impurities, such as metal oxides, within the detection limits of the PXRD method.

Low-angle powder X-ray diffraction patterns of the calcined parent—mesostructured zeolite Beta prepared using CTAB (HB-C)—and modified HB-C are shown in Figure 2. They clearly show a characteristic, yet relatively broad, low-angle peak around 1.6°, indicating the presence of mesopores in the materials. However, the breadth of the peak suggests that mesopores do not have significant order. Nevertheless, mesopores remain present after wet impregnation with the aqueous solutions of metal salts.

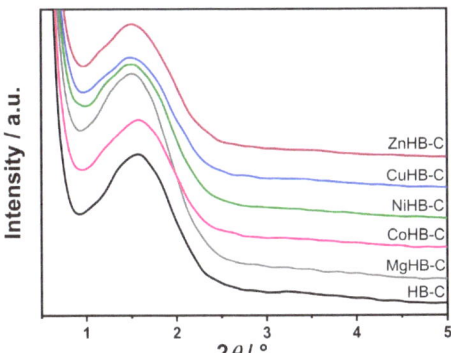

Figure 2. Low-angle diffraction patterns of samples synthesized with CTAB (parent HB-C) and modified by wet impregnation with cations (Mg^{2+}, Co^{2+}, Ni^{2+}, Cu^{2+}, and Zn^{2+}).

3.2. Scanning Electron Microscopy (SEM)

The nano-particles of zeolite Beta—HB-1 series of samples (Figure 3)—are spherical and have only one morphology. Wet impregnation followed by calcination did not alter the primary morphology of the parent sample HB-1. The particle size mainly falls within the range of 80 to 180 nm. In this series of samples, no impurities were found in terms of significantly larger-sized particles. During the drying process, the particles tend to aggregate into weak large agglomerates that can be several tens of micrometers in size as shown in Figure 4 (left). The SEM images provided show no significant influence of cation insertion on the morphology and dimensions of the MHB-1 particles. While agglomerated particles consist of even smaller crystallites, revealing a rough surface at high magnifications, their real size can only be estimated (Figure 4, right). Based on the high magnification SEM images, it can be argued that even nano-particles are polycrystalline. They are not smooth, exhibiting a rough surface, indicating a higher external surface area, which is confirmed by the BET analysis. Furthermore, relatively broad diffraction maxima of both HB-1 and HB-C series (Figure 1) point out that the crystallites, or more precisely crystalline domains, are smaller in size, corroborating findings from the SEM images.

On the other hand, the particles in the HB-C series (synthesized with CTAB, Figure 5) are much bigger, micrometer size, and contain a small amount of gel. Just a little longer time of synthesis (e.g., 8 h) results in the appearance of analcime and quartz crystals. In order to avoid it, a new procedure for synthesis of pure zeolite Beta was made, see Section 2.1. In samples impregnated with Mg(II), Ni(II), and Zn(II) acetate salts, a small quantity of material (gel) shows morphology of irregular thin sheets. So, from the diffraction patterns (Figure 1), the samples of the HB-C series show slightly lower crystallinity compared to the HB-1 series. The morphology of the parent (HB-C) sample and the samples impregnated with metal cations slightly differ. Although the rough surface of these particles is not immediately visible at lower magnifications, it becomes clear at higher magnifications (Figure 6), where agglomerates of nanometer-sized crystals are clearly visible. The particles in the HB-C series have a size in the range of 400–1300 nm (Figure 5) and are composed of much smaller crystallites whose size spans between 20–50 nm. These fine particles are better expressed in CuHB-C and CoHB-C samples (Figure 6).

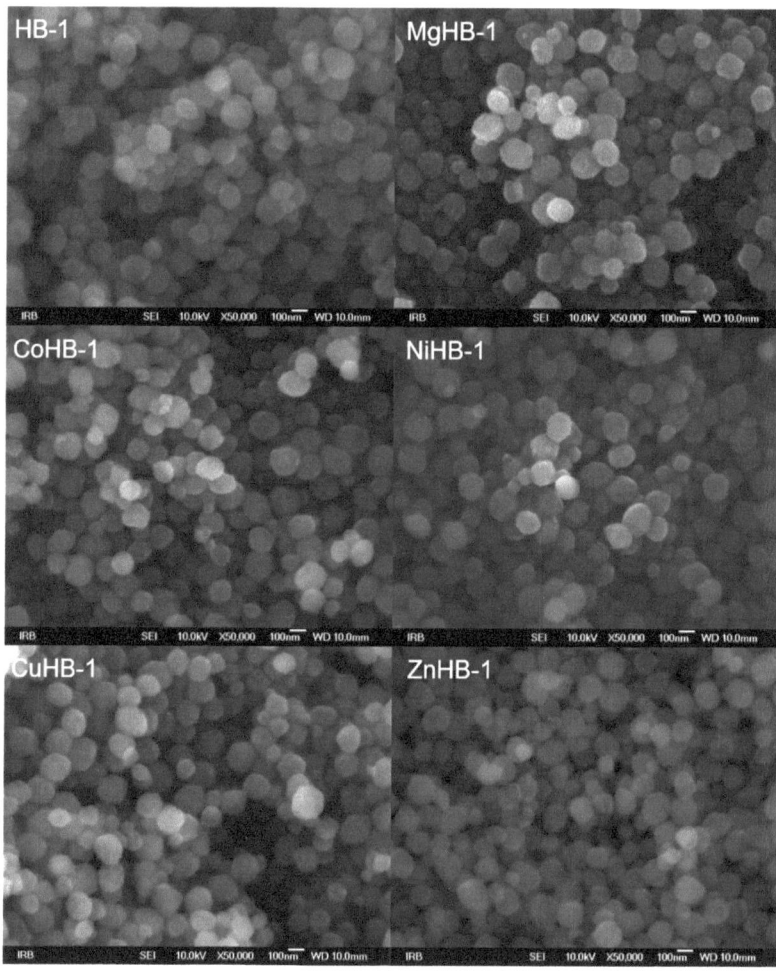

Figure 3. Photos of zeolite Beta nanocrystals modified by wet impregnation with cations (Mg^{2+}, Co^{2+}, Ni^{2+}, Cu^{2+}, and Zn^{2+}) at magnification of 50,000×.

Figure 4. Photos of zeolite Beta nanocrystals modified by wet impregnation: irregular aggregates of CoHB-1, and NiHB-1 at large magnification of 100,000×.

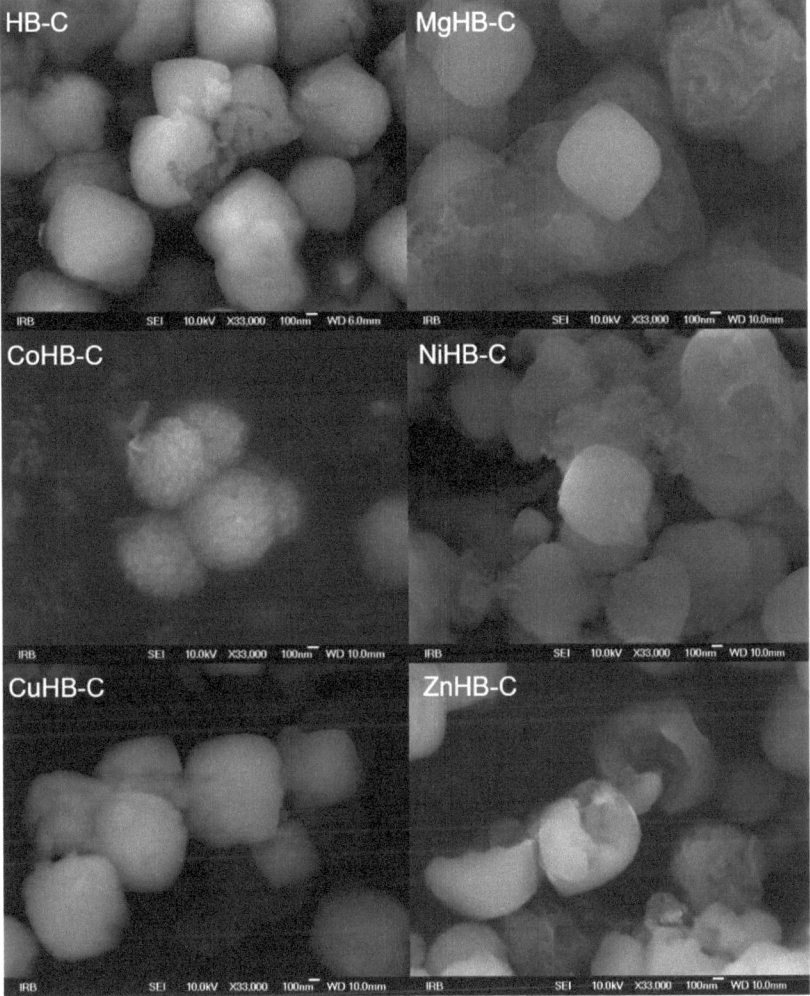

Figure 5. Photos of zeolite Beta crystals synthesized with addition of CTAB; parent (HB-C) and samples modified by wet impregnation with the following metal-acetate salts: MgHB-C, CoHB-C, NiHB-c, CuHB-C and ZnHB-C; at magnification of 33,000×.

3.3. UV Vis Spectroscopy

The DRS UV-Vis spectra of zeolite Beta samples prepared via the wet impregnation method are shown in Figure 7. The spectrum of CoHB-1 reveals a band at 515 nm which may be attributed to $[Co(H_2O)_6]^{2+}$ species in octahedral coordination. There are also broad bands around 300 nm likely related to the oxygen-to-metal charge transfer, in line with earlier reports on the framework Co^{2+} in Co-MFI and Co-APO-5 structures [32]. Moreover, the pink color of the samples also confirms the presence of octahedral Co(II) species. Sample CoHB-C contains additional bands occurring at about 525, 600, and 660 nm, and those are consistent with the formation of isolated pseudo-tetrahedral Co(II) species [32–34].

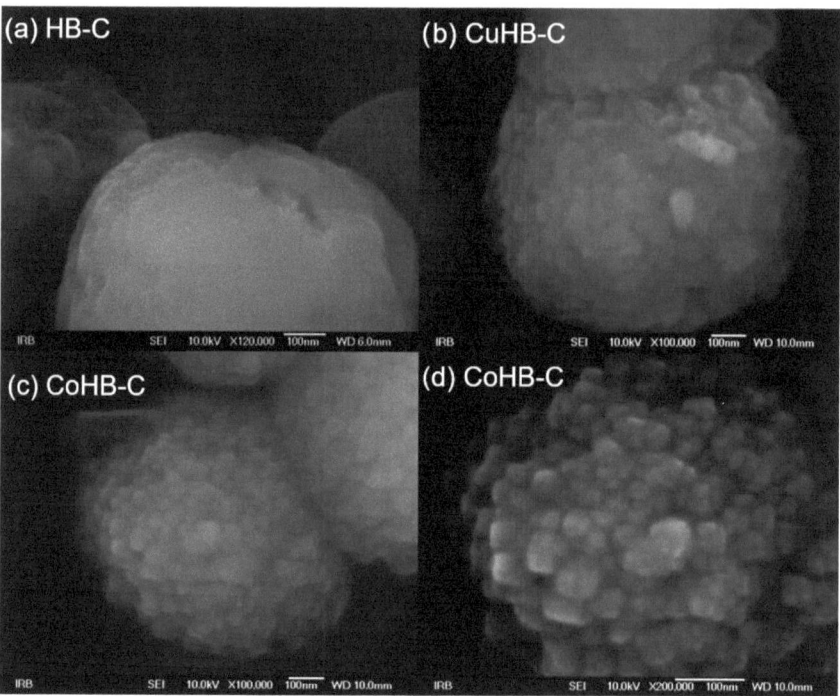

Figure 6. High magnification photos of zeolite Beta synthesized with addition of CTAB nanocrystals (**a**) and modified by Cu(II) (**b**) and Co(II) acetates (**c**,**d**).

DRS UV–Vis spectra of light green nickel-containing Beta zeolite samples look very similar and contain characteristic bands (at 405, 660, and 740 nm) related to mononuclear Ni^{2+} in octahedral coordination [35]. The shoulder visible in the spectra of sample NiHB-1 at around 300 nm indicates the presence of a small amount of NiO [35].

The DRS UV–Vis spectra, recorded for the Cu-modified Beta after calcination at 350 °C in the stream of air, reveal the presence of monomeric Cu^{2+} species. Those species are characterized with a strong absorption below 400 nm, with maxima at 230 nm [34,36]. Both samples display a broad band from 560 nm up to 1200 nm, and it is assigned to d–d transitions of isolated of Cu^{2+} ions in the distorted octahedral coordination. A weak band in a region between 380 and 600 nm can be attributed to CuO and $[Cu-O-Cu]^{2+}$ species [37]. According to PXRD patterns, there is no evidence of CuO, but the blue color of the sample also indicates that only Cu^{2+} (mainly monomeric) exists in the zeolite framework probably due to the better accessibility of the ion-exchange positions in the material [38].

Finally, the very weak band for sample CuHB-C after FTIR centered at approximately 430 nm is assigned as the O_{bridge}-Cu charge transfer transition of bis(μ-oxo)dicopper [39]. It was found that there is an optimal calcination temperature region (280–700 °C) needed to form the copper species responsible for the 440 nm band starting from a hydrated sample [40].

Samples containing Mg^{2+} and Zn^{2+} have no visible bands in the region from 200 to 1500 nm, so we can only assume that those samples would follow the same trend in metal coordination (PXRD shows that there are no metal oxides). The bands in the region between 1100 and 1430 nm are attributed to the zeolite network.

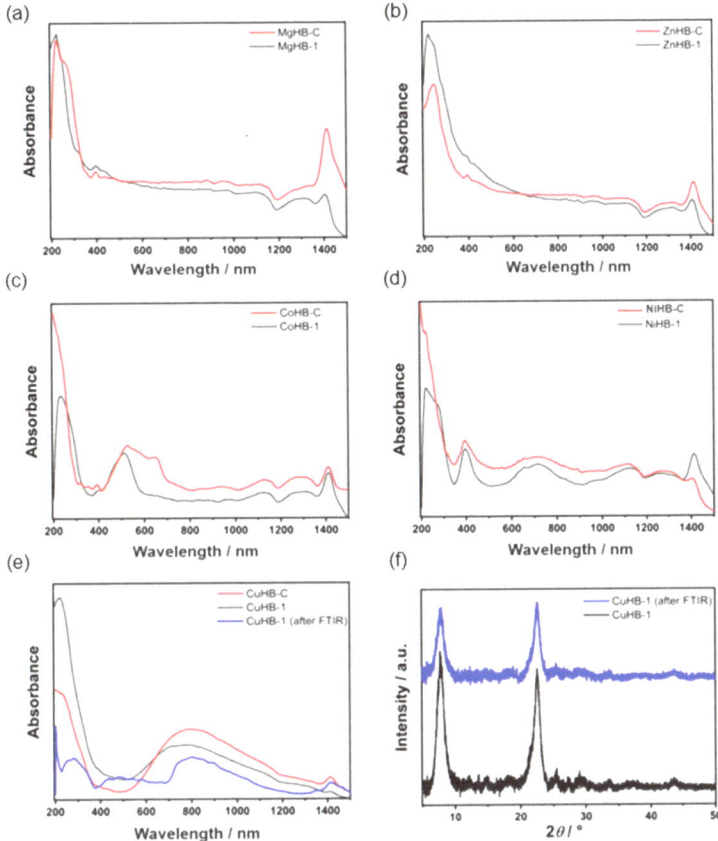

Figure 7. DRS UV-Vis spectra of metal-containing samples (**a–e**) and PXRD patterns of CuHB-1 sample before and after FTIR experiment (after desorption of CD$_3$CN at 300 °C) (**f**).

3.4. FTIR Spectroscopy

The FTIR spectra of activated HB-1 zeolite, shown in Figure 8 (top), exhibits strong bands at 3745 and 3735 cm^{-1}, associated with isolated external and isolated internal Si OH groups [41] and weak bands at 3667 cm^{-1} attributed to Al-OH groups of extra-framework aluminum (EFAL), at 3608 cm^{-1} related to bridging hydroxyls Si–O(H)–Al, and a band at 3550 cm^{-1} related to H-bonded Si-O(H)-Si groups [41–43]. The similar intensity of the bands at 3745 cm^{-1} and 3735 cm^{-1} for metal-impregnated samples in the HB-1 series, except for the sample impregnated with Mg^{2+} cations, which has a higher intensity, indicates that Mg^{2+} cations tend to generate more OH groups which directly bind to Mg.

In comparison to HB-1, the samples of HB-C have considerably weaker bands at 3745–3734 cm^{-1} and very weak bands at 3605 cm^{-1} and 3550 cm^{-1}. Generally, the smaller size of nanocrystals (HB-1) and structural defects generated by etching (desilication) are responsible for the larger number of terminal OH groups (silanol nests).

To determine the nature, number, and strength of acid sites of parents and modified, divalent metal-containing samples, the desorption FTIR spectra of adsorbed CD$_3$CN (deuterated acetonitrile), as a probe molecule, were recorded. First, the samples were activated under vacuum at 350 °C to eliminate all the water present in the materials. Then, deconvoluted spectra of CD$_3$CN desorption at different temperatures (100 °C, 150 °C, 200 °C and 300 °C) were used for calculation of the number of specific acid sites (BAS and

LAS) using Equation (1). An example of MgHB-1 sample spectra recorded at different temperatures of CD_3CN desorption and its deconvolution is shown in Figure 9.

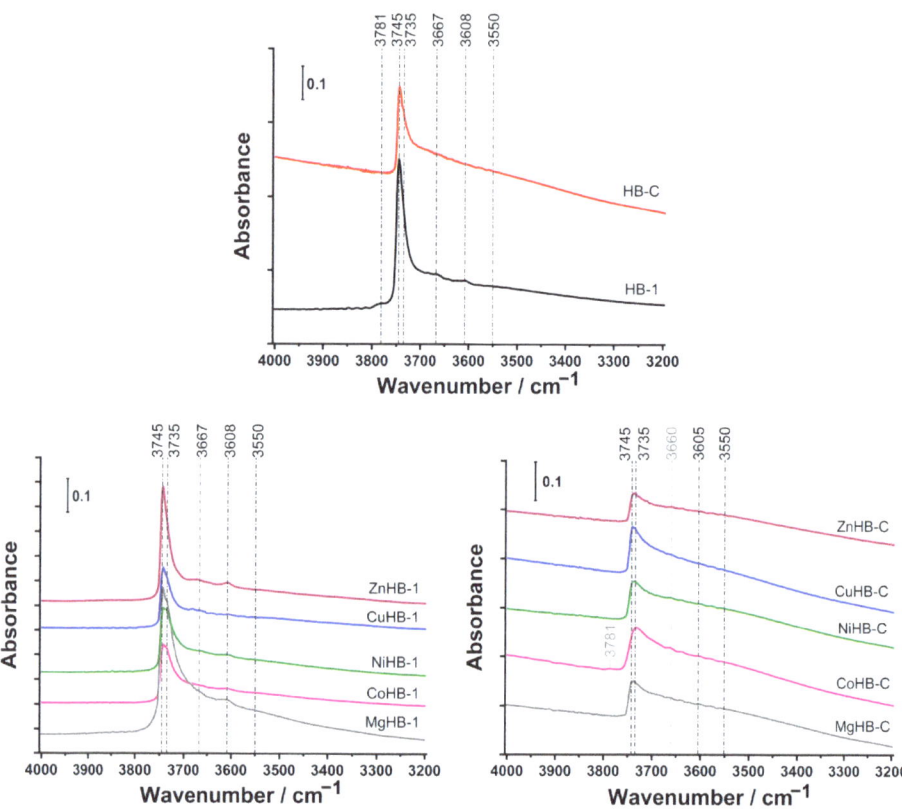

Figure 8. FTIR spectra of OH groups in the region from 4000 cm^{-1} to 3200 cm^{-1} of the samples after activation in a vacuum at 350 °C for 3 h of parents (**top**) and modified by wet impregnation with a solution of M(II) acetates: nanocrystals HB-1 series (**left**) and mesoporous crystals HB-C series (**right**).

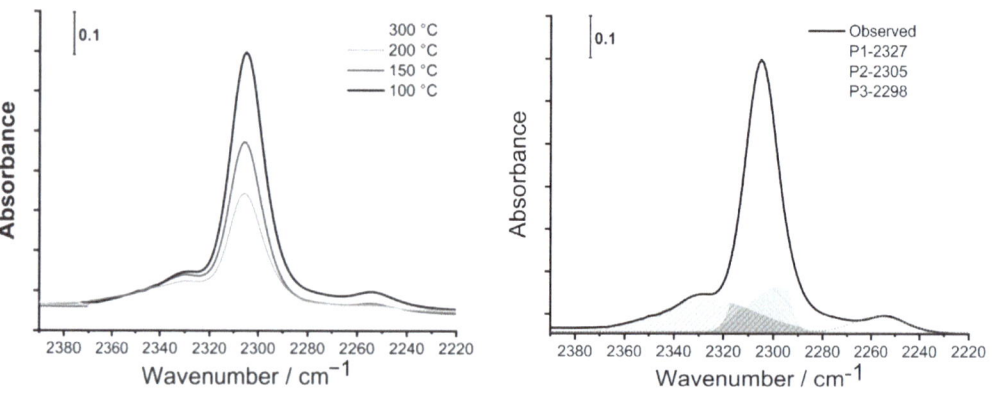

Figure 9. FTIR spectra of adsorbed CD_3CN in the region from 2390 cm^{-1} to 2220 cm^{-1} of nanocrystals impregnated with Mg^{2+} (MgHB-1) at different temperatures (**left**) and deconvolution of spectra recorded at 100 °C (**right**).

All samples have multiple bands of the adsorbed probe molecule (CD$_3$CN) in the range of 2390 cm^{-1} to 2220 cm^{-1}, covering the Lewis and Brønsted acid sites. At room temperature, the amount of adsorbed CD$_3$CN varies greatly from sample to sample due to surface adsorption. The fundamental band of ν_s(CD$_3$) and ν_{as}(CD$_3$) of CD$_3$CN in the liquid phase are at 2250 cm^{-1} and 2114 cm^{-1}, respectively. The stretching mode of ν(C≡N) is a band at 2265 cm^{-1}. Acetonitrile is a weak base and interacts with the acid sites via the lone electron pair of the C≡N group. While both modes of ν(CD$_3$) are not significantly changed after the adsorption of acetonitrile on acid sites, the strength of binding of acetonitrile to acid sites is reflected in the shift of the stretching mode of ν(C≡N) to higher frequencies [44]. The characteristic bands of adsorbed acetonitrile were attributed to acid sites of different natures: two bands of Lewis acid sites (LAS) at 2330–2320 and at 2315–2305 cm^{-1} and a band of Brønsted acid sites (BAS) at 2297 cm^{-1}, or at a slightly higher frequency (2300 cm^{-1}) due to inserted metal cations. The low-intensity bands at 2280–2275 cm^{-1}, 2265 cm^{-1}, and 2255–2245 cm^{-1} are assumed to correspond to the acetonitrile C≡N group bonded to terminal Si-OH groups or defect sites, physisorbed acetonitrile, and to C-D vibrations, respectively, as assigned by Otero Areán et al. [45].

Since the parent samples do not have a band at 2315 cm^{-1}, and the amount of BAS is larger than LAS (desorption of CD$_3$CN at 100 °C, Figure 1), it is easy to conclude that M^{2+} cations inserted to crystals were responsible for the increased amount of LAS, which prevails. The exceptions are both systems with Cu^{2+} inserted (CuHB-1 and CuHB-C, Figure 10e), where the total amount of AS is lowest.

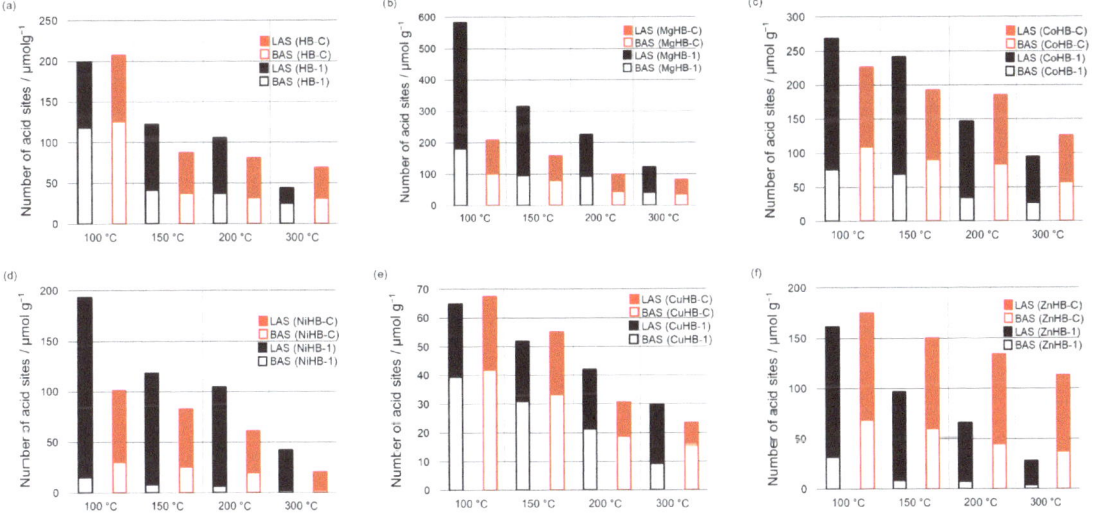

Figure 10. Relative content of specific acid sites (BAS and LAS) calculated from deconvoluted spectra of desorption of CD$_3$CN at different temperatures (100 °C, 150 °C, 200 °C, and 300 °C) for HB-1 and HB-C series (**a**–**f**).

If we compare the amount of BAS of HB-1 and the metal-impregnated HB-1 series after desorption of the probe molecule at 100 °C, all samples except MgHB-1 have less BAS than the parent HB-1. NiHB-1 has by far the smallest amount of probe molecules adsorbed on BAS at 100 °C, and only 13% of them are strong enough to bind CD$_3$CN after desorption at 300 °C; while the MgHB-1 sample has by far the largest amount of BAS and 22% of all Brønsted acid sites detected are still bound to zeolite at 300 °C, just as in the parent sample. In contrast to the Brønsted acid sites, all metal-impregnated samples have multiple bands of adsorbed probe molecules corresponding to Lewis acid sites (Figure 10). HB-1 has only one band, as mentioned above. The amount and strength of these sites differ drastically. All metal-impregnated HB-1 samples except the sample with copper, which has less BAS and LAS than the parent material, have at least 40% more LAS after desorption of D_3-acetonitrile at 100 °C. Even the sample impregnated with Cu^{2+} has a very low amount of LAS, 80% of which is strong. For the samples exchanged with Co^{2+}, Zn^{2+}, and Ni^{2+}, the number of adsorbed probe molecules at the Lewis acid sites barely differs at 150 °C and 200 °C, which is also observed for the parent samples HB-1. Finally, only the samples impregnated with Mg^{2+} and Co^{2+} have a higher total amount of acid sites than the parent material. Moreover, a large influence of the metal cation on the ratio of BAS/LAS in the sample NiHB-1 was observed.

For the HB-C series, it is obvious that all metal-impregnated samples have less BAS than the parent material. Like NiHB-1, the sample NiHB-C has by far the lowest amount of BAS. At 150 °C, only 50% of the probe molecules are still adsorbed, and after desorption at 300 °C, only 6% of them are still adsorbed.

Comparing the samples prepared by different procedures, some interesting peculiarities could be observed. For example, the HB-1 series, prepared by post-synthesis treatment with NaOH/TPAOH, have more acid sites except for the Zn^{2+} exchanged sample (Tables S1–S4). More acid sites in the HB-1 series are generated by desilication and consequentially its larger specific surface area and pore volumes. Also, there is more Al atoms in HB-1 than in the HB-C series (Table 1).

Table 1. Surface characteristics of samples determined using the N_2 adsorption–desorption method, and element ratios (Si/Al, M/Al) in the samples determined using AAS.

SAMPLE	S_{BET} (m^2g^{-1})	V_{total} [a] (cm^3g^{-1})	V_{micro} [b] (cm^3g^{-1})	S_{micro} [b] (m^2g^{-1})	S_{ext} [b] (m^2g^{-1})	V_{meso} [c] (cm^3g^{-1})	Si/Al [d]	M/Al [d]
HB-0	548	0.271	0.183	478	71	0.088	25.4	/
HB-1	674	0.527	0.118	278	397	0.409	13.9	/
MgHB-1	645	0.491	0.121	286	358	0.370	20.8	0.54
CoHB-1	658	0.525	0.123	293	365	0.402	20.6	0.58
NiHB-1	620	0.518	0.119	249	370	0.399	20.9	0.51
CuHB-1	618	0.503	0.115	275	343	0.388	21.2	0.95
ZnHB-1	621	0.489	0.122	291	330	0.367	20.2	0.60
HB-C	558	0.486	0.084	202	369	0.402	31.3	/
MgHB-C	471	0.407	0.055	131	340	0.352	31.6	0.50
CoHB-C	509	0.437	0.077	185	324	0.360	31.5	0.59
NiHB-C	519	0.443	0.084	199	320	0.359	31.9	0.50
CuHB-C	478	0.415	0.073	173	305	0.342	32.0	1.00
ZnHB-C	429	0.382	0.050	119	309	0.332	29.6	0.58

[a] total volume at $p/p_0 = 0.95$; [b] calculated using t-plot method; [c] $V_{meso} = V_{total} - V_{micro}$; [d] molar quantity of elements determined by flame atomic absorption spectroscopy.

Although the HB-1 series has a relatively large total number of LAS, only the minor part (19 to 35%) of these sites are present after desorption of CD_3CN at 300 °C, excluding the CuHB-1 sample. In contrast to HB-1, HB-C samples have a larger percentage (25 to 71%) of the strongest LAS.

When we compare the number and strength of BAS, parent and Cu^{2+} impregnated samples of HB-1 and HB-C are almost the same. However, the HB-C series has a greater number of the strongest BAS in Co^{2+}, Cu^{2+}, and Zn^{2+} impregnated samples, with MgHB-1 as an outlier.

In samples impregnated with Cu^{2+} ions, a drastic reduction of Brønsted and Lewis active sites was observed. In the FTIR spectra, a band appears at 2325 cm^{-1} as a consequence of the interaction between CD_3CN and extra-framework aluminum. In the literature, it is explained that the classical ion-exchange procedure, i.e., starting from Cu^{2+} salt, can lead to divalent oxocations, e.g., $[Cu–O–Cu]^{2+}$ which can be located only in the vicinity of a pair of nearby Al sites. Theoretically, Al pairs can be significantly present only in zeolites with a low Si/Al ratio, and calculations have shown their possible existence in ZSM-5, mordenite, and ferrierite, providing coordination sites for divalent metal oxocations [46]. For high Si/Al ratios, charge compensation must follow other ways, and can be achieved, e.g., by OH− groups, resulting in monovalent $[Cu–OH]^+$ complexes.

The mechanism for the "self-reduction" of Cu^{2+} under vacuum or inert flow is thought to start from the dehydration of two $[Cu–OH]^+$ ions located close to one framework Al [47]. There is a possibility that Cu^+ in the samples is formed via autoreduction of $[CuOH]^+$ during the vacuum pretreatment [22]. In the literature, the "blocking effect" of the active sites is explained by the formation of a strong complex between the Cu^{2+} cation and probe molecule (CD_3CN) during its introduction into the zeolite framework [47].

3.5. Textural Properties—Porosimetry (N_2 Adsorption/Desorption)

The N_2 adsorption–desorption isotherms used to estimate the textural properties of all samples studied are shown in Figure 11a,c, and the data are listed in Table 1. As can be seen in Figure 11, HB-0 shows a typical type I isotherm, reflecting its intrinsic microporous structure. Other samples obtained by etching (HB-1 series) or by using the surfactant CTAB (HB-C series) during synthesis show a composition of a type I and IV isotherm with a hysteresis loop.

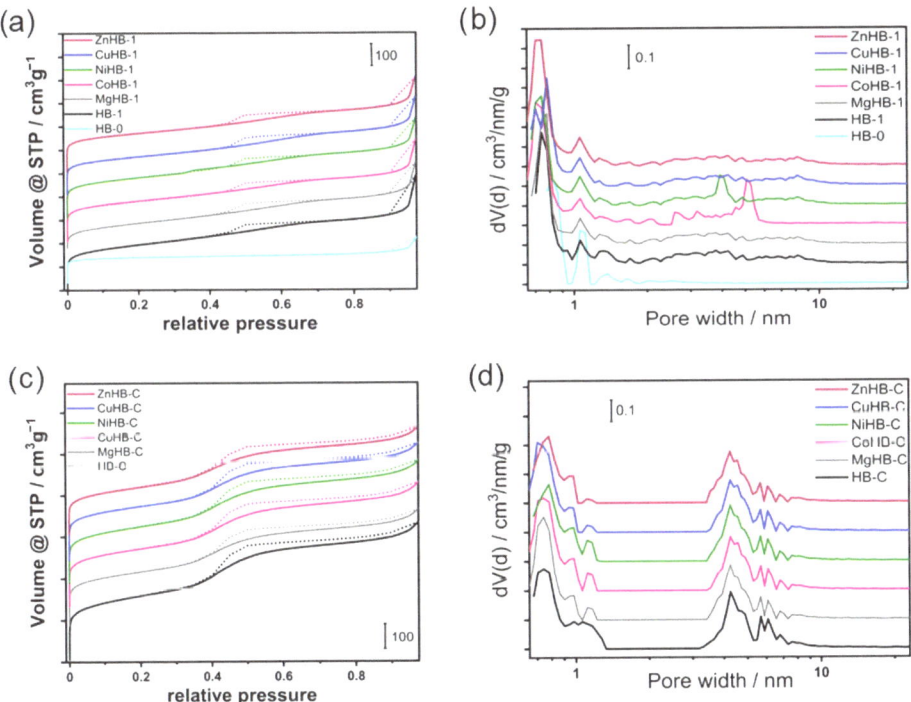

Figure 11. N_2 adsorption/desorption isotherms at −196 °C of the (**a**) nanocrystals (HB-1 series) and (**c**) microcrystals (HB-C series). The corresponding NLDFT pore size distribution of (**b**) HB-1 series (including HB-0) and (**d**) HB-C series.

These samples show gradual but continuous increasing curves after a steep increase due to the filling of the micropores in the low relative pressure region ($p/p_0 < 0.01$). At a relative pressure of 0.4–0.98 p/p_0, the desorption branch was significantly higher than the adsorption branch, along with the occurrence of capillary condensation, leading to the hysteresis loop. These analyses confirm a micro-meso hierarchical porous structural feature. The HB-1 series shows a hysteresis loop typical of a non-uniform pore size distribution (Figure 11b).

Pore size distributions were estimated using the NLDFT method on the adsorption branch of the isotherms using nitrogen on silica, sphere-cylindrical pore and show a pore size distribution where the mesopore diameters are in the range from 3 to 8 nm (Figure 11d).

The mesopores in the HB-1 series samples (Figure 11b) are widely dispersed in the range of 2–10 nm and can be attributed to inter-nanocrystals voids in the aggregates after etching (Figure 4).

The desilication of HB-0 results in an increase in the total specific surface area and substantial increase in the external surface area. The micropore volume decreases with the development of mesoporosity, with the mesopore volume almost five times higher than the microporous HB-0.

As expected, the subsequent introduction of divalent metal cations leads to a notable decrease in the total volume of both systems (HB-1 and HB-C). It is explained by the shadowing effect of inserted M^{2+} cations (column V_{total} in Table 1).

The mesopores in HB-C are well defined and open to the external surface of the crystals, which is important for improving the accessibility of the inner parts of the crystals.

It is evident that wet impregnation of HB-1 with divalent metal acetates leads to partial dealumination (an increase in the Si/Al ratio), which is not observed with the impregnation of HB-C. This is due to the presence of a certain amount of extra-framework aluminum at the surface of the etched sample which can be easily separated from the zeolite structure under acidic conditions [29].

4. Conclusions

Insertion of metal cations (Mg^{2+}, Co^{2+}, Ni^{2+}, Cu^{2+}, Zn^{2+}) into hierarchical zeolite Beta crystals strongly influence the acidity of OH groups; increasing the amount of LAS and decreasing the amount of BAS in comparison to parent crystals. Generally, etched samples (HB-1) have more structural defects including "silanol nests", and it seems that Mg^{2+} and Co^{2+} cations better fit to these defects and have more AS than the parent, confirming the findings of Medak et al. [33] in a study of zeolite FAU.

It is difficult to determine from these experiments what happens within the copper-containing samples. However, under conditions of the adsorption/desorption experiment of CD_3CN in a vacuum, an irreversible formation of the new copper species occurs, as clearly observed in the DRS UV-Vis spectra. Considering that the amount of BAS and LAS significantly deviates from the trend followed by samples impregnated with other divalent cations, it is very likely that during heating in a vacuum and saturation with CD_3CN vapor, there is an inactivation or blocking of access to the active sites which are not occupied by copper. This is likely because the metal center fills its coordination sphere with acetonitrile molecules, forming a stable complex, and the size of the Beta zeolite channels allows for the formation of such complexes.

Supplementary Materials: The following supporting information can be downloaded at: https://www.mdpi.com/article/10.3390/cryst14010053/s1, Tables S1–S4: number of acid sites calculated from deconvoluted spectra of desorption of CD_3CN for HB-1 and HB-C series at 100, 150, 200 and 300 °C.

Author Contributions: Conceptualization, J.B.; methodology, I.L.; formal analysis, I.L., M.R. and A.P.; investigation, I.L.; resources, J.B.; writing—original draft preparation, I.L.; writing—review and editing, J.B. and A.P.; visualization, I.L.; supervision, J.B. All authors have read and agreed to the published version of the manuscript.

Funding: This research was funded by the Croatian Science Foundation through project IP-2016-06-2214.

Data Availability Statement: The data presented in this study are available in the article.

Conflicts of Interest: The authors declare no conflicts of interest. The funders had no role in the design of the study; in the collection, analyses, or interpretation of data; in the writing of the manuscript; or in the decision to publish the results.

References

1. Wang, J.-J.; Chuang, Y.-Y.; Hsu, H.-Y.; Tsai, T.-C. Toward industrial catalysis of zeolite for linear alkylbenzene synthesis: A mini review. *Catal. Today* **2017**, *298*, 109–116. [CrossRef]
2. Armor, J.N. A history of industrial catalysis. *Catal. Today* **2011**, *163*, 3–9. [CrossRef]
3. Abate, S.; Barbera, K.; Centi, G.; Lanzafame, P.; Perathoner, S. Disruptive catalysis by zeolites. *Catal. Sci. Technol.* **2016**, *6*, 2485–2501. [CrossRef]
4. Hong, L.; Zang, J.; Li, B.; Liu, G.; Wang, Y.; Wu, L. Research Progress on the Synthesis of Nanosized and Hierarchical Beta Zeolites. *Inorganics* **2023**, *11*, 214. [CrossRef]
5. Valtchev, V.; Tosheva, L. Porous Nanosized Particles: Preparation, Properties, and Applications. *Chem. Rev.* **2013**, *113*, 6734–6760. [CrossRef] [PubMed]
6. Yang, L.; Wang, C.; Zhang, L.; Dai, W.; Chu, Y.; Xu, J.; Wu, G.; Gao, M.; Liu, W.; Xu, Z.; et al. Stabilizing the framework of SAPO-34 zeolite toward long-term methanol-to-olefins conversion. *Nat. Commun.* **2021**, *12*, 4661. [CrossRef]
7. Dalena, F.; Puškarić, A.; Landripet, I.; Jelić, T.A.; Bosnar, S.; Medak, G.; Marino, A.; Giordano, G.; Bronić, J.; Migliori, M. Combining hierarchization and Mg^{2+} ions insertion in ZSM-5: Acidity-modulation effect on MTO reaction. *Mol. Catal.* **2023**, *545*, 113181. [CrossRef]
8. Bordiga, S.; Lamberti, C.; Bonino, F.; Travert, A.; Thibault-Starzyk, F. Probing zeolites by vibrational spectroscopies. *Chem. Soc. Rev.* **2015**, *44*, 7262–7341. [CrossRef]
9. Hegde, S.; Abdullah, R.; Bhat; Ratnasamy, P. FTi.r. spectroscopic study of gallium beta. *Zeolites* **1992**, *15*, 951–956. [CrossRef]
10. Kefirov, R.; Penkova, A.; Hadjiivanov, K.; Dzwigaj, S.; Che, M. Stabilization of Cu^+ ions in BEA zeolite: Study by FTIR spectroscopy of adsorbed CO and TPR. *Microporous Mesoporous Mater.* **2008**, *116*, 180–187. [CrossRef]
11. Anquetil, R.; Saussey, J.; Lavalley, J.-C. Confinement effect on the interaction of hydroxy groups contained in the side pockets of H-mordenite with nitriles; a FT-IR study. *Phys. Chem. Chem. Phys.* **1999**, *1*, 555–560. [CrossRef]
12. Pelmenschikov, A.G.; van Santen, R.A.; Janchen, J.; Meijer, E. Acetonitrile-d3 as a probe of Lewis and Broensted acidity of zeolites. *J. Phys. Chem.* **1993**, *97*, 11071–11074. [CrossRef]
13. Kotrel, S.; Lunsford, J.H.; Knözinger, H. Characterizing Zeolite Acidity by Spectroscopic and Catalytic Means: A Comparison. *J. Phys. Chem. B* **2001**, *105*, 3917–3921. [CrossRef]
14. Góra-Marek, K.; Datka, J.; Dzwigaj, S.; Che, M. Influence of V Content on the Nature and Strength of Acidic Sites in VSiβ Zeolite Evidenced by IR Spectroscopy. *J. Phys. Chem. B* **2006**, *110*, 6763–6767. [CrossRef] [PubMed]
15. Bisio, C.; Massiani, P.; Fajerwerg, K.; Sordelli, L.; Stievano, L.; Silva, E.; Coluccia, S.; Martra, G. Identification of cationic and oxidic caesium species in basic Cs-overloaded BEA zeolites. *Microporous Mesoporous Mater.* **2006**, *90*, 175–187. [CrossRef]
16. Zhang, W.; Ming, W.; Hu, S.; Qin, B.; Ma, J.; Li, R. A Feasible One-Step Synthesis of Hierarchical Zeolite Beta with Uniform Nanocrystals via CTAB. *Materials* **2018**, *12*, 651. [CrossRef]
17. Verboekend, D.; Pérez-Ramírez, J. Design of hierarchical zeolite catalysts by desilication. *Catal. Sci. Technol.* **2011**, *1*, 879–890. [CrossRef]
18. De Baerdemaeker, T.; Yilmaz, B.; Müller, U.; Feyen, M.; Xiao, F.-S.; Zhang, W.; Tatsumi, T.; Gies, H.; Bao, X.; De Vos, D. Catalytic applications of OSDA-free Beta zeolite. *J. Catal.* **2013**, *308*, 73–81. [CrossRef]
19. Chaida-Chenni, F.Z.; Belhadj, F.; Casas, M.S.G.; Márquez-Álvarez, C.; Hamacha, R.; Bengueddach, A.; Pérez-Pariente, J. Synthesis of mesoporous-zeolite materials using Beta zeolite nanoparticles as precursors and their catalytic performance in m-xylene isomerization and disproportionation. *Appl. Catal. A Gen.* **2018**, *568*, 148–156. [CrossRef]
20. Wang, Y.; Yokoi, T.; Namba, S.; Tatsumi, T. Effects of Dealumination and Desilication of Beta Zeolite on Catalytic Performance in n-Hexane Cracking. *Catalysts* **2016**, *6*, 8. [CrossRef]
21. Hao, W.; Zhang, W.; Guo, Z.; Ma, J.; Li, R. Mesoporous Beta Zeolite Catalysts for Benzylation of Naphthalene: Effect of Pore Structure and Acidity. *Catalysts* **2018**, *8*, 504. [CrossRef]
22. Miao, S.; Sun, S.; Lei, Z.; Sun, Y.; Zhao, C.; Zhan, J.; Zhang, W.; Jia, M. Micron-Sized Hierarchical Beta Zeolites Templated by Mesoscale Cationic Polymers as Robust Catalysts for Acylation of Anisole with Acetic Anhydride. *Catalysts* **2023**, *13*, 1517. [CrossRef]
23. Zhao, R.; Zhao, Z.; Li, S.; Parvulescu, A.; Müller, U.; Zhang, W. Excellent Performances of Dealuminated H-Beta Zeolites from Organotemplate-Free Synthesis in Conversion of Biomass-derived 2,5-Dimethylfuran to Renewable p-Xylene. *ChemSusChem* **2018**, *11*, 3803–3811. [CrossRef] [PubMed]
24. Rutkowska, M.; Piwowarska, Z.; Micek, E.; Chmielarz, L. Hierarchical Fe-, Cu-and Co-Beta zeolites obtained by mesotemplate-free method. Part I: Synthesis and catalytic activity in N_2O decomposition. *Microporous Mesoporous Mater.* **2015**, *209*, 54–65. [CrossRef]
25. Xu, L.; Shi, C.; Zhang, Z.; Gies, H.; Xiao, F.-S.; De Vos, D.; Yokoi, T.; Bao, X.; Feyen, M.; Maurer, S.; et al. Enhancement of low-temperature activity over Cu-exchanged zeolite beta from organotemplate-free synthesis for the selective catalytic reduction of NO_x with NH_3 in exhaust gas streams. *Microporous Mesoporous Mater.* **2014**, *200*, 304–310. [CrossRef]

26. Cruz-Cabeza, A.J.; Esquivel, D.; Jiménez-Sanchidrián, C.; Romero-Salguero, F.J. Metal-Exchanged β Zeolites as Catalysts for the Conversion of Acetone to Hydrocarbons. *Materials* **2012**, *5*, 121–134. [CrossRef] [PubMed]
27. Esquivel, D.; Cruz-Cabeza, A.J.; Jiménez-Sanchidrián, C.; Romero-Salguero, F.J. Transition metal exchanged β zeolites: Characterization of the metal state and catalytic application in the methanol conversion to hydrocarbons. *Microporous Mesoporous Mater.* **2013**, *179*, 30–39. [CrossRef]
28. Atoguchi, T.; Kanougi, T. Phenol oxidation over alkaline earth metal ion exchange beta zeolite in the presence of ketone. *J. Mol. Catal. A Chem.* **2004**, *222*, 253–257. [CrossRef]
29. Essid, S.; Ayari, F.; Bulánek, R.; Vaculík, J.; Mhamdi, M.; Delahay, G.; Ghorbel, A. Improvement of the conventional preparation methods in Co/BEA zeolites: Characterization and ethane ammoxidation. *Solid State Sci.* **2019**, *93*, 13–23. [CrossRef]
30. Mintova, S.; Valtchev, V.; Onfroy, T.; Marichal, C.; Knözinger, H.; Bein, T. Variation of the Si/Al ratio in nanosized zeolite Beta crystals. *Microporous Mesoporous Mater.* **2006**, *90*, 237–245. [CrossRef]
31. Wichterlová, B.; Tvarůžková, Z.; Sobalík, Z.; Sarv, P. Determination and properties of acid sites in H-ferrierite: A comparison of ferrierite and MFI structures. *Microporous Mesoporous Mater.* **1998**, *24*, 223–233. [CrossRef]
32. Sadek, R.; Chalupka-Spiewak, K.; Krafft, J.-M.; Millot, Y.; Valentin, L.; Casale, S.; Gurgul, J.; Dzwigaj, S. The Synthesis of Different Series of Cobalt BEA Zeolite Catalysts by Post-Synthesis Methods and Their Characterization. *Catalysts* **2022**, *12*, 1644. [CrossRef]
33. Medak, G.; Puškarić, A.; Bronić, J. The Influence of Inserted Metal Ions on Acid Strength of OH Groups in Faujasite. *Crystals* **2023**, *13*, 332. [CrossRef]
34. Sobuś, N.; Michorczyk, B.; Piotrowski, M.; Kuterasiński, Ł.; Chlebda, D.K.; Łojewska, J.; Jędrzejczyk, R.J.; Jodłowski, P.; Kuśtrowski, P.; Czekaj, I. Design of Co, Cu and Fe–BEA Zeolite Catalysts for Selective Conversion of Lactic Acid into Acrylic Acid. *Catal. Lett.* **2019**, *149*, 3349–3360. [CrossRef]
35. Śrębowata, A.; Baran, R.; Łomot, D.; Lisovytskiy, D.; Onfroy, T.; Dzwigaj, S. Remarkable effect of postsynthesis preparation procedures on catalytic properties of Ni-loaded BEA zeolites in hydrodechlorination of 1,2-dichloroethane. *Appl. Catal. B Environ.* **2014**, *147*, 208–220. [CrossRef]
36. Baran, R.; Averseng, F.; Wierzbicki, D.; Chalupka, K.; Krafft, J.-M.; Grzybek, T.; Dzwigaj, S. Effect of postsynthesis preparation procedure on the state of copper in CuBEA zeolites and its catalytic properties in SCR of NO with NH_3. *Appl. Catal. A Gen.* **2016**, *523*, 332–342. [CrossRef]
37. Wang, H.; Xu, R.; Jin, Y.; Zhang, R. Zeolite structure effects on Cu active center, SCR performance and stability of Cu-zeolite catalysts. *Catal. Today* **2019**, *327*, 295–307. [CrossRef]
38. Rutkowska, M.; Díaz, U.; Palomares, A.E.; Chmielarz, L. Cu and Fe modified derivatives of 2D MWW-type zeolites (MCM-22, ITQ-2 and MCM-36) as new catalysts for DeNOx process. *Appl. Catal. B Environ.* **2015**, *168–169*, 531–539. [CrossRef]
39. Groothaert, M.H.; Smeets, P.J.; Sels, B.F.; Jacobs, P.A.; Schoonheydt, R.A. Selective Oxidation of Methane by the Bis(μ-oxo)dicopper Core Stabilized on ZSM-5 and Mordenite Zeolites. *J. Am. Chem. Soc.* **2005**, *127*, 1394–1395. [CrossRef]
40. Smeets, P.J.; Groothaert, M.H.; Schoonheydt, R.A. Cu based zeolites: A UV–vis study of the active site in the selective methane oxidation at low temperatures. *Catal. Today* **2005**, *110*, 303–309. [CrossRef]
41. Dzwigaj, S.; Massiani, P.; Davidson, A.; Che, M. Role of silanol groups in the incorporation of V in β zeolite. *J. Mol. Catal. A Chem.* **2000**, *155*, 169–182. [CrossRef]
42. Maache, M.; Janin, A.; Lavalley, J.; Joly, J.; Benazzi, E. Acidity of zeolites Beta dealuminated by acid leaching: An FTi.r. study using different probe molecules (pyridine, carbon monoxide). *Zeolites* **1993**, *13*, 419–426. [CrossRef]
43. Janin, A.; Maache, M.; Raatz, F.; Lavalley, J.C.; Joly, J.F.; Szydlowski, N. FTIR study of the silanol groups in dealuminated HY zeolites: Nature of the extraframework debris. *Zeolites* **1991**, *11*, 391–396. [CrossRef]
44. Chen, J.; Thomas, J.M.; Sankar, G. IR spectroscopic study of CD3CN adsorbed on ALPO-18 molecular sieve and the solid acid catalysts SAPO-18 and MeAPO-18. *J. Chem. Soc. Faraday Trans.* **1994**, *90*, 3455–3459. [CrossRef]
45. Areán, C.O.; Platero, E.E.; Mentruit, M.P.; Delgado, M.R.; i Xamena, F.L.; García-Raso, A.; Morterra, C. The combined use of acetonitrile and adamantane–carbonitrile as IR spectroscopic probes to discriminate between external and internal surfaces of medium pore zeolites. *Microporous Mesoporous Mater.* **2000**, *34*, 55–60. [CrossRef]
46. Giordanino, F.; Vennestrøm, P.N.R.; Lundegaard, L.F.; Stappen, F.N.; Mossin, S.; Beato, P.; Bordiga, S.; Lamberti, C. Characterization of Cu-exchanged SSZ-13: A comparative FTIR, UV-Vis, and EPR study with Cu-ZSM-5 and Cu-β with similar Si/Al and Cu/Al ratios. *Dalton Trans.* **2013**, *42*, 12741–12761. [CrossRef]
47. Deka, U.; Lezcano-Gonzalez, I.; Weckhuysen, B.M.; Beale, A.M. Local Environment and Nature of Cu Active Sites in Zeolite-Based Catalysts for the Selective Catalytic Reduction of NO_x. *ACS Catal.* **2013**, *3*, 413–427. [CrossRef]

Disclaimer/Publisher's Note: The statements, opinions and data contained in all publications are solely those of the individual author(s) and contributor(s) and not of MDPI and/or the editor(s). MDPI and/or the editor(s) disclaim responsibility for any injury to people or property resulting from any ideas, methods, instructions or products referred to in the content.

Article

Electrochemical Investigation of the OER Activity for Nickel Phosphite-Based Compositions and Its Morphology-Dependent Fluorescence Properties

Maria Poienar [1], Paula Svera [1], Bogdan-Ovidiu Taranu [1], Catalin Ianasi [2], Paula Sfirloaga [1], Gabriel Buse [3], Philippe Veber [4,*] and Paulina Vlazan [1]

[1] National Institute for Research and Development in Electrochemistry and Condensed Matter, Dr. A. Paunescu Podeanu Street, No. 144, 300569 Timisoara, Romania
[2] Coriolan Drăgulescu Institute of Chemistry, Bv. Mihai Viteazul, No. 24, 300223 Timisoara, Romania
[3] Institute for Advanced Environmental Research, West University of Timisoara (ICAM-WUT), Oituz Str., No. 4, 300086 Timisoara, Romania
[4] French National Centre for Scientific Research, Institute of Light and Matter, University Claude Bernard Lyon 1, UMR 5306 Villeurbanne, France
* Correspondence: philippe.veber2@univ-lyon1.fr; Tel.: +33-472431208

Citation: Poienar, M.; Svera, P.; Taranu, B.-O.; Ianasi, C.; Sfirloaga, P.; Buse, G.; Veber, P.; Vlazan, P. Electrochemical Investigation of the OER Activity for Nickel Phosphite-Based Compositions and Its Morphology-Dependent Fluorescence Properties. *Crystals* **2022**, *12*, 1803. https://doi.org/10.3390/cryst12121803

Academic Editors: Sanja Burazer and Lidija Androš Dubraja

Received: 31 October 2022
Accepted: 5 December 2022
Published: 12 December 2022

Publisher's Note: MDPI stays neutral with regard to jurisdictional claims in published maps and institutional affiliations.

Copyright: © 2022 by the authors. Licensee MDPI, Basel, Switzerland. This article is an open access article distributed under the terms and conditions of the Creative Commons Attribution (CC BY) license (https://creativecommons.org/licenses/by/4.0/).

Abstract: Herein, we present the investigation of catalytical and fluorescence properties for $Ni_{11}(HPO_3)_8(OH)_6$ materials obtained through a hydrothermal approach. As part of the constant search for new materials that are both cost effective and electrocatalytically active for the oxygen evolution reaction (OER) in alkaline medium, the present study involves several graphite electrodes modified with $Ni_{11}(HPO_3)_8(OH)_6$ mixed with reduced graphene oxide (rGO) and carbon black. The experimental results obtained in 0.1 mol L^{-1} KOH electrolyte solution show the electrode modified with rGO, 5 mg carbon black and 1 mg nickel phosphite as displaying the highest current density. This performance can be attributed to the synergistic effect between nickel phosphite and the carbon materials. Investigation of the electrode's OER performance in 0.1 mol L^{-1} KOH solution revealed a Tafel slope value of just 46 mV dec^{-1}. By increasing the concentration to 0.5 and 1 mol L^{-1}, this value increased as well, but there was a significant decrease in overpotential. Fluorescence properties were analyzed for the first time at the excitation length of 344 nm, and the observed strong and multiple emissions are described.

Keywords: nickel phosphite; carbon; oxygen evolution reaction; fluorescence

1. Introduction

Due to growing environmental pollution concerns from vehicles and industry, and the rapid consumption of fossil fuels, the need to satisfy the demand for sustainable and clean energy resources has attracted great interest, and it has become an increasing preoccupation of the research and development field [1]. Since hydrogen is considered a clean energy source, water splitting using a photoelectric or electric current is widely regarded as an encouraging approach to obtain eco-friendly fuel for tomorrow's energy supply [2,3]. As an example, electrocatalytic water splitting is viewed as a very convenient and environmentally friendly technology for generating oxygen and hydrogen to supply PEMFC-driven vehicles [4]. Electrochemical water splitting is a process that implies the dividing of the H_2O molecule into H_2 and O_2, with both resulting molecules being considered as fuels with zero carbon emission [5].

Basically, there are two unfolding half-cell reactions: at the anode, water splits into O_2 gas through the oxygen evolution reaction (OER), while, at the cathode, H_2 gas evolves via the hydrogen evolution reaction (HER) [6]. The HER is catalyzed with a very facile kinetics, but the OER requires more energy, since it is a four-electron reaction [7,8].

Theoretically, the electrochemical potential for O_2 evolution is 1.23 V vs. the Normal Hydrogen Electrode, but in practice, in order to overcome the energy barrier, a higher value needs to be applied [9,10]. In other words, from a kinetic point of view, the OER is difficult to control since it involves higher energy intermediates and multiple proton-coupled electron transfer steps [11,12]. There are materials that can serve as OER catalysts by lowering the reaction active energy, and they can be grouped into two categories: metal-based and carbon-based compounds [13–15].

Currently, noble metal-based electrocatalysts display the highest OER catalytic activity, especially the ones containing Ir and Ru [16,17]. However, noble metals are expensive and have a low abundance, which impedes their use in large-scale applications and impose the development of alternative strategies.

Transition metals that have been identified as efficient OER catalysts include divalent cations, such as Mn, Fe, Co, and Ni [5], that exhibit their electrocatalytic activity in the following order: $Mn^{2+} < Fe^{2+} < Co^{2+} < Ni^{2+}$ [18]. The success of eco-friendly water splitting-based technologies depends on the development of efficient and earth abundant transition metal catalysts, and this outlines the importance of directing the research focus toward metals such as Fe, Co, and Ni [19,20]. Their complexes display high OER electrocatalytic activity, and Ni-based complexes in particular have proven to be attractive due to their notable performance and relative low cost [21–23].

The present work is a continuation of previous investigations [24], and it consists in an electrochemical study involving several graphite electrodes modified with compositions containing $Ni_{11}(HPO_3)_8(OH)_6$, reduced graphene oxide (rGO), and carbon black. The catalytic material, a member of the metal phosphites class, was selected based on literature reports that indicated nickel phosphites as promising electrocatalysts for water splitting [12,23]. Considering that there are not many published studies evaluating the water splitting electrocatalytic properties of nickel phosphites, the aim of this paper is to complement the current literature by outlining the catalytic properties of the mentioned nickel phosphite-based compositions for the OER in alkaline medium. Furthermore, the compound's fluorescence properties are presented and interpreted for the first time in the case of nickel phosphite samples obtained by the hydrothermal method but in different synthesis conditions characterized by different morphology. In light of the limited research performed on the investigated material's electrocatalytic and fluorescence properties, this work relies on the proposal that the additional evaluation of these characteristics will provide the scientific community with the opportunity to better understand it.

2. Materials and Methods

Reagents and materials. $Ni_{11}(HPO_3)_8(OH)_6$ material was synthesized by the hydrothermal method at high pressure and temperature, according to [25]. Graphene PureSheets Quattro (NanoIntegris, Menlo Park, California) was concentrated up to 0.6 mg mL^{-1} by centrifugation and redispersion, Carbon Black—Vulcan XC 72 (Fuell Cell Store, Texas) and Nafion® 117 solution (Sigma-Aldrich, Saint Louis, MO, USA) were used as purchased, together with reagent grade C_2H_5OH, C_3H_7OH, KOH, KNO_3, and $K_3[Fe(CN)_6]$. The two conductive carbon materials were selected because of the important roles such compounds have been shown to play in the field of oxygen evolution catalysis [26,27]. The graphite tablets for electrode manufacturing were obtained from spectroscopic graphite rods (Ø = 6 mm), type SW. 114 (Kablo Bratislava, National Corporation "Electrocarbon Topolcany" Factory, Bratislava, Slovakia). Double-distilled water was used throughout the study.

Preparation of modified electrodes. The graphite tablets were polished using silicon carbide paper (grit size: 1200 and 2400) and felt. They were subsequently washed with double-distilled water and ethanol and dried at room temperature. The tablet's surface modification consisted of the application via the drop-casting method [28] of a 10 μL volume from suspensions having the compositions shown in Table 1. After a drying period of 24 h in air and at room temperature, twelve types of modified graphite electrodes were obtained. The codes used to identify the unmodified electrode and each of the modified ones are also presented in Table 1.

Table 1. The codes used to identify the electrodes, together with the compositions of the suspensions employed to modify the graphite tablets.

Electrode Code	$Ni_{11}(HPO_3)_8(OH)_6$ Powder (mg)	Nafion Solution (µL)	rGO Suspension (µL)	Carbon Black Powder (mg)	80% Ethanol, 15% Water, and 5% Isopropanol Solution (µL) *
G0	-	-	-	-	-
G1	-	-	83	10	500
G2	-	10	83	10	500
G3	1	-	83	10	500
G4	5	-	83	10	500
G5	1	10	83	10	500
G6	5	10	83	10	500
G7	-	-	83	5	500
G8	-	10	83	5	500
G9	1	-	83	5	500
G10	5	-	83	5	500
G11	1	10	83	5	500

* This solution was employed to lower the surface tension of the suspension and to facilitate the gradual drying of the drop-casted layer, thus improving its biding to the substrate.

Electrochemical measurements. A PGZ402 (VoltaLab 80) Universal Potentiostat from Radiometer Analytical and an electrochemical glass cell equipped with three electrodes were used for the electrochemical experiments. The counter electrode was a Pt plate (S_{geom} = 0.8 cm^2), and Ag/AgCl (sat. KCl) was the reference electrode. Each of the graphite tablets, modified and unmodified, was inserted into an electrically nonconductive support and employed as the working electrode (S_{geom} = 0.07 cm^2).

The specimens were investigated in terms of their OER electrocatalytic properties by recording linear sweep voltammograms (LSVs) in 0.1 mol L^{-1} KOH electrolyte solution (pH = 13), and the most catalytically active electrode was further studied in 0.5 and 1 mol L^{-1} KOH solutions (pH = 13.7 and 14, respectively). Cyclic voltammetry was used for electric double-layer capacitance experiments, carried out in 0.1 mol L^{-1} KOH solution, and also for electroactive surface area (EASA) and electroactive species diffusion coefficient estimations, in 1 mol L^{-1} KNO$_3$ containing 4 mmol L^{-1} K$_3$[Fe(CN)$_6$]. All electrochemical experiments were performed at 23 ± 2 °C.

Unless otherwise specified, the LSVs were recorded by applying iR compensation, and the current densities mentioned in the text are geometrical current densities. Conversion of the measured potentials to the Reversible Hydrogen Electrode (RHE) scale, calculation of the OER overpotential (η), and estimation of the EASA and diffusion coefficient were performed using the same equations that have been employed in the previous study concerning $Ni_{11}(HPO_3)_8(OH)_6$ [24] and have also been reported in the literature [29,30]. They are presented in the Supplementary Materials.

SEM and EDX investigations. Scanning electron microscopy (SEM) investigations and energy-dispersive X-ray (EDX) analyses were performed using a scanning electron microscope equipped with the EDX module, INSPECT S model from FEI Company.

An AFM analysis was performed by a Nanonics MultiView 2000 scanner (Nanonics, Jerusalem, Israel), using the intermittent mode at ambient conditions (24–25 °C). The studied materials were placed on a polished glass slide. The scanner was equipped with a silicone-type probe Cr-coated, having a tip radius of 20 nm and resonance frequency of 30–40 kHz.

A fluorescence analysis was performed in the 475–700 nm range with a Perkin Elmer LS55 Spectrofluorometer, Hamburg, Germany. The excitation scan between 300 and 500 nm revealed an excitation peak with the highest signal at approximately 344 nm. The slits used for these measurements were 15 nm for λ excitation and 15 nm for λ emission. The employed excitation length was 344 nm. No filter was used. The analysis was carried out in a suspension of water and ethyl–alcohol.

3. Results

3.1. Electrochemical Investigations

The anodic polarization curves obtained on the G0–G6 electrodes are presented in Figure 1a, while Figure 1b shows the LSVs traced on the G0 and G7–G12 electrodes. The measurements were performed in 0.1 mol L^{-1} KOH electrolyte solution at a scan rate (v) of 1 mV s^{-1}.

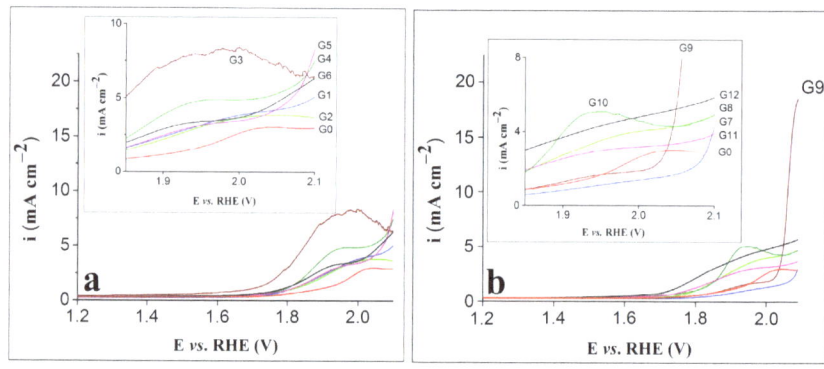

Figure 1. (**a**) LSVs recorded on the G0–G6 electrodes. (**b**) LSVs recorded on the G0 and G7–G12 electrodes. Insets in (**a**) and (**b**) correspond to enlarged regions of the LSVs.

Based on Figure 1a,b, the LSV recorded on the G0 electrode shows an oxidation feature at 2.03 V vs. RHE that is shifted towards more negative potential values in case of the modified electrodes. Furthermore, for most modified electrodes, the feature appeared at higher current densities, and the highest value was observed for G3. The LSV obtained on this electrode is not as smooth as the ones traced on the other modified electrodes, probably because of the intense bubbling effect noticed during the experiment. Basically, the OER overlapped with the process corresponding to the anodic feature [31], while the bubbles formed on its surface affected the maximum current density value by decreasing the electroactive area and hindering the reaction. Of all the differences between the LSVs recorded on the modified electrodes, the most obvious was observed for G9. The current density it exhibited was significantly higher than that of the other electrodes, and in the field of water splitting, a high-current density is an important requirement [23].

The fact that the electrode displaying the highest current density was the one modified with a composition containing rGO, 5 mg carbon black and just 1 mg nickel phosphite highlights the importance of experiments aimed at identifying the proper ratio between the materials of the compositions used to obtain the modified electrodes. The performance of the G9 electrode may be attributed to the synergistic effect of the nickel phosphite and the carbon materials. The latter ensured an improved charge transport during the OER process and, to some extent, prevented the nickel phosphite particles from aggregating into larger structures, making them better distributed into the composition and promoting the formation of catalytically active sites on the electrode surface [32].

The G9 electrode was investigated by SEM (Figure 2), and the top-view image recorded at low magnification (Figure 2a) shows that the suspension applied on the surface of the graphite tablet led to the formation of areas that differ in terms of the deposited composition's homogeneity. This observation is in agreement with other studies in which the drop-casting method was used to obtain modified electrodes [33,34]. The high-magnification image presented in Figure 2b displays the rod-shaped Ni$_{11}$(HPO$_3$)$_8$(OH)$_6$ particles with lengths between 1.2 and 10.2 μm and diameters between 0.08 and 0.6 μm. The particles investigated by Menezes et al. [23] had the same shape, but their lengths and diameters were not as big, most likely due to the different synthesis method used to obtain the samples from the current study [25].

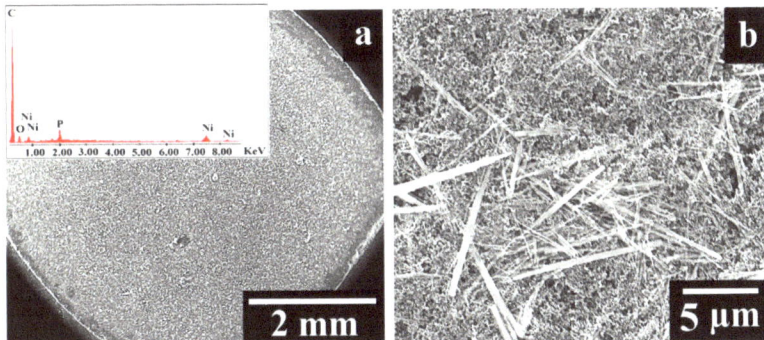

Figure 2. (a) SEM image recorded on the surface of the G9 electrode at low magnification. Inset: EDX spectrum of the G9 electrode. (b) SEM image recorded on the surface of the G9 electrode at high magnification.

The presence of nickel and phosphorus in the composition deposited on the surface of the graphite tablet was confirmed by EDX analysis (Figure 2a inset). The other identified elements (carbon and oxygen) were also expected to be present in the sample, considering the contents of the suspension used to obtain the G9 electrode.

Further characterization of the G9 electrode was performed using cyclic voltammetry, and it was aimed at determining the electric double-layer capacitance at the electrode/electrolyte solution interface and estimating both the electrode's EASA, as well as the diffusion coefficient of ferricyanide ions.

The double-layer capacitance (C_{dl}) was obtained by first recording cyclic voltammograms in 0.1 mol L^{-1} KOH solution at various scan rates (v = 0.05, 0.1, 0.15, 0.2, and 0.25 V s^{-1}) in the $-0.2 \div -0.1$ V vs. Ag/AgCl (sat. KCl) potential range, where no faradic currents were present. The capacitive current density (i_{dl}) was then calculated as the average of the absolute values of the anodic and cathodic current densities, selected at -0.15 V vs. Ag/AgCl (sat. KCl), where only double-layer adsorption and desorption features were present [35–37]. The C_{dl} value was determined as the absolute value of the slope obtained for the linear dependence between i_{dl} and the scan rate [38]. Figure 3a shows the i_{dl}-v dependence plot and the slope value of 2.2127 mF cm^{-2}.

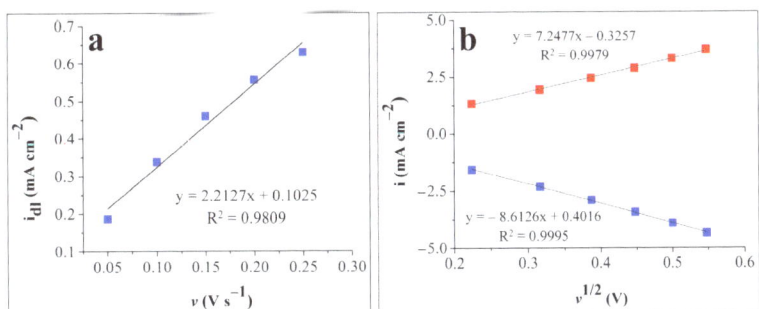

Figure 3. (a) The plot of the capacitive current density vs. the scan rate for the G9-modified electrode. (b). The graphical representations of the anodic peak current densities vs. the square root of the scan rate (red squares) and of the cathodic peak current densities vs. the square root of the scan rate (blue squares) for the G9-modified electrode.

Cyclic voltammograms were also recorded on the G9 electrode in the 1 mol L^{-1} KNO$_3$ electrolyte solution in the presence and absence of 4 mmol L^{-1} K$_3$[Fe(CN)$_6$] at various scan rate values (v = 0.05, 0.1, 0.15, 0.2, 0.25, and 0.3 V s^{-1}) and in the 0 ÷ 0.8 V vs. Ag/AgCl (sat. KCl) potential range, where the signals corresponding to the [Fe(CN)$_6$]$^{4-/3-}$ redox couple appeared (see Figure S1 from the Supplementary Materials). The experimental data were used in the Randles–Sevcik equation to estimate the EASA and diffusion coefficient of ferricyanide ions, and their values were found to be 0.13 cm^2 and 2.33 × 10^{-5} cm^2 s^{-1}, respectively. The plot of the anodic and cathodic peak current densities vs. the square root of the scan rate for the same electrode is presented in Figure 3b. As can be seen, the redox peak currents are proportional to the square root of the scan rate, which indicates a diffusion-controlled electron transfer process [24,28].

The results obtained from subsequent OER experiments performed on the G9 electrode are shown in Figure 4. The geometric area of the sample was replaced with the estimated EASA value, and the modified LSV, differing from the initial one in terms of the current density values, is presented in Figure 4a.

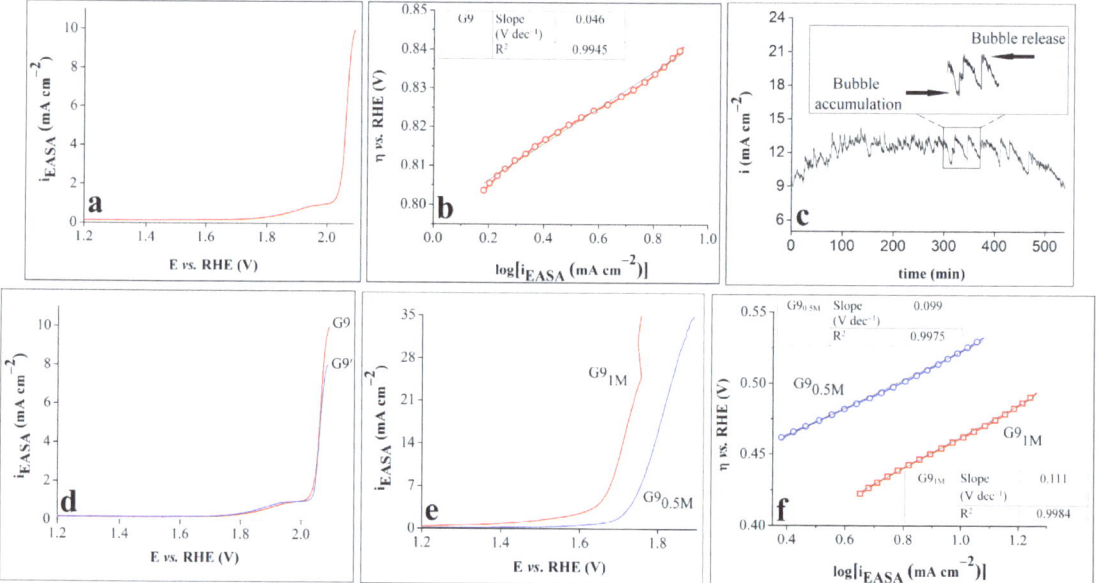

Figure 4. (**a**) LSV obtained on the G9 electrode in 0.1 mol L^{-1} KOH solution. (**b**) The Tafel curve for the G9 electrode in 0.1 mol L^{-1} KOH solution and inserted table showing the Tafel slope and R^2 values; (**c**) i-t curve recorded on the G9 electrode in 0.1 mol L^{-1} KOH solution, with inset showing an enlarged region of the curve. (**d**) LSVs traced on the G9 electrode in 0.1 mol L^{-1} KOH solution, before and after the stability test. (**e**) LSVs recorded on the G9 electrode in 0.5 mol L^{-1} KOH (G9$_{0.5M}$) and 1 mol L^{-1} KOH (G9$_{1M}$) solutions. (**f**) The Tafel curves of the G9 electrode in 0.5 mol L^{-1} KOH (G9$_{0.5M}$) and 1 mol L^{-1} KOH (G9$_{1M}$) solutions and an inserted table showing the Tafel slopes and R^2 values. All LSVs were recorded at v = 1 mV s^{-1}.

An important parameter in the study of the OER kinetics at the electrode/electrolyte interface is the Tafel slope, which reveals the relationship between the overpotential and the current density. A smaller Tafel slope usually suggests better electrocatalytic properties [27,32,39]. Figure 4b displays the Tafel curve for G9 in 0.1 mol L^{-1} KOH electrolyte solution. The value of the slope was calculated using the Tafel equation: η = b × log(i) + a, where η is the overpotential, i is the current density, and b is the Tafel slope. A low value of 46 mV dec^{-1} was obtained, indicating enhanced reaction kinetics.

The electrochemical stability of electrocatalysts is another important parameter used to evaluate their OER performance [32]. The stability of G9 in 0.1 mol L^{-1} KOH solution was studied chronoamperometrically by maintaining constant the potential value corresponding to i = 10 mA cm^{-2}. The time-dependent current density (i-t) curve was recorded for 9 h (Figure 4c), and the following observations were made: the current density increased until it reached the maximum value of 14 mA cm^{-2} after 135 min (an increase of 40%), and then, it gradually decreased to 8.9 mA cm^{-2} at the 540 min mark. The shape of the curve was affected by the alternate processes of bubble accumulation and bubble release (Figure 4c inset), and the decrease in the current density seen during the experiment can be ascribed to the reaction hindering the O$_2$ bubbles that stayed on the surface of the electrode and perhaps also to the partial detachment of the catalyst caused by the continuous bubble release [40].

The anodic polarization curves obtained on the G9 electrode before and after the stability test (denoted G9 and G9′) are presented in Figure 4d. A comparison between them reveals that they overlap almost perfectly up to the current density value of 6 mA cm^{-2}, but they gradually separate at higher values. The LSV recorded after the stability test reached 8.02 mA cm^{-2}, which was lower than the initially attained 10 mA cm^{-2}. This result outlined the stability limitations of the G9 electrode.

Lastly, the study of the electrocatalytic properties of the G9 electrode focused on the concentration of the electrolyte solution. It has been shown that, by modifying this parameter, specifically by increasing the concentration of the KOH solution, higher current densities and lower overpotential values are obtained [41]. Thus, the polarization curves were traced on G9 in 0.5 mol L^{-1} and 1 mol L^{-1} KOH electrolyte solutions (Figure 4e). By comparing Figure 4a,e, it can be seen that higher current densities were achieved when the KOH concentration of the electrolyte solution was 0.5 or 1 mol L^{-1} than when it was 0.1 mol L^{-1}. Furthermore, the OER overpotential values corresponding to the current density values became significantly smaller as the KOH concentration increased.

The Tafel curves of the G9 electrode in 0.5 mol L^{-1} and 1 mol L^{-1} KOH electrolyte solutions are presented in Figure 4f, together with the values of the Tafel slopes and R^2 values. These values are higher than the one calculated for the same sample but not in the 0.1 mol L^{-1} KOH solution. The result indicates that, when studying the OER properties of G9, even though higher KOH concentrations lead to higher anodic current densities and lower overpotential values, they also affect the OER kinetics at the electrode/electrolyte interface.

As was specified in the introduction section, the present study continues the previous work on the OER electrocatalytic properties of graphite electrodes modified with compositions containing Ni$_{11}$(HPO$_3$)$_8$(OH)$_6$. The most electrocatalytically active electrode identified in the preceding study [24] was obtained by drop-casting a suspension containing 5 mg Ni$_{11}$(HPO$_3$)$_8$(OH)$_6$ and 10 μL Nafion solution on the surface of a graphite tablet. Since the working conditions were very similar to the ones from the present study, a comparison can be made between the properties exhibited by that electrode and G9. In this sense, Table 2 shows the values of some electrochemical parameters—EASA, the OER overpotential (η) at 5 mA cm^{-2}, and the Tafel slope in 0.1 mol L^{-1} KOH electrolyte solution—for the two modified electrodes.

As can be seen in Table 2, the G9 electrode exhibits a lower Tafel slope value, which indicates faster OER kinetics. However, the electrode also exhibits a higher overpotential value.

The OER experiments from the previous study were performed only in 0.1 mol L^{-1} KOH electrolyte solution, but Menezes et al. [23] investigated the catalytic properties of Ni$_{11}$(HPO$_3$)$_8$(OH)$_6$ in 1 mol L^{-1} KOH solution as well. A comparison between the OER electrocatalytic activity in this strong alkaline medium of the best electrode identified by the researchers (nickel phosphite electrophoretically deposited on nickel foam) and that of the G9 electrode evidenced similar Tafel slope values of 0.111 V dec^{-1} for G9 and 0.091 V dec^{-1} for the reported sample. As for their OER overpotential values at 10 mA cm^{-2}, these were 0.47 V for the former and 0.232 V for the latter. Additionally, for comparative purposes, it should be pointed out that the IrO$_2$ and RuO$_2$ OER overpotential values at i = 10 mA cm^{-2}

and in alkaline medium were reported as 0.45 V and 0.42 V, while their Tafel slope values were found to be 0.083 V dec^{-1} and 0.074 V dec^{-1} [42–44].

Table 2. The values of some electrochemical parameters for the two modified electrodes.

Modified Electrode	EASA (cm^2)	η (V)	Tafel Slope (V dec^{-1})
G9: Graphite substrate modified with 1 mg nickel phosphite, 5 mg carbon black and rGO	0.13	~0.83	0.046
Graphite substrate modified with 5 mg nickel phosphite and 10 μL Nafion solution	0.105	0.55	0.081

Table S1 (see the Supplementary Materials) shows the OER electrocatalytic performance of the G9 electrode and that of other electrocatalysts reported in the scientific literature. The presented data indicate that, in most cases, the reported OER overpotential and Tafel slope values are smaller than the ones obtained for G9, and this points to a requirement for further experiments aimed at identifying a nickel phosphite-based composition that will display an improved OER activity. Regarding the more recent studies mentioned in Table S1, some of them evidenced specimens with catalytic activity surpassing that of G9 in terms of both overpotential and Tafel slope, while others outlined modified electrodes that exhibited a lower activity than the electrode identified in the current study. In the first type of investigations, the composite Co-Fe-1,4-benzenedicarboxylate catalyst displayed an OER overpotential of 0.295 V and a Tafel slope of ~0.035 V dec^{-1} [45], while for FeNi metal–organic framework nanoarrays on Ni foam, an overpotential of 0.213 V and a Tafel slope of ~0.052 V dec^{-1} were found [46]. In the second type, electrodes manufactured either with a composition containing MnO$_2$ [47] or with a Pt(II)-porphyrin [14] revealed OER overpotentials of either 0.53 V or 0.64 V. The requirement for further experiments is also indicated by the stability test result that outlines a lower degree of stability vs. that of IrO$_2$, considered highly stable in the rough oxidizing conditions of the OER [48,49].

3.2. Physicochemical Characterization and Fluorescence Study

The multifunctionality aspect of these phosphate materials [24,50–52] has led us to the study of their fluorescence properties. It is known that the fluorescence spectrum can be influenced by the morphology of the materials [53], and in this context, a new nickel phosphite sample was obtained by using a slightly modified low temperature and pressure hydrothermal synthesis [54]. The starting reactants—NaH$_2$PO$_4$•H$_2$O (0.2758 g), NiCl$_2$•6H$_2$O (1.0228 g), and CH$_3$COONa•3H$_2$O (1.8958 g)—were introduced in a Teflon steel autoclave (total volume of 65 mL) filled at 80%, and the thermal treatment occurred at 200 °C for 48 h. In the following, in order to distinguish between both materials, this sample obtained by low-temperature and low-pressure hydrothermal synthesis will be denoted S$_{LowTP}$, and the material described in the previous section, obtained by high-temperature and high-pressure synthesis, according to [24], will be denoted S$_{HighTP}$.

The XRD analysis by Rietveld refinement (Figure 5) shows that the S$_{LowTP}$ compound was well crystallized, and the calculated lattice parameters were as follows: $a = b = 12.4764$ (4) Å, $c = 4.9464$ (2) Å, and P6$_3$mc space group—parameter values in accordance with the literature [24,25,55]. The P6$_3$mc space group was used also to index the crystallized S$_{highTP}$ samples obtained for this study, as in [25].

SEM and AFM analyses performed for both samples are presented in Figures 6 and 7. SEM images revealed spherical superstructures made of platelet-like particles (S$_{LowTP}$) in Figure 6 and rods with sparse round-like formations (S$_{HighTP}$) in Figure 7. Thereby, it was difficult to obtain AFM images on certain areas of S$_{LowTP}$ where the plates are present, due to the existence of large heights (>2 μm) and abrupt differences (Figure 6a). In the S$_{HighTP}$ sample, the rod-like morphology predominated with a smaller proportion of round particles (Figure 7b), as seen in the SEM image as well. Round formations from the AFM analysis may be a result of the sample preparation process. Furthermore, individual areas were selected and measured; the height and width of some formations are shown in Figure 8.

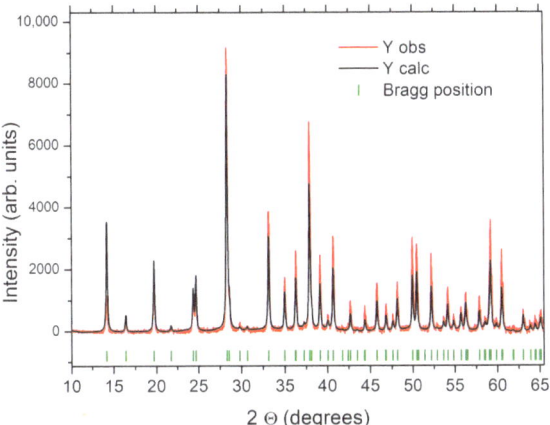

Figure 5. Rietveld refinement of the X-ray Diffractogram for the S_{LowTP} sample in the P6$_3$mc space group. The experimental XRD pattern is a red line, the calculated pattern a black line.

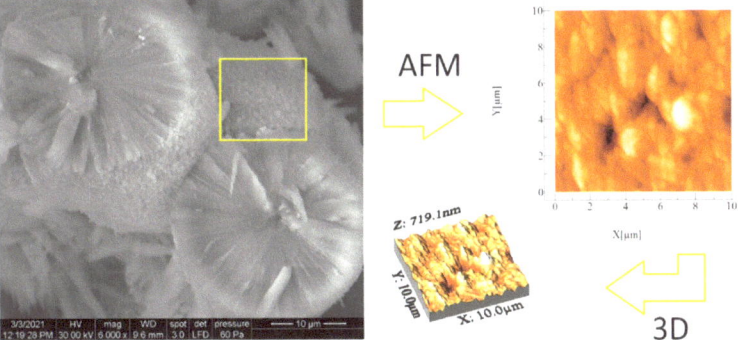

Figure 6. SEM and AFM images at different scales for the S_{LowTP} sample. The yellow square on the SEM image it's indicative of the surface area analyzed by AFM (image at right).

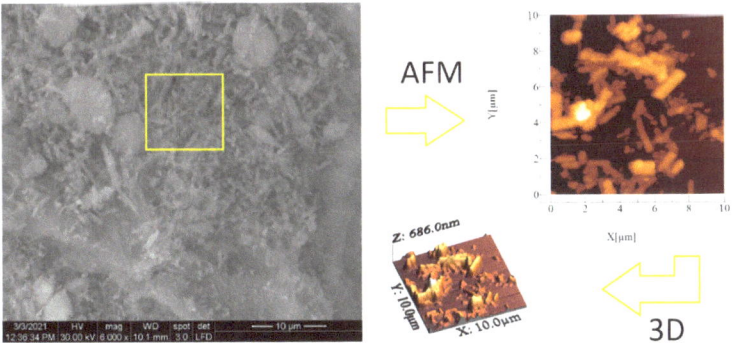

Figure 7. SEM and AFM images at different scales for the S_{HighTP} sample. The yellow square on the SEM image it's indicative of the surface area analyzed by AFM (image at right).

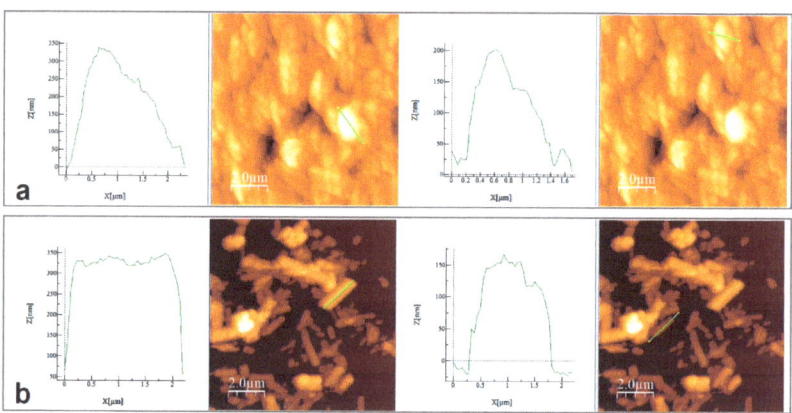

Figure 8. Height and width measurements on the selected area for S_{LowTP} (**a**) and for S_{HighTP} (**b**).

Table 3 displays the values obtained for several AFM parameters in the case of the S_{LowTP} and S_{HighTP} samples.

Table 3. Calculated values from AFM images—Average roughness (S_a), Mean square root roughness (S_q), Maximum peak height (S_p), Maximum valley depth (S_v), and Maximum peak-to-valley height (S_y)—for S_{LowTP} and S_{HighTP}.

Sample Name	Ironed Area (μm²)	S_a (μm)	S_q (μm)	S_p (μm)	S_v (μm)	S_y (μm)
S_{LowTP}	110.359	0.063	0.085	0.352	−0.366	0.719
S_{HighTP}	113.357	0.079	0.105	0.636	−0.049	0.685

The iron area in the AFM technique depends on the majority of asperities being found on the material surface, and usually, a smoother surface is characterized by a lower value for this parameter. S_p and S_v are the indicators for the highest peaks and the lowest valleys present in the analyzed area and, together, determine the S_y value. This latter parameter reveals the compaction of the deposited material or some information about the surface homogeneity. The roughness is obtained from both R_q and R_a, whereas the S_y value may indicate which sample displays a predominant tendency for valleys or heights. The results show higher roughness values for S_{HighTP} and the highest S_y value for S_{LowTP}. S_{LowTP} had lower valleys (S_v = −0.336 μm for S_{LowTP} and −0.049 μm for S_{HighTP}) in comparison to S_{HighTP} with higher S_p values (0.352 μm for S_{LowTP} and 0.636 μm for S_{HighTP}). The obtained data are explained by the fact that the rods (S_{HighTP}) were placed one on another randomly, forming angles and, therefore, higher heights, whereas the conglomerates (S_{LowTP}) were packed very closely, therefore influencing the main line calculus formula for the S_p value and S_v value, respectively [56].

A Raman analysis was performed on both S_{LowTP} and S_{HighTP} materials (Figure 9), for which the fluorescence properties are characterized in the last part of this research work. The following bands can be observed for the S_{LowTP} sample: 298, 374, 430, 500, 540, 608, 927, 1002, and 1187 cm^{-1}. There are more bands outlined in the spectrum of S_{HighTP}: 296, 374, 430, 479, 542, 613, 935, 994, 1137, 1195, and 1598 cm^{-1}. As it can be seen, some S_{HighTP} bands are slightly shifted compared to the S_{LowTP} ones, and this could be due to the different sample morphology, since it has already been shown for other materials that Raman scattering spectra can depend on the morphology of microparticles [57].

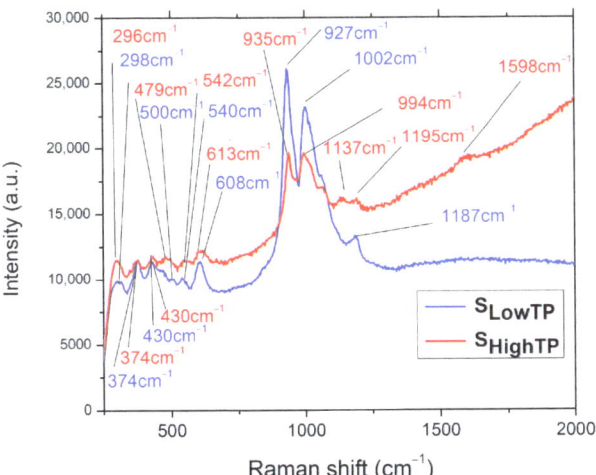

Figure 9. Raman spectra for S_{LowTP} and S_{HighTP}.

As assigned in [24], for the $Ni_{11}(HPO_3)_8(OH)_6$ compound, which was from the same batch as S_{HighTP} from this study, the $(PO_4)^{3-}$ ions are indicated by the 380 cm^{-1} band, which may also include the PO_3 deformations and rocking modes, whereas, at 296 cm^{-1}, POP deformations may take place. In the case of S_{LowTP}, a small band at 500 cm^{-1} is observed that was not registered in S_{HighTP} and was not evidenced in the previous study either. The band located at 474 cm^{-1} in [24] is also present at 479 cm^{-1} for S_{HighTP} and as a shoulder for S_{LowTP} and corresponds to the PO_3 symmetric and asymmetric deformations. In the same region (470–500 cm^{-1}), overlapped Ni-O bond bands are expected and are responsible for the peak-broadening evidenced for both S_{LowTP} and S_{HighTP} materials. When comparing the present results with the previously reported ones [24], the general tendency of the bands is to blueshift. The differences observed in the Raman spectra for the samples coming from the same batch, S_{HighTP} from this study, and the previously studied $Ni_{11}(HPO_3)_8(OH)_8$ in [24], may be due to the fact that the Raman analysis is focused on small amounts of material, whereas a large quantity of materials was obtained from the hydrothermal synthesis at a high pressure and temperature. Due to this, it could be that some inhomogeneities were present in the products from the synthesis or from the autoclave.

In comparison with the $Ni_{11}(HPO_3)_8(OH)_8$ Raman spectrum from [24], a higher intensity is noticed for the bands at 608 cm^{-1} (S_{LowTP}) and 613 cm^{-1} (S_{HighTP}), respectively. These are connected to the 474–479 cm^{-1} region referring to the symmetric and asymmetric PO_3 deformations. The highest difference is detected in the 900–1000 cm^{-1} region, where the S_{LowTP} is distinguished from the other two samples, displaying the most intense peak at 927 cm^{-1}, followed by the one at 1002 cm^{-1}. In the case of S_{HighTP}, both peaks have the same intensity and are slightly shifted in relation to S_{LowTP}. The specified bands are owed to the PO_3^- symmetric and asymmetric stretching vibrations and are the most prominent changes in the sample. In the same domain, NiO bands are expected and are less visible as a result of overlapping with the PO_3 bands, followed by the broadening of the specific peaks (994 cm^{-1} and 1002 cm^{-1}, respectively) [24].

By comparison with the previous study [24], a lower intensity is observed in the 1137–1197 cm^{-1} region. For the S_{HighTP} sample, splitting of the peak at 1137 cm^{-1} could take place due to the secondary phases and is accompanied by the red shift occurrence [58]. Compared to S_{LowTP}, the red-shifted bands of S_{HighTP} are also broader, and this may be attributed to the material synthesis, as suggested in the literature [59,60]. The differences between the Raman spectra of the S_{LowTP} and S_{HighTP} samples can be attributed to their different morphology, synthesis method, and even to the presence of any impurities. Nguyen et al. [59] confirmed the fact that phonon wavenumbers and the line shape depend strongly on impurities or doping, indicating a lower symmetry of the crystal structure. Their XRD analysis revealed no noticeable differences, as was also the case with the current study, in which impurities such as barbosalite $Fe_3(PO_4)_2(OH)_2$ could have been present in the S_{HighTP} sample obtained by high-temperature and high-pressure hydrothermal synthesis, as mentioned in [25].

The low intensity peak (1598 cm^{-1}) corresponding to the O-H stretching vibration is observed only in the case of S_{HighTP} and is red-shifted in relation to $Ni_{11}(HPO_3)_8(OH)_8$ (1580 cm^{-1}) [24].

The fluorescence measurements (Figure 10) indicate some differences between the studied S_{LowTP} and S_{HighTP} materials. Several bands can be observed in the 405–600 nm spectral range, with maxima at 426, 488, 513, 533, and 554 nm for S_{HighTP} and 429, 487, 517, 533, and 555 nm for S_{LowTP}. The differences between both materials are not only registered by the shift of the bands but also by the intensity of their maxima. For certain materials, it has been previously reported in the literature that their morphology has an influence on the obtained fluorescence results, which is usually represented by the quenching of the spectra [61,62]. Quenching may appear when the reducing of the radiative recombination of the electron and hole takes place and/or due to a lower electron–hole recombination rate [62]. Some other factor could influence the fluorescence spectra, such as the surface roughness due to the excitation of electrons from occupied d bands into states above the Fermi level [62]. In this context, the materials with higher roughness may show increased fluorescence as a consequence of the electron excitation, which is confirmed by the present results, since both AFM roughness and fluorescence intensity values showed this to be the case for S_{HighTP}. The fluorescence intensity also usually decreases with the decreasing sizes of the crystallites, but at the same time, the morphology may also play a major role, as it can cause an increase in a specific area as a result of surface defects, strain/stress effects, or other changes [53]. According to the literature, the expected bands for the Ni-O bond are in the blue emission region, more precisely centered between 424–433, 448–463, and 484–491 nm [63,64]. In the current study, as can be seen in Figure 10, the most intense bands are located in the aforementioned areas but much better outlined in the case of S_{HighTP}. Another band that is usually specific for M-O bonds is centered at 518–524 nm in the green emission area and is due to the enhanced oxygen vacancies [64]. Phosphors show broad emission bands in the 300–350 nm wavelength range in the case of $NaGd(PO_3)_4:Ce^{3+}$ materials [65], thus not possible to be put as evidence for the materials studied in this work.

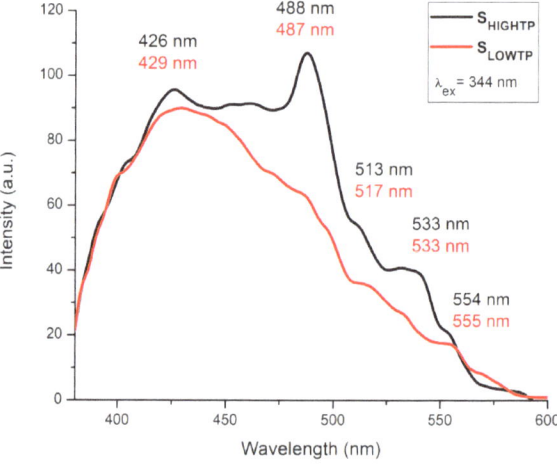

Figure 10. Fluorescence emission spectra obtained at RT for S_{LowTP} and S_{HighTP} in the 380–600 nm range, excitation at 344 nm.

4. Conclusions

The OER activity of several graphite electrodes modified with $Ni_{11}(HPO_3)_8(OH)_6$, carbon black, and rGO was investigated electrochemically in an alkaline medium. The polarization curves recorded on the electrodes show that the sample modified with 1 mg nickel phosphite, 5 mg carbon black, and rGO exhibited the highest anodic current density values. This behavior can be explained in terms of the synergistic effect of the nickel phosphite and the carbon materials. The electrochemical study outlines the OER performance of the nickel phosphite-based electrodes, as well as their limitations. Different hydrothermal synthesis conditions influenced the morphology of the obtained nickel phosphite materials, the existence of rods and plates being observed through SEM and AFM analyses. Raman spectrometry demonstrated the presence of the specific peaks for the analyzed samples, acknowledging the importance of the employed synthesis method.

The experimental data resulted from the evaluation of the nickel phosphite's OER activity, and the fluorescence properties complement the research literature, providing the scientific community with the opportunity to better understand this material and its applicative potential. Moreover, the study of the nickel phosphite's fluorescence properties increases the understanding of this material and paves the way for future investigations with potentially applicative results, highlighting the link between the roughness of the materials and the free ions.

Supplementary Materials: The following supporting information can be downloaded at: https://www.mdpi.com/article/10.3390/cryst12121803/s1: Figure S1: Cyclic voltammograms recorded on the G9 electrode in 1 mol L^{-1} KNO_3 + 4 mmol L^{-1} $K_3[Fe(CN)_6]$ electrolyte solution, at various scan rate values (v = 0.05, 0.1, 0.15, 0.2, 0.25 and 0.3 V s^{-1}); Table S1: The electrocatalytic properties for the OER of the composition containing $Ni_{11}(HPO_3)_8(OH)_6$, carbon black and rGO, applied on graphite substrate, and that of other electrocatalysts reported in the scientific literature, in 1 mol L^{-1} KOH electrolyte solution. References [66–94] are cited in the supplementary materials.

Author Contributions: Conceptualization, M.P.; sample preparation, M.P. and P.V. (Paulina Vlazan); measurements, M.P., B.-O.T., P.S. (Paula Svera), C.I. and P.S. (Paula Sfirloaga); investigation, M.P., P.V. (Philippe Veber), P.S. (Paula Svera), B.-O.T. and G.B., writing—original draft, B.-O.T., P.S. (Paula Svera) and M.P.; funding acquisition, M.P., P.S. (Paula Sfirloaga) and G.B.; and writing—review and editing, M.P., G.B., P.V. (Philippe Veber), P.S. (Paula Sfirloaga) and B.-O.T. All authors have read and agreed to the published version of the manuscript.

Funding: G.B. acknowledges the financial support for this work provided by the West University of Timisoara from overhead funding (grant PN, no.75/2020). P.S., M.P., and P.V. acknowledge the financial support for this work provided by the joint French–Romanian project ANR-UEFISCDI, no. 8 RO-Fr/01.01.2013, code PN-II-ID-JRP-2011-2-0056/ANR-12-IS08 0003, COFeIn, and the Experimental Demonstrative Project 683PED/21/06/2022.

Institutional Review Board Statement: Not applicable.

Informed Consent Statement: Not applicable.

Data Availability Statement: The data that support the findings of this study are available from the authors, upon reasonable request.

Conflicts of Interest: The authors declare no conflict of interest.

References

1. Qazi, A.; Hussain, F.; Rahim, N.; Hardaker, G.; Alghazzawi, D.; Shaban, K.; Haruna, K. Towards sustainable energy: A systematic review of renewable energy sources, technologies, and public opinions. *IEEE Access* **2019**, *7*, 63837–63851. [CrossRef]
2. Roger, I.; Shipman, M.A.; Symes, M.D. Earth-abundant catalysts for electrochemical and photoelectrochemical water splitting. *Nat. Rev. Chem.* **2017**, *1*, 0003. [CrossRef]
3. Chu, S.; Majumdar, A. Opportunities and challenges for a sustainable energy future. *Nature* **2012**, *488*, 294–303. [CrossRef] [PubMed]
4. Colmati, F.; Alonso, C.G.; Martins, T.D.; De Lima, R.B.; Ribeiro, A.C.C.; Carvalho, L.d.L.; Sampaio, A.M.B.S.; Magalhaes, M.M.; Coutinho, J.W.D.; De Souza, G.A.; et al. Chapter 4: Production of hydrogen and their use in proton exchange membrane

5. Mani, V.; Anantharaj, S.; Mishra, S.; Kalaiselvi, N.; Kundu, S. Iron hydroxyphosphate and Sn-incorporated iron hydroxyphosphate: Efficient and stable electrocatalysts for oxygen evolution reaction. *Catal. Sci. Technol.* **2017**, *7*, 5092–5104. [CrossRef]
6. Jiao, Y.; Zheng, Y.; Jaroniec, M.; Qiao, S.Z. Design of electrocatalysts for oxygen- and hydrogen-involving energy conversion reactions. *Chem. Soc. Rev.* **2015**, *44*, 2060–2086. [CrossRef]
7. Tang, Q.; Jiang, D. Mechanism of hydrogen evolution reaction on 1T-MoS2 from first principles. *ACS Catal.* **2016**, *6*, 4953–4961. [CrossRef]
8. Jamesh, M.I. Recent progress on earth abundant hydrogen evolution reaction and oxygen evolution reaction bifunctional electrocatalyst for overall water splitting in alkaline media. *J. Power Sources* **2016**, *333*, 213–236. [CrossRef]
9. Gong, M.; Dai, H. A mini review of NiFe-based materials as highly active oxygen evolution reaction electrocatalysts. *Nano Res.* **2015**, *8*, 23–39. [CrossRef]
10. Hunter, B.M.; Gray, H.B.; Muller, A.M. Earth-abundant heterogeneous water oxidation catalysts. *Chem. Rev.* **2016**, *116*, 14120–14136. [CrossRef]
11. Indra, A.; Menezes, P.W.; Driess, M. Uncovering structure–activity relationships in manganese-oxide-based heterogeneous catalysts for efficient water oxidation. *ChemSusChem* **2015**, *8*, 776–785. [CrossRef]
12. Menezes, P.W.; Indra, A.; Das, C.; Walter, C.; Gobel, C.; Gutkin, V.; Schmeiber, D.; Driess, M. Uncovering the nature of active species of nickel phosphide catalysts in high-performance electrochemical overall water splitting. *ACS Catal.* **2017**, *7*, 103–109. [CrossRef]
13. Poienar, M.; Taranu, B.O.; Svera, P.; Sfirloaga, P.; Vlazan, P. Disclosing the thermal behaviour, electrochemical and optical properties of synthetic $Fe_3(PO_4)_2(OH)_2$ materials. *J. Therm. Anal. Calorim.* **2022**, *147*, 11435. [CrossRef]
14. Fratilescu, I.; Lascu, A.; Taranu, B.O.; Epuran, C.; Birdeanu, M.; Macsim, A.-M.; Tanasa, E.; Vasile, E.; Fagadar-Cosma, E. One A3B porphyrin structure—Three successful applications. *Nanomaterials* **2022**, *12*, 1930. [CrossRef] [PubMed]
15. Taranu, B.O.; Fagadar-Cosma, E. Catalytic properties of free-base porphyrin modified graphite electrodes for electrochemical water splitting in alkaline medium. *Processes* **2022**, *10*, 611. [CrossRef]
16. Ma, Z.; Zhang, Y.; Liu, S.; Xu, W.; Wu, L.; Hsieh, Y.-C.; Liu, P.; Zhu, Y.; Sasaki, K.; Renner, J.N.; et al. Reaction mechanism for oxygen evolution on RuO_2, IrO_2, and $RuO_2@IrO_2$ core-shell nanocatalysts. *J. Electroanal. Chem.* **2018**, *819*, 296–305. [CrossRef]
17. Li, X.; Hao, X.; Abudula, A.; Guan, G. Nanostructured catalysts for electrochemical water splitting: Current state and prospects. *J. Mater. Chem. A* **2016**, *4*, 11973–12000. [CrossRef]
18. Subbaraman, R.; Tripkovic, D.; Chang, K.-C.; Strmcnik, D.; Paulikas, A.P.; Hirunsit, P.; Chan, M.; Greeley, J.; Stamenkovic, V.; Markovic, N.M. Trends in activity for the water electrolyser reactions on 3d M(Ni,Co,Fe,Mn) hydr(oxy)oxide catalysts. *Nat. Mater.* **2012**, *11*, 550–557. [CrossRef]
19. Han, L.; Dong, S.; Wang, E. Transition-metal (Co, Ni, and Fe)-based electrocatalysts for the water oxidation reaction. *Adv. Mater.* **2016**, *28*, 9266–9291. [CrossRef]
20. Osgood, H.; Devaguptapu, S.V.; Xu, H.; Cho, J.; Wu, G. Transition metal (Fe, Co, Ni, and Mn) oxides for oxygen reduction and evolution bifunctional catalysts in alkaline media. *Nano Today* **2016**, *11*, 601–625. [CrossRef]
21. Vij, V.; Sultan, S.; Harzandi, A.M.; Meena, A.; Tiwari, J.N.; Lee, W.G.; Yoon, T.; Kim, K.S. Nickel–based electrocatalysts for energy related applications: Oxygen reduction, oxygen evolution, and hydrogen evolution reactions. *ACS Catal.* **2017**, *7*, 7196–7225. [CrossRef]
22. Peugeot, A.; Creissen, C.E.; Karapinar, D.; Tran, H.N.; Schreiber, M.; Fontecave, M. Benchmarking of oxygen evolution catalysts on porous nickel supports. *Joule* **2021**, *5*, 1281–1300. [CrossRef]
23. Menezes, P.W.; Panda, C.; Loos, S.; Bunschei-Bruns, F.; Walter, C.; Schwarze, M.; Deng, X.; Dau, H.; Driess, M. A structurally versatile nickel phosphite acting as a robust bifunctional electrocatalyst for overall water splitting. *Energy Environ. Sci.* **2018**, *11*, 1287–1298. [CrossRef]
24. Taranu, B.O.; Ivanovici, M.G.; Svera, P.; Vlazan, P.; Sfirloaga, P.; Poienar, M. $Ni_{11}(HPO_3)_8(OH)_6$ multifunctional materials: Electrodes for oxygen evolution reaction and potential visible-light active photocatalysts. *J. Alloys Compd.* **2020**, *848*, 156595. [CrossRef]
25. Poienar, M.; Maignan, A.; Sfirloaga, P.; Malo, S.; Vlazan, P.; Guesdon, A.; Lainé, F.; Rouquette, J.; Martin, C. Polar space group and complex magnetism in $Ni_{11}(HPO_3)_8(OH)_6$: Towards a new multiferroic material? *Solid State Sci.* **2014**, *39*, 92–96. [CrossRef]
26. Gu, Y.; Wang, Y.; An, W.; Men, Y.; Rui, Y.; Fan, X.; Li, B. A novel strategy to boost the oxygen evolution reaction activity of NiFe-LDHs with in situ synthesized 3D porous reduced graphene oxide matrix as both the substrate and electronic carrier. *New J. Chem.* **2019**, *17*, 6555–6562. [CrossRef]
27. Li, Q.; Tang, S.; Tang, Z.; Zhang, Q.; Yang, W. Microwave-assisted synthesis of FeCoS2/XC-72 for oxygen evolution reaction. *Solid State Sci.* **2019**, *96*, 105968. [CrossRef]
28. Sebarchievici, I.; Taranu, B.O.; Birdeanu, M.; Rus, S.F.; Fagadar-Cosma, E. Electrocatalytic behaviour and application of manganese porphyrin/gold nanoparticle- surface modified glassy carbon electrodes. *Appl. Surf. Sci.* **2016**, *39*, 131–140. [CrossRef]
29. Zhao, Z.; Wu, H.; He, H.; Xu, X.; Jin, Y. Self-standing non-noble metal (Ni–Fe) oxide nanotube array anode catalysts with synergistic reactivity for high-performance water oxidation. *J. Mater. Chem. A* **2015**, *3*, 7179–7186. [CrossRef]

30. Baciu, A.; Remes, A.; Ilinoiu, E.; Manea, F.; Picken, S.J.; Schoonman, J. Carbon nanotubes composite for environmentally friendly sensing. *Environ. Eng. Manag. J.* **2012**, *11*, 239–246.
31. Wang, H.; Lee, H.-W.; Deng, Y.; Lu, Z.; Hsu, P.-C.; Liu, Y.; Lin, D.; Cui, Y. Bifunctional non-noble metal oxide nanoparticle electrocatalysts through lithium-induced conversion for overall water splitting. *Nat. Commun.* **2015**, *6*, 7261. [CrossRef]
32. Liu, C.; Ma, H.; Yuan, M.; Yu, Z.; Li, J.; Shi, K.; Liang, Z.; Yang, Y.; Zhu, T.; Sun, G.; et al. (NiFe)S_2 nanoparticles grown on graphene as an efficient electrocatalyst for oxygen evolution reaction. *Electrochim. Acta* **2018**, *286*, 195–204. [CrossRef]
33. Torres-Rivero, K.; Torralba-Cadena, L.; Espriu-Gascon, A.; Casas, I.; Bastos-Arrieta, J.; Florido, A. Strategies for surface modification with Ag-shaped nanoparticles: Electrocatalytic enhancement of screen-printed electrodes for the detection of heavy metals. *Sensors* **2019**, *19*, 4249. [CrossRef] [PubMed]
34. Bottari, D.; Pigani, L.; Zanardi, C.; Terzi, F.; Patturca, S.V.; Grigorescu, S.D.; Matei, C.; Lete, C.; Lupu, S. Electrochemical sensing of caffeic acid using gold nanoparticles embedded in poly(3,4-ethylenedioxythiophene) layer by sinusoidal voltage procedure. *Chemosensors* **2019**, *7*, 65. [CrossRef]
35. Kellenberger, A.; Ambros, D.; Plesu, N. Scan rate dependent morphology of polyaniline films electrochemically deposited on nickel. *Int. J. Electrochem. Sci.* **2014**, *9*, 6821–6833.
36. Gira, M.J.; Tkacz, K.P.; Hampton, J.R. Physical and electrochemical area determination of electrodeposited Ni, Co, and NiCo thin films. *Nano Converg.* **2016**, *3*, 1–8. [CrossRef]
37. Sebarchievici, I.; Taranu, B.-O.; Rus, S.F.; Vlazan, P.; Poienar, M.; Sfirloaga, P. Electro-Oxidation of Ascorbic Acid on Perovskite-Modified Electrodes. In Proceedings of the 25th International Symposium on Analytical and Environmental Problems, Szeged, Hungary, 7–8 October 2019; pp. 273–275.
38. Zhou, Z.; Zaman, W.Q.; Sun, W.; Cao, L.; Tariq, M.; Yang, J. Cultivating crystal lattice distortion in IrO_2 via coupling with MnO_2 to boost the oxygen evolution reaction with high intrinsic activity. *Chem. Commun.* **2018**, *54*, 4959–4962. [CrossRef]
39. Chen, S.; Qiao, S.-Z. Hierarchically porous nitrogen-doped graphene-$NiCo2O_4$ hybrid paper as an advanced electrocatalytic water splitting material. *ACS Nano* **2013**, *7*, 10190–10196. [CrossRef]
40. Li, Y.H.; Liu, P.F.; Pan, L.F.; Wang, H.F.; Yang, Z.Z.; Zheng, L.R.; Hu, P.; Zhao, H.J.; Gu, L.; Yang, H.G. Local atomic structure modulations activate metal oxide as electrocatalyst for hydrogen evolution in acidic water. *Nat. Commun.* **2015**, *6*, 8064. [CrossRef]
41. Zhao, Y.; Chen, S.; Sun, B.; Su, D.; Huang, X.; Liu, H.; Yan, Y.; Sun, K.; Wang, G. Graphene-Co_3O_4 nanocomposite as electrocatalyst with high performance for oxygen evolution reaction. *Sci. Rep.* **2015**, *5*, 7629. [CrossRef]
42. Hona, R.K.; Karki, S.B.; Ramezanipour, F. Oxide electrocatalysts based on earth-abundant metals for both hydrogen- and oxygen-evolution reactions. *ACS Sustain. Chem. Eng.* **2020**, *8*, 11549–11557. [CrossRef]
43. Zhu, Y.; Zhou, W.; Chen, Z.-G.; Chen, Y.; Su, C.; Tad, M.O.; Shao, Z. $SrNb_{0.1}Co_{0.7}Fe_{0.2}O_{3-\delta}$ perovskite as a next-generation electrocatalyst for oxygen evolution in alkaline solution. *Angew. Chem. Int. Ed.* **2015**, *54*, 3897–3901. [CrossRef] [PubMed]
44. Das, D.; Das, A.; Reghunath, M.; Nanda, K.K. Phosphine-free avenue to Co_2P nanoparticle encapsulated N,P co-doped CNTs: A novel non-enzymatic glucose sensor and an efficient electrocatalyst for oxygen evolution reaction. *Green Chem.* **2017**, *19*, 1327–1335. [CrossRef]
45. Li, F.; Li, J.; Zhou, L.; Dai, S. Enhanced OER performance of composite Co–Fe–based MOF catalysts via a one-pot ultrasonic-assisted synthetic approach. *Sustain. Energy Fuels* **2021**, *5*, 1095–1102. [CrossRef]
46. Wang, C.-P.; Feng, Y.; Sun, H.; Wang, Y.; Yin, J.; Yao, Z.; Bu, X.-H.; Zhu, J. Self-optimized metal–organic framework electrocatalysts with structural stability and high current tolerance for water oxidation. *ACS Catal.* **2021**, *11*, 7132–7143. [CrossRef]
47. Taranu, B.-O.; Vlazan, P.; Racu, A. Water splitting studies in alkaline medium using graphite electrodes modified with transition metal oxides and compositions containing them. *Stud. UBB Chem.* **2022**, *67*, 79–95. [CrossRef]
48. Trasatti, S. Electrocatalysis in the anodic evolution of oxygen and chlorine. *Electrochim. Acta* **1984**, *29*, 1503–1512. [CrossRef]
49. Cherevko, S.; Geiger, S.; Kasian, O.; Kulyk, N.; Grote, J.-P.; Savan, A.; Shrestha, B.R.; Merzlikin, S.; Breitbach, B.; Ludwig, A.; et al. Oxygen and hydrogen evolution reactions on Ru, RuO_2, Ir, and IrO_2 thin film electrodes in acidic and alkaline electrolytes: A comparative study on activity and stability. *Catal. Today* **2016**, *262*, 170–180. [CrossRef]
50. Zhang, D.; Zhang, Y.; Luo, Y.; Zhang, Y.; Li, X.; Yu, X.; Ding, H.; Chu, P.; Sun, L. High-performance asymmetrical supercapacitor composed of rGO-enveloped nickel phosphite hollow spheres and N/S co-doped rGO aerogel. *Nano Res.* **2018**, *11*, 1651–1663. [CrossRef]
51. Li, B.; Shi, Y.; Huang, K.; Zhao, M.; Qiu, J.; Xue, H.; Pang, H. Cobalt-doped nickel phosphite for high performance of electrochemical energy storage. *Small* **2018**, *14*, 1703811. [CrossRef]
52. Tu, J.; Lei, H.; Wang, M.; Yu, Z.; Jiao, S. Facile synthesis of $Ni_{11}(HPO_3)_8(OH)_6$/rGO nanorods with enhanced electrochemical performance for aluminum-ion batteries. *Nanoscale* **2018**, *10*, 21284–21291. [CrossRef]
53. Wang, X.; Xu, J.; Yu, J.; Bu, Y.; Marques-Hueso, J.; Yan, X. Morphology control, spectrum modification and extended optical applications of rare earth ion doped phosphors. *Phys. Chem. Chem. Phys.* **2020**, *22*, 15120–15162. [CrossRef] [PubMed]
54. Liao, K.; Ni, Y. Synthesis of hierarchical $Ni_{11}(HPO_3)_8(OH)_6$ superstructures based on nanorods through a soft hydrothermal route. *Mater. Res. Bull.* **2010**, *45*, 205–209. [CrossRef]
55. Marcos, M.D.; Amoros, P.; Beltran-Porter, A.; Martinez-Manez, R.; Attfield, J.P. Novel crystalline microporous transition-metal phosphites $M_{11}(HPO_3)_8(OH)_6$ (M = Zn, Co, Ni). X-ray powder diffraction structure determination of the cobalt and nickel derivatives. *Chem. Mater.* **1993**, *5*, 121–128. [CrossRef]

56. Bhushan, B. *Modern Tribology Handbook, Chapter 2: Surface Roughness Analysis and Measurement Techniques*, 1st ed.; CRC Press: Boca Raton, FL, USA, 2001; pp. 49–120. [CrossRef]
57. Owen, J.F.; Chang, R.K.; Barber, P.W. Morphology–dependent resonances in Raman scattering, fluorescence emission, and elastic scattering from microparticles. *Aerosol Sci. Technol.* **1982**, *1*, 293–302. [CrossRef]
58. Lan, Y.; Zondode, M.; Deng, H.; Yan, J.A.; Ndaw, M.; Lisfi, A.; Wang, C.; Pan, Y.-L. Basic concepts and recent advances of crystallographic orientation determination of graphene by Raman spectroscopy. *Crystals* **2018**, *8*, 375. [CrossRef]
59. Nguyen, T.H.; Nguyen, T.M.H.; Kang, B.; Cho, B.; Han, M.; Choi, H.J.; Kong, M.; Lee, Y.; Yang, I. Raman spectroscopic evidence of impurity-induced structural distortion in SmB6. *J. Raman Spectrosc.* **2019**, *50*, 1661–1671. [CrossRef]
60. Oliver, S.M.; Beams, R.; Krylyuk, S.; Kalish, I.; Singh, A.K.; Bruma, A.; Tavazza, F.; Joshi, J.; Stone, I.R.; Stranick, S.J.; et al. The structural phases and vibrational properties of $Mo_{1-x}W_xTe_2$ alloys. *2D Mater.* **2017**, *4*, 045008. [CrossRef]
61. Van Stam, J.; Lindqvist, C.; Hansson, R.; Ericsson, L.; Moons, E. Fluorescence and UV/VIS absorption spectroscopy studies on polymer blend films for photovoltaics. *Proc. SPIE Int. Soc. Opt. Eng.* **2015**, *9549*, 95490L1-9. [CrossRef]
62. Hamzah, M.; Khenfouch, M.; Srinivasu, V.V. The quenching of silver nanoparticles photoluminescence by graphene oxide: Spectroscopic and morphological investigations. *J. Mater. Sci. Mater. Electron.* **2017**, *28*, 1804–1811. [CrossRef]
63. Gangwar, J.; Dey, K.K.; Tripathi, S.K.; Wan, M.; Yadav, R.R.; Singh, R.K.; Srivastava, A.K. NiO-based nanostructures with efficient optical and electrochemical properties for high-performance nanofluids. *Nanotechnology* **2013**, *24*, 415705. [CrossRef]
64. Vijayaprasath, G.; Sakthivel, P.; Murugan, R.; Mahalingam, T.; Ravi, G. Deposition and characterization of ZnO/NiO thin films. *AIP Conf. Proc.* **2016**, *1731*, 080033. [CrossRef]
65. Zhong, J.; Liang, H.; Su, Q.; Dorenbos, P.; Danang Birowosuto, M. Luminescence of $NaGd(PO_3)_4:Ce^{3+}$ and its potential application as a scintillator material. *Chem. Phys. Lett.* **2007**, *445*, 32–36. [CrossRef]
66. Bard, A.J.; Faulkner, L.R. *Electrochemical Methods: Fundamentals and Applications*, 2nd ed.; John Wiley & Sons: New York, NY, USA, 2001; pp. 186–191.
67. Yang, M.; Yang, Y.; Liu, Y.; Shen, G.; Yu, R. Platinum nanoparticles-doped sol–gel/carbon nanotubes composite electrochemical sensors and biosensors. *Biosens. Bioelectron.* **2006**, *21*, 1125–1131. [CrossRef] [PubMed]
68. Hrapovic, S.; Liu, Y.; Male, K.B.; Luong, J.H.T. Electrochemical biosensing platforms using platinum nanoparticles and carbon nanotubes. *Anal. Chem.* **2004**, *76*, 1083–1088. [CrossRef]
69. Xu, H.; Zhang, W.; Zhang, J.; Wu, Z.; Sheng, T.; Gao, F. An Fe-doped $Co11(HPO3)8(OH)6$ nanosheets array for high-performance water electrolysis. *Electrochim. Acta* **2020**, *334*, 135616. [CrossRef]
70. Lu, W.-X.; Wang, B.; Chen, W.-J.; Xie, J.-L.; Huang, Z.-Q.; Jin, W.; Song, J.-L. Nanosheet-like $Co_3(OH)_2(HPO_4)_2$ as a highly efficient and stable electrocatalyst for oxygen evolution reaction. *ACS Sustain. Chem. Eng.* **2019**, *7*, 3083–3091. [CrossRef]
71. Sial, M.; Lin, H.; Wang, X. Microporous 2D NiCoFe phosphate nanosheets supported on Ni foam for efficient overall water splitting in alkaline media. *Nanoscale* **2018**, *10*, 12975–12980. [CrossRef]
72. Zhang, Q.; Li, T.; Liang, J.; Wang, N.; Kong, X.; Wang, J.; Qian, H.; Zhou, Y.; Liu, F.; Wei, C.; et al. Highly wettable and metallic NiFe-phosphate/phosphide catalyst synthesized by plasma for highly efficient oxygen evolution reaction. *J. Mater. Chem. A* **2018**, *6*, 7509–7516. [CrossRef]
73. Lei, Z.; Bai, J.; Li, Y.; Wang, Z.; Zhao, C. Fabrication of nanoporous nickel-iron hydroxylphosphate composite as bifunctional and reversible catalyst for highly efficient intermittent water splitting. *ACS Appl. Mater. Interfaces* **2017**, *9*, 35837–35846. [CrossRef]
74. Zhou, J.; Dou, Y.B.; Zhou, A.; Guo, R.M.; Zhao, M.J.; Li, J.R. MOF template-directed fabrication of hierarchically structured electrocatalysts for efficient oxygen evolution reaction. *Adv. Energy Mater.* **2017**, *7*, 1602643. [CrossRef]
75. Zhou, H.Q.; Yu, F.; Sun, J.Y.; He, R.; Chen, S.; Chu, C.W.; Ren, Z.F. Highly active catalyst derived from a 3D foam of $Fe(PO3)2/Ni2P$ for extremely efficient water oxidation. *Proc. Natl. Acad. Sci. USA* **2017**, *114*, 5607–5611. [CrossRef] [PubMed]
76. Chang, J.; Lv, Q.; Li, G.; Ge, J.; Liu, C.; Xing, W. Core-shell structured $Ni_{12}P_5/Ni_3(PO_4)_2$ hollow spheres as difunctional and efficient electrocatalysts for overall water electrolysis. *Appl. Catal. B Environ.* **2017**, *204*, 486–496. [CrossRef]
77. Jin, Y.S.; Wang, H.T.; Li, J.J.; Yue, X.; Han, Y.J.; Shen, P.K.; Cui, Y. Porous MoO_2 nanosheets as non-noble bifunctional electrocatalysts for overall water splitting. *Adv. Mater.* **2016**, *28*, 3785–3790. [CrossRef] [PubMed]
78. Masud, J.; Umapathi, S.; Ashokaan, N.; Nath, M. Iron phosphide nanoparticles as an efficient electrocatalyst for OER in alkaline solution. *J. Mater. Chem. A* **2016**, *4*, 9750–9754. [CrossRef]
79. Rao, Y.; Wang, Y.; Ning, H.; Li, P.; Wu, M.B. Hydrotalcite-like $Ni(OH)_2$ nanosheets in situ grown on nickel foam for overall water splitting. *ACS Appl. Mater. Interfaces* **2016**, *8*, 33601–33607. [CrossRef]
80. You, B.; Jiang, N.; Sheng, M.L.; Bhushan, M.W.; Sun, Y.J. Hierarchically porous urchin-like Ni_2P superstructures supported on nickel foam as efficient bifunctional electrocatalysts for overall water splitting. *ACS Catal.* **2016**, *6*, 714–721. [CrossRef]
81. Yu, X.Y.; Feng, Y.; Guan, B.Y.; Lou, X.W.; Paik, U. Carbon coated porous nickel phosphides nanoplates for highly efficient oxygen evolution reaction. *Energy Environ. Sci.* **2016**, *9*, 1246–1250. [CrossRef]
82. Zhao, S.L.; Wang, Y.; Dong, J.C.; He, C.T.; Yin, H.J.; An, P.F.; Zhao, K.; Zhang, X.F.; Gao, C.; Zhang, L.J.; et al. Ultrathin metal–organic framework nanosheets for electrocatalytic oxygen evolution. *Nat. Energy* **2016**, *1*, 1–10. [CrossRef]
83. Li, J.; Yan, M.; Zhou, X.; Huang, Z.-Q.; Xia, Z.; Chang, C.-R.; Ma, Y.; Qu, Y. Mechanistic insights on ternary $Ni_2-xCoxP$ for hydrogen evolution and their hybrids with graphene as highly efficient and robust catalysts for overall water splitting. *Adv. Funct. Mater.* **2016**, *26*, 6785–6796. [CrossRef]

84. Liang, H.; Gandi, A.N.; Anjum, D.H.; Wang, X.; Schwingenschlogl, U.; Alshareef, H.N. Plasma-assisted synthesis of NiCoP for efficient overall water splitting. *Nano Lett.* **2016**, *16*, 7718–7725. [CrossRef]
85. Yuan, C.-Z.; Jiang, Y.-F.; Wang, Z.; Xie, X.; Yang, Z.-K.; Yousaf, A.B.; Xu, A.-W. Cobalt phosphate nanoparticles decorated with nitrogen-doped carbon layers as highly active and stable electrocatalysts for the oxygen evolution reaction. *J. Mater. Chem. A* **2016**, *4*, 8155–8160. [CrossRef]
86. Feng, L.L.; Yu, G.T.; Wu, Y.Y.; Li, G.D.; Li, H.; Sun, Y.H.; Asefa, T.; Chen, W.; Zou, X.X. High-index faceted Ni_3S_2 nanosheet arrays as highly active and ultrastable electrocatalysts for water splitting. *J. Am. Chem. Soc.* **2015**, *137*, 14023–14026. [CrossRef] [PubMed]
87. Han, A.; Chen, H.L.; Sun, Z.J.; Xu, J.; Du, P.W. High catalytic activity for water oxidation based on nanostructured nickel phosphide precursors. *Chem. Commun.* **2015**, *51*, 11626–11629. [CrossRef] [PubMed]
88. Hu, H.; Guan, B.Y.; Xia, B.Y.; Lou, X.W. Designed formation of $Co_3O_4/NiCo_2O_4$ double-shelled nanocages with enhanced pseudocapacitive and electrocatalytic properties. *J. Am. Chem. Soc.* **2015**, *137*, 5590–5595. [CrossRef]
89. Jiang, N.; You, B.; Sheng, M.L.; Sun, Y.J. Electrodeposited cobalt-phosphorous-derived films as competent bifunctional catalysts for overall water splitting. *Angew. Chem. Int. Ed.* **2015**, *54*, 6251–6254. [CrossRef]
90. Jin, H.Y.; Wang, J.; Su, D.F.; Wei, Z.Z.; Pang, Z.F.; Wang, Y. In-situ cobalt-cobalt oxide/N-doped carbon hybrids as superior bifunctional electrocatalysts for hydrogen and oxygen evolution. *J. Am. Chem. Soc.* **2015**, *137*, 2688–2694. [CrossRef]
91. Liang, H.F.; Meng, F.; Caban-Acevedo, M.; Li, L.S.; Forticaux, A.; Xiu, L.C.; Wang, Z.C.; Jin, S. Hydrothermal continuous flow synthesis and exfoliation of NiCo layered double hydroxide nanosheets for enhanced oxygen evolution catalysis. *Nano Lett.* **2015**, *15*, 1421–1427. [CrossRef] [PubMed]
92. Ledendecker, M.; Calderon, S.K.; Papp, C.; Steinruck, H.P.; Antonietti, M.; Shalom, M. The synthesis of nanostructured Ni_5P_4 films and their use as a non-noble bifunctional electrocatalyst for full water splitting. *Angew. Chem. Int. Ed.* **2015**, *54*, 12361–12365. [CrossRef] [PubMed]
93. Etesami, M.; Khezri, R.; Abbasi, A.; Nguyen, M.T.; Yonezawa, T.; Kheawhom, S.; Somwangthanaroj, A. Ball mill-assisted synthesis of NiFeCo-NC as bifunctional oxygen electrocatalysts for rechargeable zinc-air batteries. *J. Alloys Compd.* **2022**, *922*, 166287. [CrossRef]
94. Etesami, M.; Mohamad, A.A.; Nguyen, M.T.; Yonezawa, T.; Pornprasertsuk, R.; Somwangthanaroj, A.; Kheawhom, S. Benchmarking superfast electrodeposited bimetallic (Ni, Fe, Co, and Cu) hydroxides for oxygen evolution reaction. *J. Alloys Compd.* **2012**, *889*, 161738. [CrossRef]

Article

Investigating the Formation of Different (NH$_4$)$_2$[M(H$_2$O)$_5$(NH$_3$CH$_2$CH$_2$COO)]$_2$[V$_{10}$O$_{28}$]·nH$_2$O (M = Co^{II}, Ni^{II}, Zn^{II}, n = 4; M = Cd^{II}, Mn^{II}, n = 2) Crystallohydrates

Jana Chrappová [1,*], Yogeswara Rao Pateda [1,*], Lenka Bartošová [2] and Erik Rakovský [1]

[1] Department of Inorganic Chemistry, Faculty of Natural Sciences, Comenius University in Bratislava, Ilkovičova 6, 842 15 Bratislava, Slovakia; erik.rakovsky@uniba.sk
[2] Department of Food Databases, Food Research Institute, National Agricultural and Food Centre, Priemyselná 4, 824 75 Bratislava, Slovakia; lenka.bartosova@nppc.sk
* Correspondence: jana.chrappova@uniba.sk (J.C.); rao2@uniba.sk (Y.R.P.)

Abstract: Three hybrid compounds based on decavanadates, i.e., (NH$_4$)$_2$[Co(H$_2$O)$_5$(β-HAla)]$_2$[V$_{10}$O$_{28}$]·4H$_2$O (**1**), (NH$_4$)$_2$[Ni(H$_2$O)$_5$(β-HAla)]$_2$[V$_{10}$O$_{28}$]·4H$_2$O (**2**), and (NH$_4$)$_2$[Cd(H$_2$O)$_5$(β-HAla)]$_2$[V$_{10}$O$_{28}$]·2H$_2$O (**3**), (where β-Hala = zwitterionic form of β-alanine) were prepared by reactions in mildly acidic conditions (pH ~ 4) at room temperature. These compounds crystallise in two structure types, both crystallising in monoclinic $P2_1/n$ space group but with dissimilar cell packing, i.e., as tetrahydrates (**1** and **2**) and as a dihydrate (**3**). An influence of crystal radii and spin state of the central atom in [M(H$_2$O)$_5$(β-HAla)]$^{2+}$ complex cations on the crystal packing leading to the formation of different crystallohydrate forms was investigated together with previously prepared (NH$_4$)$_2$[Zn(H$_2$O)$_5$(β-HAla)]$_2$[V$_{10}$O$_{28}$]·4H$_2$O (**4**) and (NH$_4$)$_2$[Mn(H$_2$O)$_5$(β-HAla)]$_2$[V$_{10}$O$_{28}$]·2H$_2$O (**5**) and spin states of [M(H$_2$O)$_5$(β-HAla)]$^{2+}$ (M = Co^{2+}, Ni^{2+}, and Mn^{2+}) cations in solution were confirmed by ^{1}H-NMR paramagnetic effects. FT-IR and FT-Raman spectra for **1**–**5** are in agreement with the X-ray structure analysis results.

Keywords: hybrid decavanadates; β-alanine; crystallohydrates; crystal structure; vibrational spectroscopy; paramagnetic NMR

Citation: Chrappová, J.; Pateda, Y.R.; Bartošová, L.; Rakovský, E. Investigating the Formation of Different (NH$_4$)$_2$[M(H$_2$O)$_5$ (NH$_3$CH$_2$CH$_2$COO)]$_2$[V$_{10}$O$_{28}$]·nH$_2$O (M = Co^{II}, Ni^{II}, Zn^{II}, n = 4; M = Cd^{II}, Mn^{II}, n = 2) Crystallohydrates. *Crystals* **2024**, *14*, 685. https://doi.org/10.3390/cryst14080685

Academic Editor: Pedro Marques De Almeida

Received: 4 July 2024
Revised: 23 July 2024
Accepted: 24 July 2024
Published: 27 July 2024

Copyright: © 2024 by the authors. Licensee MDPI, Basel, Switzerland. This article is an open access article distributed under the terms and conditions of the Creative Commons Attribution (CC BY) license (https://creativecommons.org/licenses/by/4.0/).

1. Introduction

Decavanadate anion is the predominant and most stable V(V) oxo-anion found in the acidic aqueous solutions [1]. It is known in different protonation states, [H$_n$V$_{10}$O$_{28}$]$^{(6-n)-}$ (n = 0–4) [2–4].

The decavanadate anion consists of ten edge-sharing VO$_6$ octahedra. Six of them form a rectangular 2 × 3 arrangement, with two VO$_6$ octahedra joining the arrangement from the top and two from the bottom via edge sharing with the six octahedra lying in the central plane. The ideal symmetry of [V$_{10}$O$_{28}$]$^{6-}$ anion is given by the D_{2h} point group symmetry. In most crystal structures, [V$_{10}$O$_{28}$]$^{6-}$ anion usually occupies special positions, mostly inversion centres, and its symmetry is changed correspondingly; however, its geometry remains close to the ideal D_{2h} symmetry.

In crystal structures, decavanadates act as acceptors of protons in supramolecular arrangements; protonated decavanadate species can also act as proton donors. Typically, if decavanadate anion is not protonated on mutually centrosymmetrically arranged sites, such decavanadate anions will likely form a centrosymmetric dimer interconnected via anion-anion hydrogen bonds [5] or via hydrogen bonding to counter-cations lying on a centre of symmetry [6]. Considering the sterical effect of the decavanadate anion and the influence of size, shape, and nature of counter-cations, a substantial part of decavanadates include water molecules as a stabilising element of the crystal structure. Formation of different crystallohydrates depends on reaction conditions [7] or on temperature changes, even if the reaction conditions remain the same [8]. As a result, decavanadates are known to form a large variety of supramolecular assemblies [9].

The oxovanadates (V), including decavanadates and peroxovanadium compounds, are of great interest in bioinorganic chemistry and biochemistry because of their antidiabetic, antibacterial, antiprotozoal, antiviral, and anticancer properties [10,11]. Studying of non-covalent interactions based on electrostatic attractive forces between decavanadate anion and appropriate molecules (organic cations, hybrid inorganic–organic complex cations, and biomacromolecules) could provide important information on the cooperative effects of decavanadate ions in biological systems [12–14].

Apart from the biological importance of decavanadates, there are other emerging application possibilities. Some decavanadates exhibit water oxidation activity [15] or heterogeneous bifunctional catalytic properties in the selective oxidation of sulphides and Mannich reaction [16]. Decavanadates with superior proton conductivity [17] and photoluminescent sensing properties for the detection of Zn^{2+} and Co^{2+} [18] were also prepared. Ammonium decavanadate nanodots—holey reduced graphene oxide nanoribbons [19] and α-Co(OH)$_2$ nanoplates with decavanadate anion [20] can be used as electrodes for supercapacitors.

In this work, we report the synthesis, crystal structure determination, and properties of three decavanadates, i.e., $(NH_4)_2[Co(H_2O)_5(\beta\text{-HAla})]_2[V_{10}O_{28}]\cdot 4H_2O$ (**1**), $(NH_4)_2[Ni(H_2O)_5(\beta\text{-HAla})]_2[V_{10}O_{28}]\cdot 4H_2O$ (**2**), and $(NH_4)_2[Cd(H_2O)_5(\beta\text{-HAla})]_2[V_{10}O_{28}]\cdot 2H_2O$ (**3**), prepared by the same procedure we used previously for the preparation of two decavanadates containing the same $[M(H_2O)_5(\beta\text{-HAla})]^{2+}$ cations in two different crystallohydrate forms, i.e., tetrahydrate $(NH_4)_2[Zn(H_2O)_5(\beta\text{-HAla})]_2[V_{10}O_{28}]\cdot 4H_2O$ (**4**) and dihydrate $(NH_4)_2[Mn(H_2O)_5(\beta\text{-HAla})]_2[V_{10}O_{28}]\cdot 2H_2O$ (**5**) [21]. The compounds **1** and **2** are isostructural with **4**, while **3** is isostructural with **5**. In [22], compound **2** was prepared by using an alternative hydrothermal "Direct Synthesis" approach by the reaction of metal powder in anion-deficient conditions with V_2O_5 in an aqueous solution of β-alanine and ammonium acetate. Subsequently, compound **2** was used as a precursor for the preparation of V_2O_5/MnV_2O_6 mixed oxide, and its ability to act as a water oxidation catalyst for oxygen production was studied. Later, **1** was prepared by the same "Direct Synthesis" method and used as a precursor for the preparation of solid mixed oxides CoV_2O_6/V_2O_5 and $Co_2V_2O_7/V_2O_5$ by thermal decomposition; both these mixed oxides and **1** can act as catalysts for photoinduced water oxidation [23]. The aim of the present study is to obtain more complete insight into the formation of different crystallohydrates with the general formula $(NH_4)_2[M(H_2O)_5(\beta\text{-HAla})]_2[V_{10}O_{28}]\cdot nH_2O$, where M is a divalent metal ion, from the same reaction conditions.

2. Materials and Methods

2.1. General

All chemicals were of reagent or better grade, obtained from commercial sources (Mikrochem (Pezinok, Slovakia), Sigma-Aldrich (Saint Louis, MI, USA), Slavus (Donja Stubica, Croatia), Lachema (Brno, Czech Republic)). Infrared spectra were obtained from KBr pellets on a Thermo Scientific Nicolet 6700 FTIR spectrometer in the 400–4000 cm^{-1} range (Waltham, MA, USA). The Raman spectra were registered with a Raman microspectrometer Senterra (Bruker Optik, Ettlingen, Germany), using a laser with the wavelength 532 nm, with a maximum power of 2 mW, in the range 64–4467 cm^{-1}, with 10× objective lenses, and 2 scans for 10 s each. ^{1}H-NMR spectra were recorded on a Bruker AVANCE Neo 400 MHz (operating at 9.37 T, 400 MHz) using D_2O as a reference. Chemical shifts are reported in Hz. Vanadium (V) was determined volumetrically by titration with $FeSO_4$ ($c = 0.1$ M) using diphenylamine as the indicator.

2.2. Synthesis and Crystallisation

2.2.1. Synthesis of $(NH_4)_2[Co(H_2O)_5(\beta\text{-HAla})]_2[V_{10}O_{28}]\cdot 4H_2O$ (1)

$CoSO_4\cdot 7H_2O$ (1.4055 g; 5 mmol) was added to a solution of β-alanine (0.89 g; 10 mmol) in water (20 mL). After stirring for 15 min, a solution of NH_4VO_3 (1.170 g; 10 mmol) in water (40 mL) was added under the immediate formation of a fine precipitate. The solution with precipitate was stirred for 30 min and filtered. The pH of the filtrate was adjusted to 4.0 with 2M H_2SO_4. To the reddish-brown solution obtained, ethanol (10 mL) was added. Reddish brown crystals were isolated after standing for 22 days at 4 °C in the refrigerator.

Yield: 0.8 g/52.1% (calc. for vanadium). Anal. Calc. for $C_6H_{50}N_4O_{46}Co_2V_{10}$ (MW = 1541.76 g/mol) V, 34.10%. Found: V, 33.96%.

2.2.2. Synthesis of $(NH_4)_2[Ni(H_2O)_5(\beta\text{-HAla})]_2[V_{10}O_{28}]\cdot 4H_2O$ (2)

$NiCl_2\cdot 6H_2O$ (1.188 g; 5 mmol) was added to a solution of β-alanine (0.89 g; 10 mmol) in water (20 mL). After stirring for 15 min, a solution of NH_4VO_3 (1.170 g; 10 mmol) in water (60 mL) was added, followed by the immediate formation of a fine precipitate. The solution with precipitate was stirred for 30 min and filtered. The pH of the filtrate was adjusted to 4.0 with 4M HCl. To the yellow solution obtained, ethanol (10 mL) was added. Dark orange crystals with an intense green tone were isolated after standing for 20 days at 4 °C in the refrigerator.

Yield: 1.343 g/87.5% (calc. for vanadium). Anal. Calc. for $C_6H_{50}N_4O_{46}Ni_2V_{10}$ (MW = 1541.32 g/mol) V, 34.11%. Found: V, 33.69%.

2.2.3. Synthesis of $(NH_4)_2[Cd(H_2O)_5(\beta\text{-HAla})]_2[V_{10}O_{28}]\cdot 2H_2O$ (3)

$Cd(NO_3)_2\cdot 4H_2O$ (1.542 g; 5 mmol) was added to a solution of β-alanine (0.89 g; 10 mmol) in water (20 mL). After stirring for 15 min, a solution of NH_4VO_3 (1.170 g; 10 mmol) in water (40 mL) was added, followed by the immediate formation of a fine precipitate. The solution with precipitate was stirred for 30 min and filtered. The pH of the filtrate was adjusted to 4.25 with 4M HNO_3. To the yellow solution obtained, ethanol (10 mL) was added. Orange crystals were isolated after standing for 15 days at 4 °C in the refrigerator.

Yield: 0.702 g/43.7% (calc. for vanadium). Anal. Calc. for $C_6H_{46}N_4O_{44}Cd_2V_{10}$ (MW = 1612.67 g/mol) V, 31.59%. Found: V, 31.44%.

$(NH_4)_2[Zn(H_2O)_5(\beta\text{-HAla})]_2[V_{10}O_{28}]\cdot 4H_2O$ (4) and $(NH_4)_2[Mn(H_2O)_5(\beta\text{-HAla})]_2[V_{10}O_{28}]\cdot 2H_2O$ (5) samples were prepared for physical measurements according to [22].

2.3. X-ray Data Collection and Structure Determination

Intensity data for the compounds 1–3 were collected on a Kuma KM–4 CCD diffractometer using graphite monochromated MoKα radiation (0.71073 Å) by the ω- and φ-scan techniques at room temperature. Data collection, data reduction, and finalisation were carried out using *CrysAlis Pro-Version 1.171.43.128a* software [24]. Intensity data for 4 and 5 were reprocessed from original images collected under the conditions in [21] by using the above-mentioned up-to-date version of *CrysAlis Pro* software to take advantage of improved data processing, especially regarding twin data reduction handling.

The structures were solved by *SHELXT* [25] and refined by the full matrix least-squares method on all F^2 data using *SHELXL-2018/3* [26]. All non-hydrogen atoms were refined anisotropically. Hydrogen atoms of methylene groups were refined using a riding model, and the ammonium group was refined as a free rotor, with C–H and N–H distances free to refine. Hydrogen atoms of water molecules were refined with all O–H and all H...H distances restrained to be equal; similarly, the tetrahedral shape of ammonium cation was retained by restraining all N–H and all H...H distances to be equal. Thermal parameters of the H atoms were constrained to $U_{iso}(H) = 1.2U_{eq}(C)$ and $U_{iso}(H) = 1.5U_{eq}(N, O)$.

Both compounds 3 and 5 crystallise as non-merohedral twins. Two domains related to the twofold rotation about the c-axis were found using *Ewald Explorer* in the *CrysAlis Pro* software, and data were refinalised using internal programme features. The twin domain volume ratios were refined to 0.6515(7):0.3485(7) for **3** and 0.8436(4):0.1564(4) for **5**.

Geometrical calculations were performed using *SHELXL-2018/3, Olex2 1.5* [27], and *PARST* [28]. *Olex2 1.5* and *DIAMOND* [29] were used for molecular graphics. Octahedral distortion parameters ζ [30], Δ [31], Σ [32], and Θ [33] were calculated using *OctaDist 3.1.0* [34]. Hydrogen bonding geometries were normalised to neutron distances following a literature procedure [35,36].

Crystal data, conditions of data collection, and refinement results for the compounds **1–5** are reported in Table S1.

2.4. Paramagnetic ^{1}H-NMR Measurements

For the Evans method [37–39], approx. 2 mM solutions of **1**, **2**, **4**, and **5** using 3% (v/v) *t-Bu*OH solution in H$_2$O were prepared. The inner n.m.r. tube (o.d. 2 mm) was loaded with 3% (v/v) *t-Bu*OH solution in D$_2$O as a reference. The outer n.m.r. tube (o.d. 5 mm) was filled with 500 µL of the sample solution.

The Δ*f* value, defined as the difference between the chemical shift of the ^{1}H-NMR *t-Bu* signal in the sample solution and that of the *t-Bu*OH reference solution, was used to calculate the molar susceptibility of the complex using the following equation:

$$\chi_m = \frac{3\Delta f}{4\pi c f} \quad (1)$$

where *c* is the concentration of the paramagnetic complex in the solution in mmol.L^{-1} and *f* is the spectrometer frequency (400 MHz). Considering there are 2 [M(H$_2$O)$_5$(β-HAla)]$^{2+}$ cations per formula unit, the value of *c* is twice the concentration of the compound in the solution.

The χ_p values [cm^{3}mol^{-1}] are molar susceptibilities corrected for diamagnetic contribution to the susceptibility in the sample.

$$\chi_p = \chi_m - \chi_d \quad (2)$$

where χ_d represents diamagnetic contributions from the ligands, ions, inner-core electrons, etc. [40]. The following values were used to calculate the effective magnetic moment:

$$\mu_{eff} = \sqrt{\frac{3kT\chi_p}{N_A \mu_B^2}} \quad (3)$$

where *k* is the Boltzmann constant, *T* is temperature [K], N_A is Avogadro's number, and μ_B is the Bohr magneton. Effective magnetic moment is compared to the calculated value for the central atom in question as follows [41]:

$$\mu_{cal} = 2\sqrt{S(S+1)}\mu_B \quad (4)$$

3. Results and Discussion

Decavanadates with complex cations, i.e., (NH$_4$)$_2$[Co(H$_2$O)$_5$(β-HAla)]$_2$[V$_{10}$O$_{28}$]·4H$_2$O (**1**), (NH$_4$)$_2$[Ni(H$_2$O)$_5$(β-HAla)]$_2$[V$_{10}$O$_{28}$]·4H$_2$O (**2**), and (NH$_4$)$_2$[Cd(H$_2$O)$_5$(β-HAla)]$_2$[V$_{10}$O$_{28}$]·2H$_2$O (**3**), were obtained by crystallisation from the β-alanine—CoSO$_4$—NH$_4$VO$_3$—H$_2$SO$_4$—H$_2$O—ethanol (**1**), β-alanine—NiCl$_2$—NH$_4$VO$_3$—HCl—H$_2$O—ethanol (**2**), and β-alanine—Cd(NO$_3$)$_2$—NH$_4$VO$_3$—HNO$_3$—H$_2$O—ethanol (**3**) reaction systems in mildly acidic (pH~4) conditions. Attempts to prepare compounds of Mg^{2+}, Sr^{2+}, Ba^{2+}, Pb^{2+}, and Hg^{2+} from the same reaction system were unsuccessful, obtaining ammonium decavanadate as a product.

The preparation of compound **1** by the hydrothermal reaction of cobalt powder with V$_2$O$_5$ in an aqueous solution containing β-alanine and ammonium acetate was reported earlier [23]. However, from the described green colour of the solution and products obtained by hydrothermal synthesis when compared with the orange product we obtained, it can be inferred that a partial reduction of V(V) to V(IV) has taken place. This partial reduction can be avoided by using our preparation method. Although the synthesis was targeted towards using the prepared compound as a precursor for the preparation of mixed Co/V oxides by thermal decomposition, to avoid the presence of lower oxidation states of vanadium in the products, not only during the reactions in solutions but even after thermal decomposition in the air atmosphere, it is a good practise to use purified V(V) precursors [42].

3.1. Crystallographic Characterisation of Prepared Compounds

All prepared compounds belong to one of two monoclinic (space group $P2_1/n$) structure types with substantially different cell parameters and cell packing—tetrahydrates (**1**, **2**, and **4**) (Figure 1a) and dihydrates (**3** and **5**) (Figure 1b).

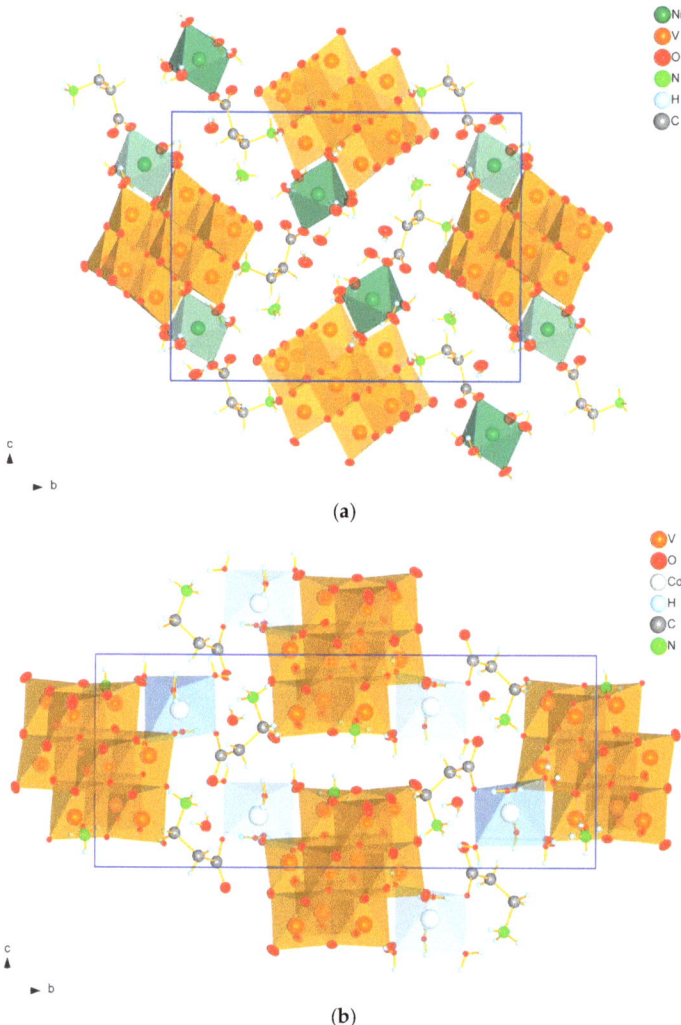

Figure 1. A view of the cell packing of tetrahydrates (**2** as an example) (**a**) and dihydrates (**3** as an example) (**b**) along the *a*-axis. Dashed lines indicate hydrogen bonds, blue rectangle denote unit cell boundaries.

The asymmetric unit of all prepared dihydrates consists of one half of the $[V_{10}O_{28}]^{6-}$ anion with C_i symmetry lying on an inversion centre, one $[M(H_2O)_5(\beta\text{-HAla})]^{2+}$ cation, one NH_4^+ cation, and two water molecules in general positions (Figure 2a).

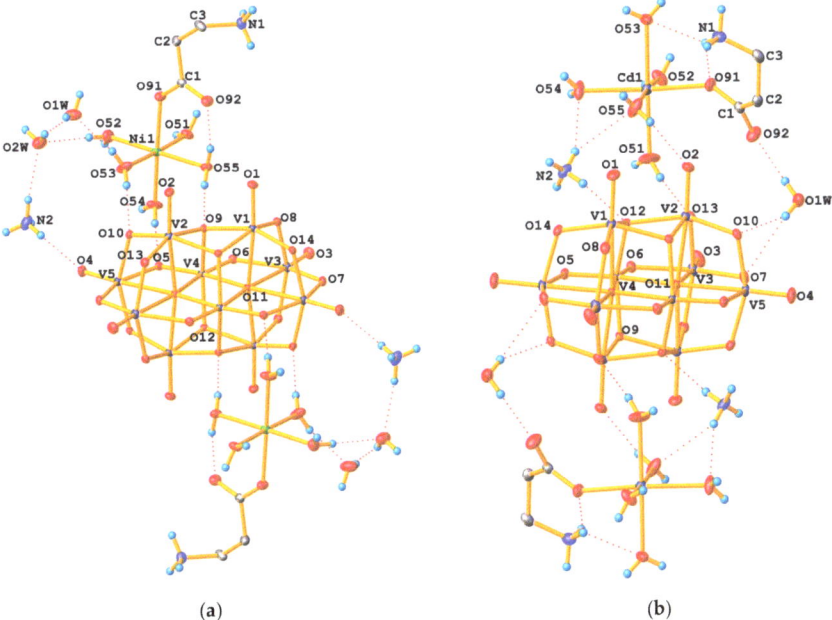

Figure 2. ADP representations of the crystal structure of (**a**) tetrahydrates (**2** as an example) and (**b**) dihydrates (**3** as an example) with numbering scheme (methylene H atoms are excluded). Labelled atoms are related to the unlabelled ones by the centre of symmetry. Displacement ellipsoids are drawn at 30% probability level. Dashed lines indicate hydrogen bonds.

The asymmetric unit of tetrahydrates consists of one half of the $[V_{10}O_{28}]^{6-}$ anion with C_i symmetry lying on an inversion centre, one $[M(H_2O)_5(\beta\text{-HAla})]^{2+}$ cation, one NH_4^+ cation, and one water molecule in general positions (Figure 2b).

The quality of X-ray diffraction data allowed us to find all hydrogen atoms from the ifference electron density map and refine them semi-freely using a set of suitable restraints [43]. As a result, the accurate orientations of water molecules, $-NH_3^+$ groups, and NH_4^+ cations were successfully determined.

The following four types of oxygen atoms can be distinguished in the decavanadate anions: terminal O_T bonded to only one vanadium atom, and $O(\mu_2)$, $O(\mu_3)$, and $O(\mu_6)$ bridging atoms, connecting 2, 3, and 6 vanadium atoms, respectively. Table S2 contains V–O and M–O bond lengths and bond valences/bond valence sums (BVS) calculated using the following equation [44,45]:

$$s = \exp\left(\frac{R_0 - R}{B}\right) \quad (5)$$

where R is bond length and R_0 and B are empirical parameters for given bond type, to confirm protonation state of the decavanadate anion, oxidation number of V, and central atoms of $[M(H_2O)_5(\beta\text{-HAla})]^{2+}$ complex cations. For V^V–O bonds, the values R_0 = 1.803 Å and B = 0.37 were used [46]. The BVS values obtained are in the range of 1.61–1.76 for V = O_T bonds, 1.764–1.893 for V–$O(\mu_2)$ bonds, 1.88–1.92 for V–$O(\mu_3)$ bonds, and 1.975–1.998 for V–$O(\mu_6)$ bonds. A comparison of BVS values obtained for particular bonds also confirms the rigidity of the $[V_{10}O_{28}]^{6-}$ anion, only marginally influenced by adjacent molecules. For protonated O atoms, expected BVS values are $\Sigma s < 1.5$; thus, in all prepared compounds, the presence of an unprotonated $[V_{10}O_{28}]^{6-}$ anion was confirmed. BVS values for all V atoms are in the range of 5.02–5.11, confirming oxidation state V(V).

For the BVS calculations for central atoms M in $[M(H_2O)_5(\beta\text{-HAla})]^{2+}$ cations, the following values of the bond parameters were used: $R_0 = 1.685$ Å and $B = 0.37$ for Co^{2+} in **1** [47], $R_0 = 1.675$ Å and $B = 0.37$ for Ni^{2+} in **2** [45], $R_0 = 1.904$ Å and $B = 0.37$ for Cd^{2+} in **3** [46], $R_0 = 1.704$ Å and $B = 0.37$ for Zn^{2+} in **4** [46], and $R_0 = 1.762$ Å and $B = 0.4$ for Mn^{2+} in **5** [48]. In all $[M(H_2O)_5(\beta\text{-HAla})]^{2+}$ cations in the compounds **1–5**, oxidation state II for the central atom was confirmed.

All the molecules present in both dihydrates and tetrahydrates are involved in the creation of an extensive hydrogen bonding network (Table S3). Most of these interactions are medium hydrogen bonds with bond distances d(H$\cdots$A) between 1.5 to 2.2 Å and bond angles belonging to the range 130–180° after normalisation to neutron distances [49].

The values of average M–O distances $\bar{d}(M\text{–O})$ in $[M(H_2O)_5(\beta\text{-HAla})]^{2+}$ complex cations of the compounds **1–5** (Table 1) are in good agreement with the values of the crystal radii for $r(Co^{2+}$ hs$) = 85.5$ pm, $r(Ni^{2+}) = 83$ pm, $r(Cd^{2+}) = 109$ pm, $r(Zn^{2+}) = 88$ pm, $r(Mn^{2+}$ hs$) = 97$ pm, and $r(O^{2-}) = 121$ pm, respectively; crystal radii corresponding with low spin states are significantly smaller ($r(Co^{2+}$ ls$) = 79$ pm, $r(Mn^{2+}$ ls$) = 81$ pm) [50]. The M–O_{water} distances are in the range 2.070(2)–2.1195(19) Å for **1**, 2.038(2)–2.0729(19) Å for **2**, 2.232(3)–2.327(3) Å for **3**, 2.0665(11)–2.1152(10) Å for **4**, and 2.1369(19)–2.2197(18) for **5**. The M–$O_{\beta\text{-HAla}}$ distances $d(M\text{–O91})$ (Table S2) are in the range 2.0614(18)–2.200(3) Å in the order of increasing crystal radii. Octahedral distortion parameters in $[M(H_2O)_5(\beta\text{-HAla})]^{2+}$ complex cations (Table 1) also show significant differences between particular values for tetrahydrates and for dihydrates.

Table 1. Average M–O distances, octahedral distortion parameters, and octahedron volumes in $[M(H_2O)_5(\beta\text{-HAla})]^{2+}$ complex cations of the compounds **1–5**.

Compound	$\bar{d}(M\text{–O})$ [Å]	ζ	Δ	Σ	Θ	V [Å^3]
1	2.090(15)	0.065267	5.1×10^{-5}	21.9313	51.4217	12.14
2	2.055(11)	0.050418	2.7×10^{-5}	23.2468	50.2841	11.54
3	2.26(4)	0.195609	3.10×10^{-4}	53.3874	144.5566	15.30
4	2.090(15)	0.064040	4.8×10^{-5}	28.7336	66.9865	12.14
5	2.17(4)	0.198905	3.38×10^{-4}	43.3304	113.7109	13.47

The zwitterionic form of β-alanine is characterised by the formation of an intramolecular salt bridge between the ammonium group and the carboxylic group, forming a six-membered S(6) ring [51,52]. In tetrahydrates (Figure 3a), smaller Co^{2+}, Ni^{2+}, and Zn^{2+} ions prefer the coordination of the less bulky oxygen atom of the carboxylate group not involved in the intramolecular hydrogen bond, while larger Mn^{2+} and Cd^{2+} cations in dihydrates (Figure 3b) prefer the coordination of the oxygen atom involved in the formation of the intramolecular hydrogen bond. This different behaviour influences not only the strength of the intermolecular hydrogen bond in β-alanine but also the entire hydrogen bond network. As a result, the $[M(H_2O)_5(\beta\text{-HAla})]^{2+}$ cation in tetrahydrates contains two intramolecular hydrogen bonds, i.e., the above-mentioned salt bridge N1–H1$\cdots$O92 between –NH_3^+ hydrogen and non-coordinated carboxylate oxygen, and O55–H55B$\cdots$O92 hydrogen bond between coordinated water molecules and the same non-coordinated carboxylate oxygen, thus forming another S(6) supramolecular ring. In dihydrates, the $[M(H_2O)_5(\beta\text{-HAla})]^{2+}$ cation contains one three-centred chelated intramolecular hydrogen bond, i.e., N1–H1B$\cdots$O53 and N1–H1B$\cdots$O91, where H1B belongs to the –NH_3^+ group, O53 to the coordinated water molecule, and O91 carboxylate oxygen coordinated to the metal atom. Alongside the intramolecular S(6) ring formed by β-alanine, the S(4) ring involving the coordinated oxygen atom of the carboxylate group, the hydrogen atom of the ammonium group, and the oxygen atom of the coordinated water molecule is formed.

These changes are also reflected as a change in corresponding torsion angles (Table 2) and in the presence of different numbers of cocrystallised water molecules in both crystal forms.

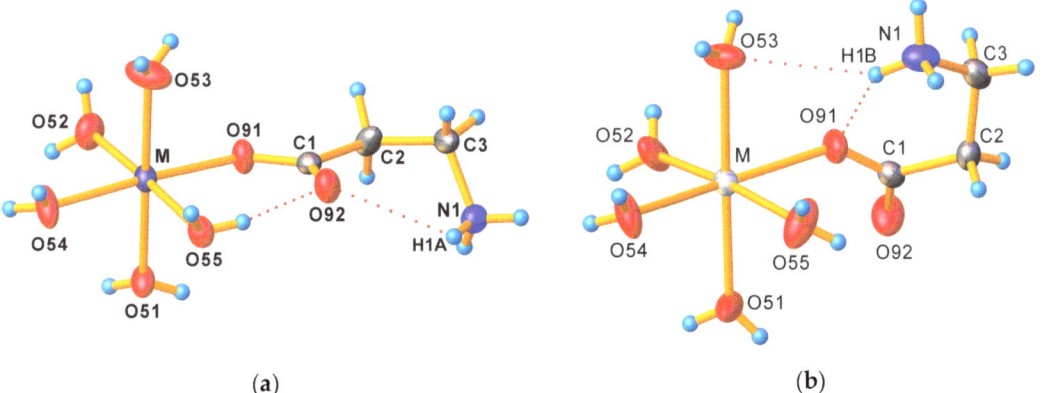

Figure 3. A detailed view of $[M(H_2O)_5(\beta\text{-HAla})]^{2+}$ geometry and intramolecular hydrogen bonds (**a**) in tetrahydrates and (**b**) in dihydrates. Displacement ellipsoids are drawn at 50% probability level.

Table 2. Selected torsion angles in $[M(H_2O)_5(\beta\text{-HAla})]^{2+}$ complex cations of the compounds **1**–**5**.

	1	2	3	4	5
M–O91–C1–O92	3.8(4)	0.8(4)	51.7(7)	4.3(2)	59.8(4)
M–O91–C1–C2	−176.41(18)	−179.03(17)	−128.4(4)	−175.70(9)	−119.9(3)
O91–C1–C2–C3	174.9(2)	173.3(2)	−36.8(5)	174.61(12)	−39.3(3)
O92–C1–C2–C3	−5.3(4)	−6.5(4)	143.1(4)	−5.41(18)	141.0(2)
C1–C2–C3–N1	70.5(3)	71.4(3)	59.0(5)	70.03(16)	60.9(3)

3.2. Molar Susceptibility Determination of $[M(H_2O)_5(\beta\text{-HAla})]^{2+}$ Ions in Solution by Paramagnetic ^{1}H-NMR

To determine the spin state of paramagnetic central atoms in the $[M(H_2O)_5(\beta\text{-HAla})]^{2+}$ (M = Co^{II}, Ni^{II}, Mn^{II}) complex cations, molar susceptibilities of the complex cations in the solutions of **1**, **2**, and **5** were determined based on the paramagnetic shift of the ^{1}H-NMR signal of the *t-Bu* group in the paramagnetic solution against the diamagnetic reference (3% v/v *t*BuOH solution in D$_2$O). For the assessment of diamagnetic contribution, the ^{1}H-NMR chemical shifts of the signal belonging to the *t-Bu* group in the solution of **4** and in a blank sample containing *t*-BuOH in D$_2$O were also recorded (Figure 4).

At the concentrations given, there was no significant diamagnetic contribution χ_d from the ligands and other ions in the solution based on a comparison of **4** and blank *t*-BuOH ^{1}H-NMR spectra, thus μ_{eff} values were calculated directly from χ_m values (Table 3).

Table 3. Experimental data, molar magnetic susceptibilities, and effective magnetic moments of the $[M(H_2O)_5(\beta\text{-HAla})]^{2+}$ complex cations in the compounds **1**, **2**, and **5**.

Compound	Δf [Hz]	c [mmol.L^{-1}]	T [K]	χ_m [cm^{3}mol^{-1}]	μ_{eff} [μ_B]
1	67.576	4.011	302.35	0.010052	4.95
2	30.2619	4.004	302.25	0.004509	3.32
5	95.534	4.000	300.85	0.014235	5.88

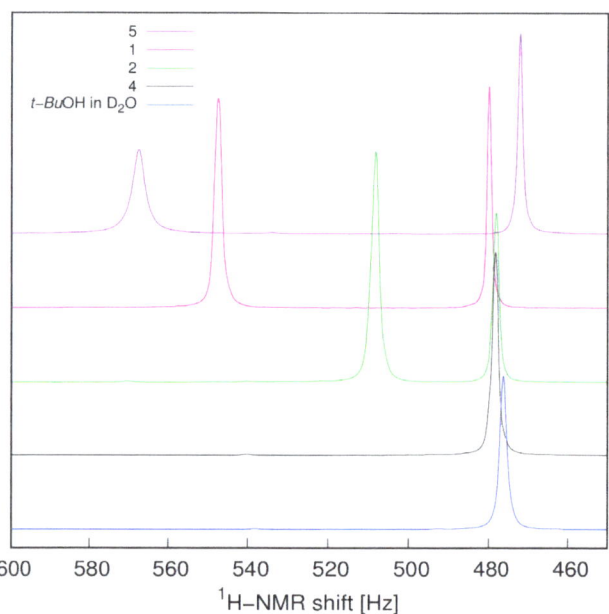

Figure 4. Paramagnetic shifts of the *t-Bu* ^{1}H-NMR signal in the solutions of the compounds **1**, **2**, **4**, and **5** and a blank sample against the reference solution of *t-Bu*OH in D$_2$O.

The results are in good agreement with the calculated μ_{cal} values by using Equation (4) for three unpaired electrons of Co^{2+}(hs) in **1**, two unpaired electrons of Ni^{2+} in **2**, and five unpaired electrons of Mn^{2+}(hs) in **5**. These findings agree with the assignment of spin states based on average $\bar{d}$(M–O) distances.

3.3. Vibrational Spectroscopy

The FT-IR (Figure 5a) and FT-Raman (Figure 5b) spectra of crystalline **1–5** are quite similar because of the same chemical components present. The FT-IR and FT-Raman spectra of the individual compounds are mutually compared on Figure S1a–e. The assignments of the bands are summarised in Table S4 [7,21,53].

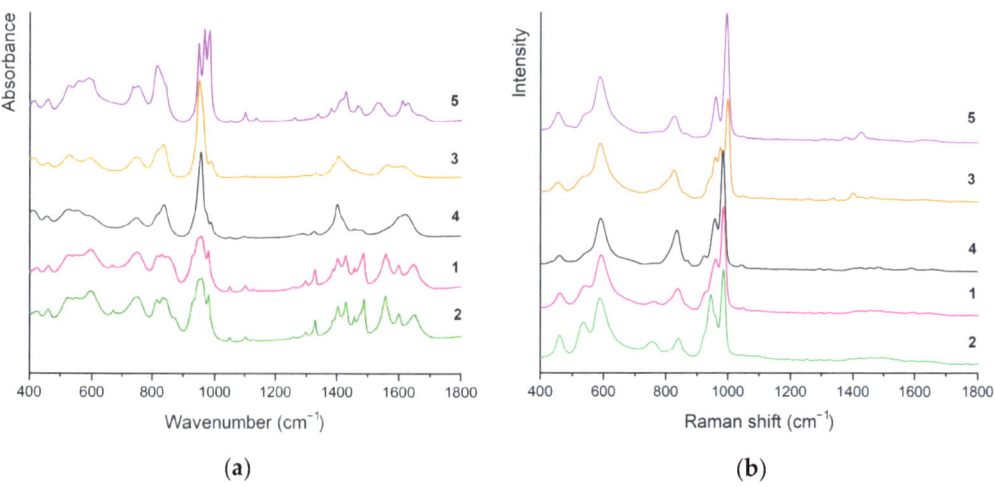

Figure 5. Stacked FT-IR (**a**) and FT-Raman (**b**) spectra of the compounds **1–5**.

The common features of decavanadate IR spectra involve the bands with highest intensity in the range 920–1000 cm^{-1} corresponding to valence vibrations of terminal V=O bonds and two broad series of bands related to asymmetric and symmetric bridging O–V$_2$ vibration modes in the range of 843–748 cm^{-1} (ν_{as}, IR)/866–823 cm^{-1} (ν_{as}, Raman) and 596–524 cm^{-1} (ν_s, IR)/591–533 cm^{-1} (ν_s, Raman), respectively.

The assignment of the bands corresponding to $\delta_d(NH_3^+)$, $\delta_d(NH_4^+)$, and both ν_s and $\nu_{as}(COO^-)$, is based on a comparison of partially deuterated samples [21].

To distinguish between binding modes of carboxylate group in complexes, it is possible to use the differences between frequencies of asymmetric and symmetric stretching vibrations, $\Delta = \nu_{as}(COO^-) - \nu_s(COO^-)$. Carboxylato complexes exhibiting Δ values that are significantly greater than ionic values, typically with $\Delta \geq 200$ cm^{-1} (cf. 164 and 184 cm^{-1} for zwitterionic β-HAla [53]), have usually unidentate coordination [54]. The Δ values of studied compounds ($\Delta = 230$ for **1, 2,** and **3**; $\Delta = 233$ for **4**; and $\Delta = 199$ for **5**) are typical for monodentate mode of carboxylate group bonding, thus confirming the results of X-ray structure analysis.

The presence of water molecules is confirmed by the bands in the region 3484–3380 cm^{-1}. These bands usually appear in the 3600–3400 cm^{-1} range, and their shift towards lower wavenumbers is caused by the occurrence of O–H...O hydrogen bonds.

4. Conclusions

We prepared $(NH_4)_2[Co(H_2O)_5(β-HAla)]_2[V_{10}O_{28}]\cdot 4H_2O$, $(NH_4)_2[Ni(H_2O)_5(β-HAla)]_2[V_{10}O_{28}]\cdot 4H_2O$, and $(NH_4)_2[Cd(H_2O)_5(β-HAla)]_2[V_{10}O_{28}]\cdot 2H_2O$ by synthesising them in a mildly acidic aqueous solution, which, in comparison with hydrothermal direct synthesis from metallic powder and V_2O_5, prevented partial reduction of V(V) to V(IV) during the preparation of $(NH_4)_2[Co(H_2O)_5(β-HAla)]_2[V_{10}O_{28}]\cdot 4H_2O$. The prepared compounds were characterised by X-ray structure analysis and vibration spectroscopy. The FT-IR and FT-Raman spectra confirmed the presence of the $[V_{10}O_{28}]^{6-}$ anion and monodentate coordination mode of β-HAla in complex cation in accordance with crystallographic findings. To confirm the spin state of central atoms in $[M(H_2O)_5(β-HAla)]^{2+}$ cations, including previously prepared $(NH_4)_2[Mn(H_2O)_5(β-HAla)]_2[V_{10}O_{28}]\cdot 2H_2O$, the Evans method was used to confirm the assignment of spin states based on crystallographic data. The increasing crystal radius of central atoms and their different preferences for donor atoms resulting in rebuilding of the hydrogen bonding network is the main cause of the existence of two different crystallohydrate forms.

Supplementary Materials: The following supporting information can be downloaded at: https://www.mdpi.com/article/10.3390/cryst14080685/s1. Table S1: Crystallographic data for the compounds **1**–**5**; Table S2: Selected bond lengths, bond valences and bond valence sums in the compounds **1**–**5**; Table S3: Hydrogen bonds in the structures of **1**–**5**; Figure S1a: A comparison of IR and Raman spectra of **1**; Figure S1b: A comparison of IR and Raman spectra of **2**; Figure S1c: A comparison of IR and Raman spectra of **3**; Figure S1d: A comparison of IR and Raman spectra of **4**; Figure S1e: A comparison of IR and Raman spectra of **5**. Table S4: Assignments of the IR and Raman absorption bands for compounds **1**–**5** [7,21,53].

Author Contributions: Conceptualisation, L.B. and E.R.; methodology, J.C. and E.R.; validation, J.C. and E.R.; formal analysis, Y.R.P. and E.R.; investigation, L.B., J.C., Y.R.P. and E.R.; resources, L.B.; data curation, E.R.; writing—original draft preparation, E.R. and Y.R.P.; writing—review and editing, E.R. and J.C.; visualisation, E.R. and J.C.; project administration, E.R. and Y.R.P.; funding acquisition, Y.R.P. and E.R. All authors have read and agreed to the published version of the manuscript.

Funding: This work was supported by the Grant of Comenius University no. UK/3049/2024, Scientific Grant Agency of the Ministry of Education of the Slovak Republic and of Slovak Academy of Sciences VEGA 1/0669/22, and Slovak Research and Development Agency under Contract no. APVV-21-0503.

Data Availability Statement: The deposition numbers CCDC 2366040—2366044 contain the supplementary crystallographic data for this article, including structure factors. These data can be obtained free of charge at https://www.ccdc.cam.ac.uk/structures/ (accessed date 23 July 2024). Crystallographic data can also be obtained from the Crystallography Open Database (COD) under COD ID 3000552—3000556.

Acknowledgments: The authors thank Aleksandra Cyganiuk from Nicolaus Copernicus University in Toruń (Poland) for the Raman spectra measurements.

Conflicts of Interest: The authors declare no conflicts of interest.

References

1. Hayashi, Y. Hetero and Lacunary Polyoxovanadate Chemistry: Synthesis, Reactivity and Structural Aspects. *Coord. Chem. Rev.* **2011**, *255*, 2270–2280. [CrossRef]
2. Kempf, J.Y.; Rohmer, M.M.; Poblet, J.M.; Bo, C.; Benard, M. Relative basicities of the oxygen sites in $[V_{10}O_{28}]^{6-}$. An analysis of the ab initio determined distributions of the electrostatic potential and of the Laplacian of charge density. *J. Am. Chem. Soc.* **1992**, *114*, 1136–1146. [CrossRef]
3. Day, V.W.; Klemperer, W.G.; Maltbie, D.J. Where are the protons in $H_3V_{10}O_{28}{}^{3-}$? *J. Am. Chem. Soc.* **1987**, *109*, 2991–3002. [CrossRef]
4. Biagioli, M.; Strinna-Erre, L.; Micera, G.; Panzanelli, A.; Zema, M. Tetrahydrogendecavanadate(V) and Its Binding to Glycylglycine. *Inorg. Chem. Commun.* **1999**, *2*, 214–217. [CrossRef]
5. Rakovský, E.; Joniaková, D.; Gyepes, R.; Schwendt, P.; Mička, Z. Synthesis and Crystal Structure of $[CuCl(phen)_2]_3H_3V_{10}O_{28}\cdot 7H_2O$. *Cryst. Res. Technol.* **2005**, *40*, 719–722. [CrossRef]
6. Kaziev, G.Z.; Oreshkina, A.V.; Stepnova, A.F.; Holguin Quinones, S.; Stash, A.I.; Morales Sanchez, L.A. Synthesis and Study of the Physicochemical Properties of Ammonium Hydrogen Hexaaquacobaltate(III) Isopolyvanadate $[(NH_4)_2][Co(H_2O)_6]\cdot H[V_{10}O_{28}]\cdot 8H_2O$. *Russ. J. Coord. Chem.* **2011**, *37*, 766–771. [CrossRef]
7. Sánchez-Lara, E.; Pérez-Benítez, A.; Treviño, S.; Mendoza, A.; Meléndez, F.J.; Sánchez-Mora, E.; Bernès, S.; González-Vergara, E. Synthesis and 3D Network Architecture of 1- and 16-Hydrated Salts of 4-Dimethylaminopyridinium Decavanadate, $(DMAPH)_6[V_{10}O_{28}]\cdot nH_2O$. *Crystals* **2016**, *6*, 65. [CrossRef]
8. Lv, Y.-K.; Jiang, Z.-G.; Gan, L.-H.; Liu, M.-X.; Feng, Y.-L. Three Novel Organic-Inorganic Hybrid Materials Based on Decaoxovanadates Obtained from a New Liquid Phase Reaction. *CrystEngComm* **2012**, *14*, 314–322. [CrossRef]
9. Ferreira da Silva, J.L.; Fátima Minas da Piedade, M.; Teresa Duarte, M. Decavanadates: A Building-Block for Supramolecular Assemblies. *Inorganica Chim. Acta* **2003**, *356*, 222–242. [CrossRef]
10. Crans, D.C.; Smee, J.J.; Gaidamauskas, E.; Yang, L. The Chemistry and Biochemistry of Vanadium and the Biological Activities Exerted by Vanadium Compounds. *Chem. Rev.* **2004**, *104*, 849–902. [CrossRef]
11. Aureliano, M.; Gumerova, N.I.; Sciortino, G.; Garribba, E.; Rompel, A.; Crans, D.C. Polyoxovanadates with Emerging Biomedical Activities. *Coord. Chem. Rev.* **2021**, *447*, 214143. [CrossRef]
12. Aureliano, M.; Gumerova, N.I.; Sciortino, G.; Garribba, E.; McLauchlan, C.C.; Rompel, A.; Crans, D.C. Polyoxidovanadates' Interactions with Proteins: An Overview. *Coord. Chem. Rev.* **2022**, *454*, 214344. [CrossRef]
13. Samart, N.; Saeger, J.; Haller, K.J.; Aureliano, M.; Crans, D.C. Interaction of Decavanadate With Interfaces and Biological Model Membrane Systems: Characterization of Soft Oxometalate Systems. *J. Mol. Eng. Mater.* **2014**, *02*, 1440007. [CrossRef]
14. Aureliano, M.; Gândara, R.M.C. Decavanadate Effects in Biological Systems. *J. Inorg. Biochem.* **2005**, *99*, 979–985. [CrossRef] [PubMed]
15. Buvailo, H.I.; Pavliuk, M.V.; Makhankova, V.G.; Kokozay, V.N.; Bon, V.; Mijangos, E.; Shylin, S.I.; Jezierska, J. Facile One-Pot Synthesis of Hybrid Compounds Based on Decavanadate Showing Water Oxidation Activity. *Inorg. Chem. Commun.* **2020**, *119*, 108111. [CrossRef]
16. Huang, X.; Gu, X.; Qi, Y.; Zhang, Y.; Shen, G.; Yang, B.; Duan, W.; Gong, S.; Xue, Z.; Chen, Y. Decavanadate-Based Transition Metal Hybrids as Bifunctional Catalysts for Sulfide Oxidation and C—C Bond Construction. *Chin. J. Chem.* **2021**, *39*, 2495–2503. [CrossRef]
17. Cao, J.-P.; Shen, F.-C.; Luo, X.-M.; Cui, C.-H.; Lan, Y.-Q.; Xu, Y. Proton Conductivity Resulting from Different Triazole-Based Ligands in Two New Bifunctional Decavanadates. *RSC Adv.* **2018**, *8*, 18560–18566. [CrossRef] [PubMed]
18. Kang, R.; Cao, J.; Han, Y.; Hong, Y.; Yang, M.; Xu, Y. Three New Ln-Decavanadates Materials: Synthesis, Structure, and Photoluminescent Sensing for Detection of Zn^{2+} and Co^{2+}. *Zeitschrift Anorg. Allg. Chem.* **2020**, *646*, 1315–1323. [CrossRef]
19. Kumar, D.; Tomar, A.K.; Singal, S.; Singh, G.; Sharma, R.K. Ammonium Decavanadate Nanodots/Reduced Graphene Oxide Nanoribbon as "Inorganic-Organic" Hybrid Electrode for High Potential Aqueous Symmetric Supercapacitors. *J. Power Sources* **2020**, *462*, 228173. [CrossRef]
20. Kumar, D.; Tomar, A.K.; Singh, G.; Sharma, R.K. Interlayer Gap Widened 2D α-Co(OH)$_2$ Nanoplates with Decavanadate Anion for High Potential Aqueous Supercapacitor. *Electrochim. Acta* **2020**, *363*, 137238. [CrossRef]
21. Klištincová, L.; Rakovský, E.; Schwendt, P. Decavanadates with Complex Cations: Synthesis and Structure of $(NH_4)_2[M(H_2O)_5(NH_3CH_2CH_2COO)]_2V_{10}O_{28}\cdot nH_2O$ ($M. = Zn^{II}$, $n = 4$; $M. = Mn^{II}$, $n = 2$). *Transit. Met. Chem.* **2010**, *35*, 229–236. [CrossRef]

22. Pavliuk, M.V.; Makhankova, V.G.; Kokozay, V.N.; Omelchenko, I.V.; Jezierska, J.; Thapper, A.; Styring, S. Structural, Magnetic, Thermal and Visible Light-Driven Water Oxidation Studies of Heterometallic Mn/V Complexes. *Polyhedron* **2015**, *88*, 81–89. [CrossRef]
23. Pavliuk, M.V.; Mijangos, E.; Makhankova, V.G.; Kokozay, V.N.; Pullen, S.; Liu, J.; Zhu, J.; Styring, S.; Thapper, A. Homogeneous Cobalt/Vanadium Complexes as Precursors for Functionalized Mixed Oxides in Visible-Light-Driven Water Oxidation. *ChemSusChem* **2016**, *9*, 2957–2966. [CrossRef] [PubMed]
24. Rigaku Oxford Diffraction. *CrysAlisPro Software System, Version 1.171.43.128a*; Rigaku Corporation: Wroclaw, Poland, 2024.
25. Sheldrick, G.M. SHELXT—Integrated Space-Group and Crystal-Structure Determination. *Acta Crystallogr. Sect. A Found. Adv.* **2015**, *71*, 3–8. [CrossRef]
26. Sheldrick, G.M. Crystal Structure Refinement with SHELXL. *Acta Crystallogr. Sect. C Struct. Chem.* **2015**, *71*, 3–8. [CrossRef] [PubMed]
27. Bourhis, L.J.; Dolomanov, O.V.; Gildea, R.J.; Howard, J.A.K.; Puschmann, H. The Anatomy of a Comprehensive Constrained, Restrained Refinement Program for the Modern Computing Environment—Olex2 Dissected. *Acta Crystallogr. Sect. A Found. Adv.* **2015**, *71*, 59–75. [CrossRef] [PubMed]
28. Nardelli, M. PARST 95—An Update to PARST: A System of Fortran Routines for Calculating Molecular Structure Parameters from the Results of Crystal Structure Analyses. *J. Appl. Crystallogr.* **1995**, *28*, 659. [CrossRef]
29. Brandenburg, K. *Diamond. Release 3.2k*; Crystal Impact GbR: Bonn, Germany, 2014.
30. Buron-Le Cointe, M.; Hébert, J.; Baldé, C.; Moisan, N.; Toupet, L.; Guionneau, P.; Létard, J.F.; Freysz, E.; Cailleau, H.; Collet, E. Intermolecular Control of Thermoswitching and Photoswitching Phenomena in Two Spin-Crossover Polymorphs. *Phys. Rev. B Condens. Matter Mater. Phys.* **2012**, *85*, 064114. [CrossRef]
31. Lufaso, M.W.; Woodward, P.M. Jahn-Teller Distortions, Cation Ordering and Octahedral Tilting in Perovskites. *Acta Crystallogr. Sect. B Struct. Sci.* **2004**, *60*, 10–20. [CrossRef]
32. Marchivie, M.; Guionneau, P.; Létard, J.F.; Chasseau, D. Photo-Induced Spin-Transition: The Role of the Iron(II) Environment Distortion. *Acta Crystallogr. Sect. B Struct. Sci.* **2005**, *61*, 25–28. [CrossRef]
33. McCusker, J.K.; Rheingold, A.L.; Hendrickson, D.N. Variable-Temperature Studies of Laser-Initiated $^5T_2 \rightarrow {}^1A_1$ Intersystem Crossing in Spin-Crossover Complexes: Empirical Correlations between Activation Parameters and Ligand Structure in a Series of Polypyridyl Ferrous Complexes. *Inorg. Chem.* **1996**, *35*, 2100–2112. [CrossRef]
34. Ketkaew, R.; Tantirungrotechai, Y.; Harding, P.; Chastanet, G.; Guionneau, P.; Marchivie, M.; Harding, D.J. OctaDist: A Tool for Calculating Distortion Parameters in Spin Crossover and Coordination Complexes. *J. Chem. Soc. Dalt. Trans.* **2021**, *50*, 1086–1096. [CrossRef] [PubMed]
35. Jeffrey, G.A.; Lewis, L. Cooperative Aspects of Hydrogen Bonding in Carbohydrates. *Carbohydr. Res.* **1978**, *60*, 179–182. [CrossRef]
36. Taylor, R.; Kennard, O. Comparison of X-ray and Neutron Diffraction Results for the N-H···O=C Hydrogen Bond. *Acta Crystallogr. Sect. B* **1983**, *39*, 133–138. [CrossRef]
37. Evans, D.F. The Determination of the Paramagnetic Susceptibility of Substances in Solution by Nuclear Magnetic Resonance. *J. Chem. Soc.* **1959**, 2003–2005. [CrossRef]
38. Evans, D.F.; Fazakerley, G.V.; Phillips, R.F. Organometallic Compounds of Bivalent Europium, Ytterbium, and Samarium. *J. Chem. Soc. A Inorg. Phys. Theor. Chem.* **1971**, 1931–1934. [CrossRef]
39. Schubert, E.M. Utilizing the Evans Method with a Superconducting NMR Spectrometer in the Undergraduate Laboratory. *J. Chem. Educ.* **1992**, *69*, 62. [CrossRef]
40. Bain, G.A.; Berry, J.F. Diamagnetic Corrections and Pascal's Constants. *J. Chem. Educ.* **2008**, *85*, 532–536. [CrossRef]
41. Mugiraneza, S.; Hallas, A.M. Tutorial: A Beginner's Guide to Interpreting Magnetic Susceptibility Data with the Curie-Weiss Law. *Commun. Phys.* **2022**, *5*, 95. [CrossRef]
42. Rakovský, E.; Krivosudsky, L. Tetrakis(2,6-Dimethylpyridinium) Dihydrogen Decavanadate Dihydrate. *Acta Crystallogr. Sect. E Struct. Rep. Online* **2014**, *70*, m225–m226. [CrossRef]
43. Cooper, R.I.; Thompson, A.L.; Watkin, D.J. CRYSTALS Enhancements: Dealing with Hydrogen Atoms in Refinement. *J. Appl. Crystallogr.* **2010**, *43*, 1100–1107. [CrossRef]
44. Brown, I.D. *The Chemical bond in Inorganic Chemistry*; IUCr Monographs on Crystallography; Oxford University Press: New York, NY, USA, 2002; Volume 12.
45. (IUCr) Bond Valence Parameters. Available online: https://www.iucr.org/__data/assets/file/0011/150779/bvparm2020.cif (accessed on 23 July 2024).
46. Brown, I.D.; Altermatt, D. Bond-Valence Parameters Obtained from a Systematic Analysis of the Inorganic Crystal Structure Database. *Acta Crystallogr. Sect. B Struct. Sci.* **1985**, *41*, 244–247. [CrossRef]
47. Wood, R.M.; Palenik, G.J. Bond Valence Sums in Coordination Chemistry. A Simple Method for Calculating the Oxidation State of Cobalt in Complexes Containing Only Co−O Bonds. *Inorg. Chem.* **1998**, *37*, 4149–4151. [CrossRef] [PubMed]
48. Urusov, V.S. Problem of Optimization of Bond Valence Model Parameters (as Exemplified by Manganese in Different Oxidation States). *Dokl. Phys. Chem.* **2006**, *408*, 152–155. [CrossRef]
49. Jeffrey, G.A. *An Introduction to Hydrogen Bonding*, 1st ed.; Oxford University Press: New York, NY, USA, 1997; p. 12.
50. Shannon, R.D. Revised Effective Ionic Radii and Systematic Studies of Interatomic Distances in Halides and Chalcogenides. *Acta Crystallogr. Sect. A* **1976**, *32*, 751–767. [CrossRef]

51. Etter, M.C.; MacDonald, J.C.; Bernstein, J. Graph-Set Analysis of Hydrogen-Bond Patterns in Organic Crystals. *Acta Crystallogr. Sect. B Struct. Sci.* **1990**, *46*, 256–262. [CrossRef] [PubMed]
52. Bernstein, J.; Davis, R.E.; Shimoni, L.; Chang, N.-L. Patterns in Hydrogen Bonding: Functionality and Graph Set Analysis in Crystals. *Angew. Chemie Int. Ed. Engl.* **1995**, *34*, 1555–1573. [CrossRef]
53. Berezhinsky, L.I.; Dovbeshko, G.I.; Lisitsa, M.P.; Litvinov, G.S. Vibrational Spectra of Crystalline β-Alanine. *Spectrochim. Acta Part A Mol. Biomol. Spectrosc.* **1998**, *54*, 349–358. [CrossRef]
54. Deacon, G.B.; Phillips, R.J. Relationships between the carbon-oxygen stretching frequencies of carboxylato complexes and the type of carboxylate coordination. *Coord. Chem. Rev.* **1980**, *33*, 227–250. [CrossRef]

Disclaimer/Publisher's Note: The statements, opinions and data contained in all publications are solely those of the individual author(s) and contributor(s) and not of MDPI and/or the editor(s). MDPI and/or the editor(s) disclaim responsibility for any injury to people or property resulting from any ideas, methods, instructions or products referred to in the content.

Article

Supramolecular Structure, Hirshfeld Surface Analysis, Morphological Study and DFT Calculations of the Triphenyltetrazolium Cobalt Thiocyanate Complex

Essam A. Ali [1], Rim Bechaieb [2], Rashad Al-Salahi [1], Ahmed S. M. Al-Janabi [3], Mohamed W. Attwa [1] and Gamal A. E. Mostafa [1,*]

[1] Department of Pharmaceutical Chemistry, College of Pharmacy, King Saud University, P.O. Box 2457, Riyadh 11451, Saudi Arabia; ralsalahi@ksu.edu.sa (R.A.-S.); mzeidan@ksu.edu.sa (M.W.A.)
[2] Laboratoire de Chimie Théorique, Sorbonne Universités, UPMC Univ Paris 06, UMR 7616, F-75005 Paris, France
[3] Department of Chemistry, College of Science, Tikrit University, Tikrit 34001, Iraq; dr.ahmed.chem@tu.edu.iq
* Correspondence: gmostafa@ksu.edu.sa

Abstract: Polymorphism is a prevalent occurrence in pharmaceutical solids and demands thorough investigation during product development. This paper delves into the crystal growth and structure of a newly synthesized polymorph (TPT)$_2$[Co^{II}(NCS)$_4$], (1), where TPT is triphenyl tetrazolium. The study combines experimental and theoretical approaches to elucidate the 3D framework of the crystal structure, characterized by hydrogen-bonded interactions between (TPT)$^+$ cations and [Co(NCS)$_4$]$^{2-}$ anions. Hirshfeld surface analysis, along with associated two-dimensional fingerprints, is employed to comprehensively investigate and quantify intermolecular interactions within the structure. The enrichment ratio is calculated for non-covalent contacts, providing insight into their propensity to influence crystal packing interactions. Void analysis is conducted to predict the mechanical behavior of the compound. Utilizing Bravais-Friedel, Donnay-Harker (BFDH), and growth morphology (GM) techniques, the external morphology of (TPT)$_2$[Co^{II}(NCS)$_4$] is predicted. Experimental observations align well with BFDH predictions, with slight deviations from the GM model. Quantum computational calculations of the synthesized compounds is performed in the ground state using the DFT/UB3LYP level of theory. These calculations assess the molecule's stability and chemical reactivity, including the computation of the HOMO-LUMO energy difference and other chemical descriptors. The study provides a comprehensive exploration of the newly synthesized polymorph, shedding light on its crystal structure, intermolecular interactions, mechanical behavior, and external morphology, supported by both experimental and computational analyses.

Keywords: supramolecular structure; crystal structure; Hirshfeld surface analysis; void analysis; morphologies; DFT calculations; cobalt thiocyanate; triphenyltetrazolium

Citation: Ali, E.A.; Bechaieb, R.; Al-Salahi, R.; Al-Janabi, A.S.M.; Attwa, M.W.; Mostafa, G.A.E. Supramolecular Structure, Hirshfeld Surface Analysis, Morphological Study and DFT Calculations of the Triphenyltetrazolium Cobalt Thiocyanate Complex. *Crystals* 2023, 13, 1598. https://doi.org/10.3390/cryst13111598

Academic Editors: Sanja Burazer and Lidija Androš Dubraja

Received: 16 October 2023
Revised: 12 November 2023
Accepted: 15 November 2023
Published: 19 November 2023

Copyright: © 2023 by the authors. Licensee MDPI, Basel, Switzerland. This article is an open access article distributed under the terms and conditions of the Creative Commons Attribution (CC BY) license (https://creativecommons.org/licenses/by/4.0/).

1. Introduction

The chemistry of cobalt complexes is gaining interest in several inorganic chemistry groups due to the distinct reactivity of the produced complexes and the variety of ligands that influence the characteristics of such complexes [1]. The Co(II) cations having a d^9 configuration are found in most organometallic compounds in square planar, square-pyramidal, or square-bipyramidal geometries [2]. Thiocyanate (SCN)$^-$ is an ambidentate ligand that has been significant in the generation of bonding models for the mechanisms that influence linkage isomerism in transition metal complexes, resulting in a variety of structures of varying dimensions [3,4]. Increasing the diversity of the paramagnetic metal centers used can lead to more exciting magnetic and optic properties [5]. Thiocyanate coordination modes have been the subject of numerous studies. Thiocyanates' linkage isomerism preferences as terminal or bridging modes were investigated across a range of experimental conditions. Notable experimental factors include electronic and steric effects, central metal type, solvent type, organic molecule type, and position of substituents in N donor ligands [6,7].

Tetrazole derivatives are commonly employed as color markers for detecting enzyme systems that produce reduction equivalents. They are incredibly important tools in academic and clinical research, as well as for numerous diagnostic applications, because of this property [8]. Aromatic derivatives of 1,2,3,4-tetrazole (substitution at positions 2, 3, and 5) are the most commonly utilized tetrazole derivatives in biochemistry and cell biology. The 2,3,5-triphenyl-2H-tetrazolium salt has garnered a lot of attention so far [9]. It is a heterocyclic compound composed of four nitrogen atoms, one of which is positively charged, and a five-member ring. The use of triphenyl tetrazolium salt for element and ion extraction, spectrophotometric, and potentiometric determination has been reported [5,10]. Triphenyl tetrazolium salt was recently used as an ion-pair reagent for the detection of different analytes as PVC membrane sensors [11–13]. It could also be employed in salt synthesis experiments where the size of the cationic group is connected to the kind and structure of the matching anionic units.

The work presents the synthesis, characterization, and determination of the single-crystal structure and DFT calculations of the tetraphenyltetrazolium thiocyanate cobalt (II) complex. The molecular structure was stabilized by hydrogen bonds and non-covalent interactions studied by Hirshfeld surface analysis. The characterization of this complex was performed using differential spectroscopic techniques such as UV spectrometry, infrared, mass spectrometry, elemental analysis, and NMR. Additionally, DFT calculations for the compound was carried out.

2. Experimental
2.1. General

A Gallenkamp melting point instrument was used to calculate the melting point. A Perkin-Elmer FTIR spectrometer was used to record the IR spectra. The experiments involved the utilization of Bruker 500 and 700 MHz instruments to acquire 1H NMR and 13C NMR spectra in DMSOd6. TMS was employed as an internal standard, and the chemical shifts were reported in δ-ppm. The complex spectrum was scanned using a Shimadzu double-beam spectrophotometer (1800 UV) with a quartz cell. Perkin Elmer's 2400 series II, CHNS/O elemental analysis was used for the elemental analysis.

2.2. Synthesis

Twenty-five milliliters of 1 mmol (0.291 mg) of cobalt (II) nitrate hexahydrate in methanol was added to fifty milliliters of potassium thiocyanate of 1 mmol (0.0972 g), then the cobalt thiocyanate complex was formed $[Co(SCN)_4]^{-2}$. Twenty-five milliliters of triphenyl tetrazolium chloride solution in methanol 1 mmol (0.3348 g) was added to the previously formed cobalt thiocyanate complex. A blue precipitate was produced, thereafter separated by filtration, and finally washed with methanol. The resulting precipitate was subjected to vacuum drying in order to obtain the desired ion-pair complex. The titled compound (74% yield, m.p.225 °C) was obtained by recrystallization in acetonitrile. The formed ion-pair complex was confirmed by different spectroscopic and instrumentation analyses. IR (KBr cm^{-1}): 2000.85 cm^{-1} for CN and ^{1}H-NMR (DMSO-d$_6$) δ: 7.63–8.26 ppm (all Ar H of triphenyl rings). ^{13}C-NMR (DMSO-d$_6$) δ: 123.23, 126.73, 127.66, 130.63, 130.83, 133.28, 134.05, 134.55, 164.45.

2.3. Single Crystal X-ray Diffraction Measurement

We measured single-crystal X-ray diffraction at 293 K using a STOE IPDS 2 diffractometer with a sealed X-ray tube and a graphite monochromator. Cell refinement and data reduction were completed by APX3 [14]. The structure was solved and refined using the ShelXT and ShelXL programs, respectively [15], using Olex2 software [16]. Non-hydrogen atoms were refined with anisotropic displacement parameters. Geometrically, hydrogen atoms were positioned (C-H = 0.93 Å) and refined using a riding model, with Uiso(H) = 1.2 Ueq(C). The crystal structure was visualized using the Mercury software 4.0 [17]. The crystallographic data have been stored in the Cambridge Crystallographic Data Centre (CCDC) as CCDC-2259943. A summary of crystallographic and structure refinement data is given in Table 1.

Table 1. Crystal data and structure refinement parameters for (**1**).

Chemical Formula	$C_{42}H_{30}CoN_{12}S_4$
CCDC number	2,259,943
Formula weight	889.95
Crystal system	Triclinic
space group	P-1
Temperature (K)	293
a (Å)	9.7110 (16)
b (Å)	12.892 (2)
c (Å)	18.753 (3)
α/β/γ (°)	87.215 (14)/79.122 (14)/74.971 (13)
V (Å^3)	2226.6 (7)
Z	2
μ (mm^{-1})	0.62
Crystal size (mm)	0.32 × 0.26 × 0.12
θ$_{min}$/θ$_{max}$ (°)	1.6/25.1
No. of measured, independent and observed [I > 2σ(I)] reflections	20,928/7891/4056
R$_{int}$	0.094
R[F^2 > 2σ(F^2)]/ wR(F^2), S	0.049/0.095/0.92
Data/restraints/parameters	7891/0/532
Δρ$_{max}$/Δρ$_{min}$ (e Å^{-3})	0.37/−0.34

2.4. Hirshfeld Surface Analysis and Enrichment Ratio Calculations

Hirshfeld surface analysis (HS) and generation of 2D fingerprint plots [18,19] were calculated using the CrystalExplorer 17.5 software [20] as a highly useful tool for evaluating and illustrating all non-covalent interactions in multi-component crystal structures. For quantifying and decoding the intercontacts in the crystal packing, the normalized contact distance d$_{norm}$ [20] based on Bondi's van der Waals radii [21] and 2D fingerprint plots were employed. The Hirshfeld surface is defined by the distances de and d$_i$, which represent the distance from the surface to the nearest atoms outside and inside, respectively. The normalized contact distance, d$_{norm}$, is calculated based on these distances. The deep red color corresponds to the interactions that are shorter than the sum of van der Waals radii, as shown by negative d$_{norm}$ values. Other intermolecular distances near van der Waals interactions with d$_{norm}$ = 0 appear as light-red circles. On the other hand, contacts with positive d$_{norm}$ values that are longer than the total van der Waals radii are colored blue.

The enrichment ratio E$_{XY}$ for an element pair (X, Y) is defined as the ratio of the actual percentage of random contacts in the crystal (C$_{XY}$) to the percentage of theoretically equivalently distributed random contacts (R$_{XY}$) (E$_{XY}$ = C$_{XY}$/R$_{XY}$) [22]. If the enrichment ratio of two elements is greater than 1, they are more likely to make contacts in the crystal, whereas contacts with E$_{XY}$ values less than 1 are less likely to create contacts.

2.5. Growth Morphology Prediction

There are good relationships between an investigational pharmaceutical compound's crystal structure, crystal morphology, and physicochemical characteristics [23,24]. The theoretical crystal morphology of (**1**) was simulated using Materials Studio 7.0 [25] utilizing the Bravais-Friedel-Donnay-Harker (BFDH) model and growth morphology (GM) models. The BFDH approach is capable of independently predicting the relative growth rates of faces, without considering factors such as atoms, partial charges, bond types, or interatomic forces. Nevertheless, this approach remains the most straightforward and expedient technique for identifying potential crystal development facets and providing an approximate initial assessment of a crystal's structure. The determination relies solely on the lattice parameters and the crystal's symmetry. This is done based on the fact that the facets with the largest interplanar spacing (d_{hkl}) are likely to be morphologically important (MI) [26].

The concept of attachment energy, denoted as E_{hkl}^{att}, refers to the energy that is released when a building unit, also known as a growth slice, is added to the surface of a growing crystal. The variable in question holds considerable importance in comprehending the characteristics of chemical bonding inside solid materials. In this technique, it is assumed the attachment energy E_{hkl} is directly proportional to the growth rate R_{hkl} of a specific crystal face (hkl). The AE method additionally posits the Miller index MI_{hkl} exhibits an inverse relationship with the face's E_{hkl}^{att}. Consequently, this particular approach posits facial features exhibiting the lowest E_{hkl}^{att} values are associated with the highest MI_{hkl} values and exhibit the slowest growth rates, as represented by the subsequent equations:

$$R_{hkl} \sim |E_{hkl}^{att}| \quad (1)$$

$$MI_{hkl} \sim 1/|E_{hkl}^{att}| \quad (2)$$

2.6. DFT Calculations

2.6.1. Computational Details

In this study, we conducted comprehensive calculations of geometric and electronic properties for the [(TPT)$_2$[Co(NCS)$_4$] crystal using the Gaussian 16 program [27]. Our computational methodology of geometry optimization, frequency calculations, and electronic properties was initiated by the crystal's asymmetric unit with a charge of zero and a doublet spin. The geometry was fully optimized in the C1 symmetry group. All structures, analyzed using different levels of theory, exhibit a real minimum on their potential energy surfaces, as indicated by the absence of negative values in the calculated wavenumbers (imaginary frequencies).

To explore the optical behavior and chemical reactivity descriptors of the studied compound, we utilized density functional theory (DFT) and time-dependent density functional theory (TD-DFT) methodologies. As we have an open shell system, the unrestricted formalism UB3LYP functional [28,29], was used for the system's characteristics. To accurately represent the electronic behavior of the cobalt atoms (Co) within the crystal, we adopted the LANL2DZ pseudopotential and basis set [30]. Additionally, for the constituent atoms of nitrogen (N), carbon (C), hydrogen (H), and sulfur (S), we used the 6–31+G* basis set as implemented in Gaussian 16 [27]. This choice of basis set effectively captured the electron distribution of these elements, enabling precise calculations of their properties within the compound. This analysis holds particular significance in the context of biology and the study of molecular structures. It sheds light on the studied crystal's properties, which can be very important in various biological processes. Understanding these properties can have far-reaching implications, from drug design to understanding biomolecular interactions, ultimately contributing to advancements in the field of molecular biology.

2.6.2. Global Reactivity Descriptors

DFT method and frontier molecular orbitals (FMOs) analysis are now essential instruments employed by both theorists and experimentalists to explore, comprehend, and forecast chemical properties of organic and inorganic compounds [31–34]. This is chiefly attributable to the exceptional balance struck between precision and computational efficiency.

In the forthcoming phase of our research, our focus extends beyond predicting the energy gap (Eg), this term refers to the disparity between the energy levels of the highest occupied molecular orbital (HOMO) and the lowest unoccupied molecular orbital (LUMO). We are committed to conducting an in-depth exploration of chemical reactivity descriptors to gain a comprehensive understanding of the compound's chemical behavior under scrutiny. To accomplish this, we draw upon a range of well-established chemical reactivity descriptors, extensively discussed in our previous works [35–37]. These descriptors, which are integral to understanding the electronic properties of our compound, are rigorously evaluated and summarized as follows [38,39]:

$$\text{Energy Gap (Eg): Eg} = E_{(LUMO)} - E_{(HOMO)} \quad (3)$$

This term denotes the disparity in energy levels between the highest occupied molecular orbital (HOMO) and the lowest unoccupied molecular orbital (LUMO). It provides insights into the compound's electronic transitions.

$$\text{First Ionization energy (I), I} = -E_{(HOMO)} \quad (4)$$

The symbol I denotes the energy that is necessary for the removal of an electron from the highest occupied molecular orbital (HOMO), indicating the compound's electron-donating capacity.

$$\text{Affinity (A), A} = -E_{(LUMO)} \quad (5)$$

Affinity refers to the quantification of the energy variation that occurs upon the addition of an electron to the lowest unoccupied molecular orbital (LUMO), indicating the compound's electron-accepting propensity.

$$\text{Chemical Hardness } (\eta), \eta = \frac{(E_{LUMO} - E_{HOMO})}{2} \quad (6)$$

This descriptor reflects the stability of the compound, providing information about its resistance to electron exchange.

$$\text{Chemical potential } (\mu), \mu = \frac{(E_{HOMO} + E_{LUMO})}{2} \quad (7)$$

μ offers insights into the compound's ability to exchange electrons with its environment.

$$\text{Electrophilicity } (\omega), \omega = \frac{\mu^2}{2\eta} \quad (8)$$

Electrophilicity combines information about the chemical potential and hardness, offering valuable insights into the compound's reactivity.

$$\text{Electronegativity } (\chi), \chi = \frac{(I + A)}{2} \quad (9)$$

Electronegativity indicates the compound's tendency to attract electrons in chemical bonds, reflecting its polarity and reactivity.

$$\text{Softness (S), S} = \frac{1}{2\eta} \quad (10)$$

Softness describes the compound's response to electron addition or removal, offering further insights into its reactivity and stability.

3. Results and Discussion

3.1. Chemistry

The reaction of triphenyl tetrazolium chloride with cobalt thiocyanate in a methanolic solution at ambient temperature afforded the title compound of cobalt (II) thiocyanate-tetraphenyl tetrazolium complex in 78% yield through an anion exchange reaction. The elemental analysis results show C = 55.63%, H 3.45%, and N 18.74%, which agree with theoretical values of C 56.69%, H 3.37%, and N 18.89%. The UV-visible spectra show the complex has a maximum wavelength of 623 nm.

Several resonances in the ^{1}H-NMR spectrum are caused by chemical shifts in the aromatic zone. This is strong evidence that the triphenyl tetrazolium cobalt thiocyanate complex has been formed. When comparing the NMR spectra of the triphenyl tetrazolium ion (Figure 1A) and the triphenyl tetrazolium cobalt thiocyanate, we observe in the aromatic zone of the former more peaks, such as 7.68–7.80 ppm, assignable to particular protons in the triphenyl rings (a); 7.94–8.00 ppm also assignable to other different protons in the two phenyl rings (b); and 8.30–8.35 ppm assignable for protons in one phenyl ring (c). However, the ^{1}H-NMR spectra exhibit only four peaks (7.63, 7.69, 7.77, and 8.26 ppm) for the protons of triphenyl groups, confirming the establishment of the triphenyl-tetrazolium complex. A prominent absorption band at 2000.85 cm^{-1} for CN can be seen in the IR spectrum of the triphenyl tetrazolium cobalt thiocyanate complex, which is not present in the IR spectrum of triphenyl-tetrazole. This is also another evidence for the complex. All carbons for the triphenyl group appear in the 13C-NMR for both the triphenyl tetrazolium and the triphenyl tetrazolium cobalt thiocyanate complex but with slightly distinct chemical shifts.

3.2. Structural Description

The structure of (**1**), (TPT)$_2$[Co^{II}(NCS)$_4$], is determined by single crystal X-ray analysis. It is a polymorph of the previously published structure by Nakashima et al. [40]. The asymmetric unit of (**1**) consists of one tetrathiocyanatecobaltate [Co^{II}(NCS)$_4$]$^{2-}$ anionic complex and two triphenyltetrazolium (TPT)$^+$ organic cations (Figure 1B). The tetrathiocyanate-cobaltate [Co(NCS)$_4$]$^{2-}$ anion consists of the Co^{II} ion, which is tetrahedrally coordinated by four nitrogen atoms of thiocyanate ligands with averaged Co-N bond lengths of 1.94–1.95 Å (Table 2). The thiocyanate ligands are bound through nitrogen atoms and are quasi-linear [N-C-S= 179.5 (4) Å], while the Co-NCS linkages are bent [C-N-Co = 157.0 (4) Å] (Table 2). These structural characteristics have been reported by its previously described polymorph (**2**) [40] and other similar compounds containing [M(NCS)$_4$]$^{2-}$ anion (M is a transition metal) [2,41]. Regarding polymorphs, it is not surprising the structure determination for (**2**) revealed a network that differed from that of (**1**). In our case, the structure crystallizes in the chiral space group, P-1, though the starting material in both structures, 2,3,5-triphenyltetrazolium, is achiral. There are two crystallographically different molecules of bis(2,3,5-triphenyltetrazolium) tetrathiocyanatecobaltate in the asymmetric unit of (**2**). The coordination environments exhibited by complexes (**1**) and (**2**) are comparable; however, there are modest variations in the bond angles observed in each complex [40]. This difference may also be due to the intermolecular interactions between anions and cations. In (**1**), all five atoms in the tetrazolium rings are coplanar (r.m.s deviations are 0.0046 (Å) and 0.0028 (Å) for each tetrazolium ring, respectively). Delocalization in the tetrazolium ring is supported by N-C and N-N lengths (Table 2). The tetrazolium rings are close to being planar with one of the phenyl rings (r.m.s deviations = 0.052 and 0.031 Å) and form dihedral angles of 56.72(4)° and 54.99(5)° with the planes of the two other benzene rings. These values are different from its polymorphic form (**2**) and 2,3,5-triphenyltetrazolium hexachloroantimonate(V) [40,41], 2,3,5-triphenyltetrazolium perrhenate(vii) [9] and agree well with those reported for 2,3,5-triphenyltetrazolium chloride acetonitrile solvate [6], 2,3,5-triphenyltetrazolium chloride monohydrate [6], and 2,3,5-triphenyltetrazolium chloride ethanol solvate [6].

The cohesion and stability of (**1**) are ensured by C-H$\cdots$N hydrogen bonds between aromatic H atoms of [TPT] cations and N-atoms of thiocyanate groups of [Co(NCS)$_4$]$^{2-}$ anion (Figure 2, Table 3) and X$\cdots\pi$ interactions (Figure 3, Table 4). All the hydrogen bonds in (**1**) are found to be electrostatic in nature and weak with respect to donor-acceptor bond lengths (D$\cdots$A > 3 Å) [41,42]. The X$\cdots\pi$ interactions are established between the sulfur atoms of the thiocyanate groups and the tetrazolium rings (Figure 3, Table 4). In this structure, no π-stacking interactions is observed, compared to those reported for 2,3,5-triphenyltetrazolium cations [6,9,41]. All these non-covalent interactions give rise to a three-dimensional supramolecular framework (Figure S1).

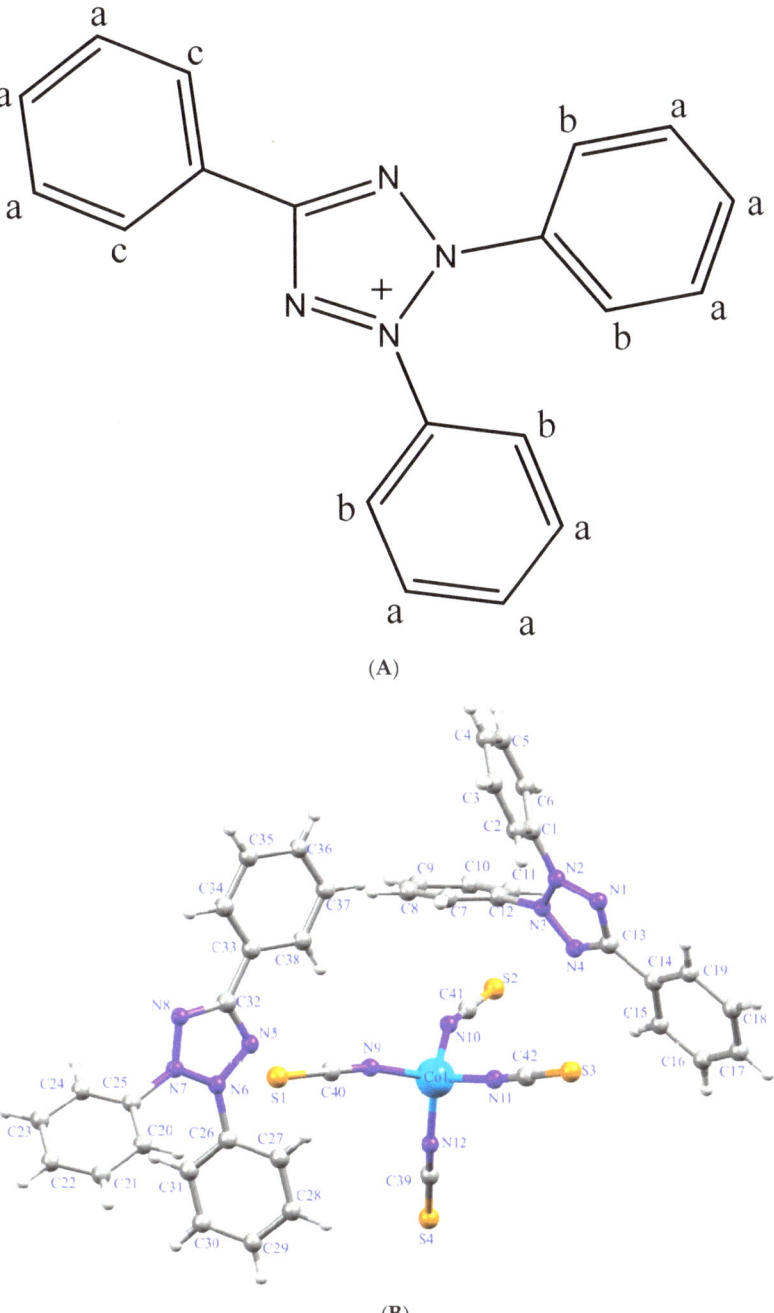

Figure 1. (**A**) Chemical structure of triphenyltetrazolium. (**B**) The asymmetric unit of (**1**) with atom labeling scheme. Thermal ellipsoids are drawn at the 50% probability level.

Table 2. Selected bond lengths (Å) and angles (°) for (**1**).

Bond Lengths (Å)			
Co1-N12	1.946 (4)	C27-C28	1.379 (5)
Co1-N9	1.947 (3)	C31-C30	1.381 (5)
Co1-N11	1.950 (3)	C7-C8	1.382 (5)
Co1-N10	1.950 (3)	C29-C28	1.367 (5)
S3-C42	1.618 (4)	C29-C30	1.375 (5)
S1-C40	1.605 (4)	C6-C5	1.403 (5)
S2-C41	1.600 (4)	C2-C3	1.397 (5)
S4-C39	1.618 (5)	C37-C36	1.373 (5)
N3-N4	1.315 (3)	C17-C18	1.368 (5)
N3-N2	1.343 (3)	C17-C16	1.368 (5)
N3-C12	1.449 (4)	C21-C22	1.363 (5)
N7-N8	1.321 (3)	C11-C10	1.384 (5)
N7-N6	1.340 (3)	C22-C23	1.364 (5)
N7-C25	1.451 (4)	C34-C35	1.383 (5)
N2-N1	1.317 (3)	C12-C7	1.355 (4)
N2-C1	1.455 (4)	C12-C11	1.370 (4)
N1-C13	1.346 (4)	C33-C38	1.376 (4)
N5-N6	1.317 (3)	C33-C34	1.380 (5)
N5-C32	1.348 (4)	C33-C32	1.461 (4)
N8-C32	1.350 (4)	C14-C19	1.367 (4)
N4-C13	1.349 (4)	C14-C15	1.380 (4)
N6-C26	1.451 (4)	C25-C24	1.368 (4)
N9-C40	1.164 (4)	C25-C20	1.373 (4)
C13-C14	1.468 (4)	N12-C39	1.149 (5)
N10-C41	1.157 (4)	N11-C42	1.161 (4)
C26-C31	1.366 (4)	C20-C21	1.377 (4)
C26-C27	1.373 (4)	C15-C16	1.391 (5)
C1-C2	1.357 (4)	C19-C18	1.395 (5)
C1-C6	1.357 (4)	C38-C37	1.388 (5)
C4-C5	1.356 (5)	C9-C10	1.381 (6)
C4-C3	1.369 (5)	C24-C23	1.378 (5)
C9-C8	1.362 (6)	C36-C35	1.358 (5)
Bond Angles (°)			
N12-Co1-N9	109.84 (14)	C12-C7-C8	117.5 (4)
N12-Co1-N11	111.90 (14)	C28-C29-C30	120.0 (4)
N9-Co1-N11	107.90 (13)	C1-C6-C5	117.6 (4)
N12-Co1-N10	101.52 (14)	C1-C2-C3	119.2 (4)
N9-Co1-N10	116.21 (14)	C36-C37-C38	120.4 (4)
N11-Co1-N10	109.46 (13)	C18-C17-C16	119.5 (4)
N4-N3-N2	109.4 (2)	C29-C28-C27	120.5 (3)
N4-N3-C12	122.7 (3)	C22-C21-C20	120.3 (3)
N2-N3-C12	127.8 (3)	C12-C11-C10	118.0 (4)
N8-N7-N6	110.1 (2)	C31-C26-C27	122.3 (3)
N8-N7-C25	122.9 (3)	C31-C26-N6	120.6 (3)
N6-N7-C25	126.8 (3)	C27-C26-N6	117.0 (3)
N1-N2-N3	110.2 (3)	C2-C1-C6	122.4 (3)
N1-N2-C1	123.7 (3)	C2-C1-N2	117.1 (3)
N3-N2-C1	126.1 (2)	C6-C1-N2	120.3 (3)
N2-N1-C13	103.9 (2)	C7-C12-C11	123.3 (3)
N6-N5-C32	103.7 (3)	C7-C12-N3	120.0 (3)
N7-N8-C32	103.4 (3)	C11-C12-N3	116.4 (3)
N3-N4-C13	104.4 (3)	N9-C40-S1	179.4 (4)
N5-N6-N7	110.0 (3)	C38-C33-C34	119.1 (3)
N5-N6-C26	122.9 (3)	C38-C33-C32	119.6 (3)
N7-N6-C26	126.9 (3)	C34-C33-C32	121.4 (3)
C40-N9-Co1	167.8 (3)	N5-C32-N8	112.7 (3)
N1-C13-N4	112.1 (3)	N5-C32-C33	123.5 (3)
N1-C13-C14	124.9 (3)	N8-C32-C33	123.7 (3)

Table 2. Cont.

Bond Angles (°)			
N4-C13-C14	123.0 (3)	C19-C14-C15	120.0 (3)
C41-N10-Co1	154.4 (4)	C31-C26-C27	122.3 (3)
C19-C14-C13	120.6 (3)	C17-C16-C15	119.9 (4)
C15-C14-C13	119.4 (3)	C21-C22-C23	120.6 (3)
C24-C25-C20	122.5 (3)	C33-C34-C35	120.9 (4)
C24-C25-N7	118.0 (3)	C5-C4-C3	120.0 (4)
C20-C25-N7	119.5 (3)	C8-C9-C10	119.5 (4)
N10-C41-S2	178.8 (4)	C25-C24-C23	117.9 (3)
C39-N12-Co1	162.5 (4)	C17-C18-C19	121.1 (4)
C42-N11-Co1	172.1 (3)	C29-C30-C31	120.6 (3)
C25-C20-C21	118.2 (3)	C35-C36-C37	120.2 (4)
N11-C42-S3	179.7 (4)	C9-C8-C7	121.5 (4)
C14-C15-C16	120.3 (3)	C4-C3-C2	119.6 (4)
C14-C19-C18	119.2 (4)	C2-C3-H3	120.2
N12-C39-S4	179.4 (4)	C22-C23-C24	120.6 (4)
C33-C38-C37	119.7 (4)	C36-C35-C34	119.7 (4)
C26-C27-C28	118.4 (3)	C9-C10-C11	120.2 (4)
C26-C31-C30	118.1 (3)	C4-C5-C6	121.2 (4)

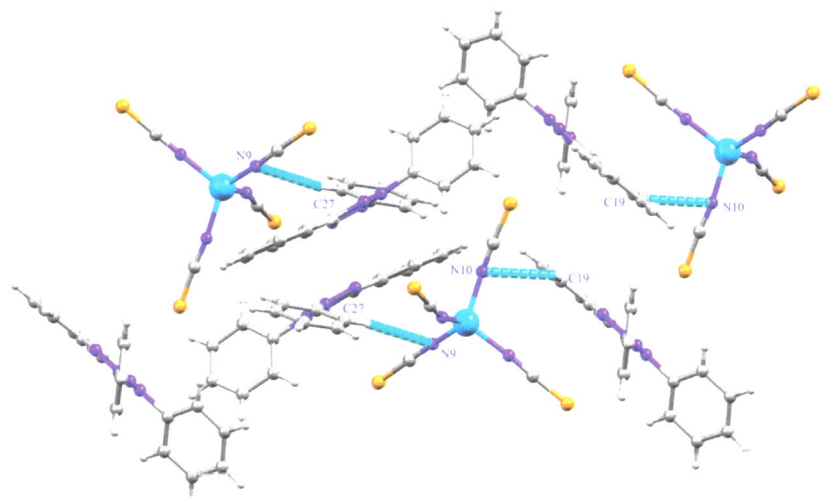

Figure 2. C-H···N Hydrogen bonds in (**1**).

Table 3. Hydrogen-bond geometry in (**1**).

D-H···A	D-H (Å)	H···A (Å)	D···A (Å)	D-H···A (°)
C27-H27···N9	0.93	2.83	3.748 (4)	167
C19-H19···N10 [(i)]	0.93	2.80	3.413 (5)	125

Symmetry code: [(i)] 1-x, 1-y, -z.

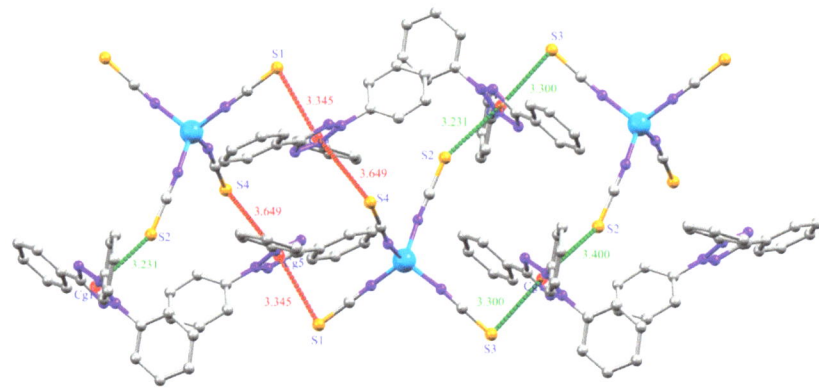

Figure 3. S···π interactions in (**1**).

Table 4. Analysis of Y-X···Cg (Pi-Ring) interactions (X···Cg < 4.0 Å).

	d(X···Cg) (Å)	d(Y···Cg) (Å)	Y-X···Cg(°)
C40-S1···Cg5 [(ii)]	3.345(2)	3.647(4)	87.53(14)
C41-S2···Cg1 [(ii)]	3.231(2)	4.713(4)	153.09(14)
C42-S3···Cg1 [(i)]	3.299(2)	3.718(4)	91.69(14)
C39-S4···Cg5 [(iii)]	3.649(2)	4.490(5)	111.00(18)

Symmetry code: [(i)] 1-x,1-y,-z, [(ii)] x,y,z, [(iii)] 1-x,1-y,1-z. (Where, Cg1 and Cg5 are centroids of rings N(1)-N(2)-N(3)-N(4)-C(13) and N(5)-N(6)-(7)-N(8)-C(32), respectively).

3.3. Hirshfeld Surface Analysis and Enrichment Ratio Calculations

Figure 4 shows the Hirshfeld surface generated over a d_{norm} range of -0.0794 to 1.5355 a.u. The red spot on the views of the d_{norm} surfaces near nitrogen and ortho C-H of the phenyl ring indicates these atoms are involved in the C-H···N hydrogen-bonding contacts. Two-dimensional (2D) fingerprint plot analysis can be used to quantify contacts present in the structure [26,35]. The overall 2D fingerprint plot (Figure 5) and those delineated into H···C/C···H, H···H, H···S/S···H, and N···H/H···N, are given in Figure 5, respectively, together with their relative contributions to the HS. The most significant contribution from H···C/C···H contacts is 30.2% (Table 3), shown in the 2D fingerprint plot by a pair of sharp spikes pointed at $d_e + d_i = 2.7$ Å (Figure 5). Furthermore, because of the abundance of carbon (%S_C) and hydrogen (%S_H) on the molecular surface (%S_C = 17.25% and %S_H = 59.5%, respectively), rather than the presence of C-H···π interactions and an enrichment ratio greater than the unit $E_{H···C}$ = 1.47 (Table 5), these types of contacts are the most prominent interactions. Hydrogen-hydrogen (H···H) interactions are prominently observed in the two-dimensional fingerprint plot, exhibiting a substantial presence within the central region at a distance of $d_e + d_i = 2.4$ Å. These interactions account for approximately 28.7% of the entire area encompassed by the hydrogen bond surface (Figure 5); the second most frequent connections occur as a result of the high abundance of hydrogen on the molecular surface, with a hydrogen content of 59.5%. However, these contacts are significantly less prevalent, with an enrichment ratio of approximately 0.81. The H···S contacts, occur as a pair of sharp spikes pointed at $d_e + d_i = 3$ Å, and refer to S···π interactions, the third most significant interactions on the surface account for approximately 18.7% of the Hirshfeld surfaces (Figure 5) and have an enrichment ratio higher than unit $E_{H···S}$ = 1.38 (Table 5). In addition, H···N contacts present the fourth most abundant interactions on the HS (12.8%), corresponding to the presence of weak C-H···N hydrogen bonding, which is represented by two large spikes pointed at $d_e + d_i = 3.2$ Å (Figure 5). These contacts are over-represented with an enrichment ratio $E_{N···H}$ = 1.22. Finally, C···C, N···C, N···S, C···S, and Co···H contacts all make minor contributions to the overall HS (3.2,

2.6, 2.3, 1.7, and 0.2% to the total HS, respectively). However, C···C, N···S, and Co···H contacts are more enriched at $E_{C···C} = 1.07$, $E_{N···S} = 1.16$ and $E_{Co···H} = 1.66$ (Table 5).

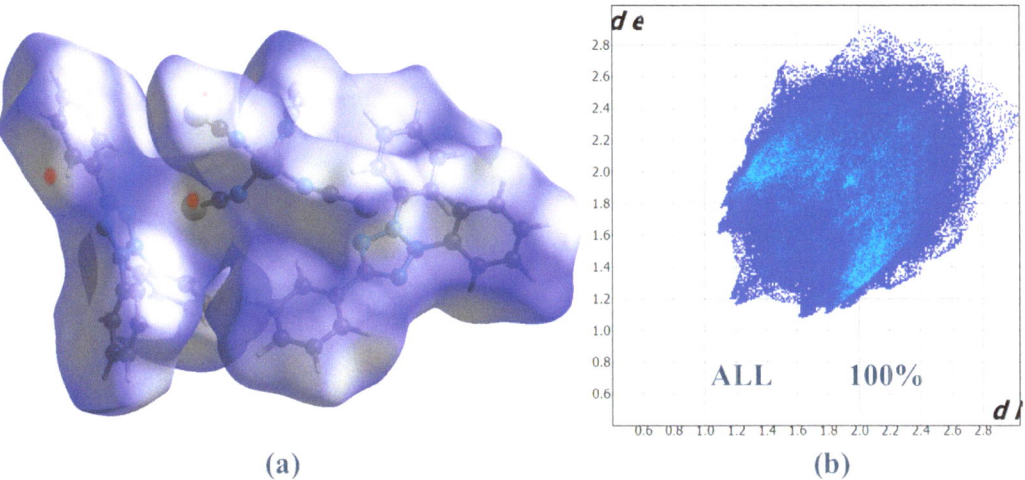

(a) (b)

Figure 4. (a) HS plotted over d_{norm} for (**1**) and (b) Fingerprint plot illustrating the total percentage contribution of various interactions to HS area.

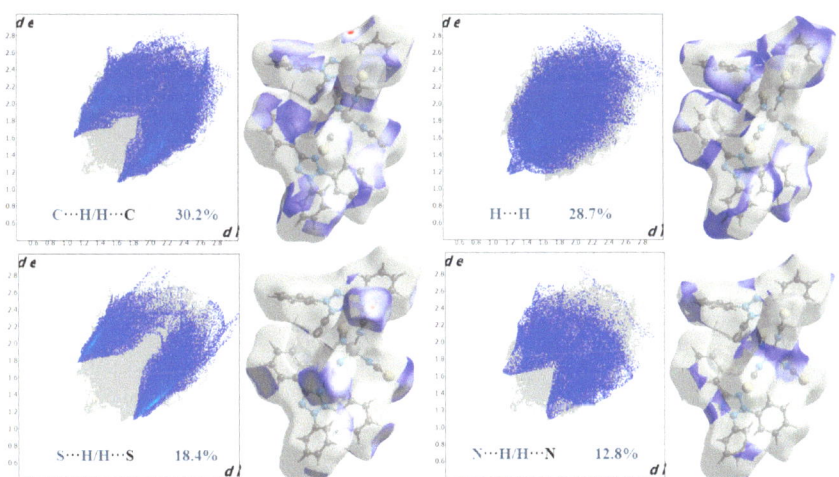

Figure 5. Two-dimensional fingerprints made with standard HS surfaces of C-H, H-H, S-H, and N-H contacts in (**1**).

3.4. Void Analysis

The mechanical characteristics of single crystals are strongly linked to the voids. If a single crystal has a very small percentage of voids, it has good mechanical properties such as stress response, melting temperature, and so on [43,44] In this light, we calculated voids in (**1**) by considering all atoms were spherically symmetric and by combining the electronic density of all atoms in the crystal structure. The volume of the crystal voids (Figure 6) and the percentage of free spaces in the unit cell are calculated as 352.86 Å^3 and 15.84%, respectively. Therefore, the molecules are closely packed, and there are no large voids in the crystal packing. This higher value of volume occupied by voids reveals (**1**) is a hard crystal with good mechanical properties.

Table 5. Hirshfeld contact surfaces and derived random contact and enrichment ratios calculations for (**1**).

	Contacts (%)				
Atoms	Co	N	C	S	H
Co	0	-	-	-	-
N	0	0	-	-	-
C	0	2.6	3.2	-	-
S	0	2.3	1.7	0	-
H	0.2	12.8	30.2	18.4	28.7
Surface%	0.1	8.85	17.25	11.2	59.5
	Random Contacts (%)				
Co	0	-	-	-	-
N	0.02	0.78	-	-	-
C	0.03	3.05	2.98	-	-
S	0.02	1.98	3.86	1.25	-
H	0.12	10.53	20.53	13.33	35.40
	Enrichment rations E_{XY}				
Co	0	-	-	-	-
N	0	0	-	-	-
C	0	0.85	1.07	-	-
S	0	1.16	0.44	0	-
H	1.66	1.22	1.47	1.38	0.81

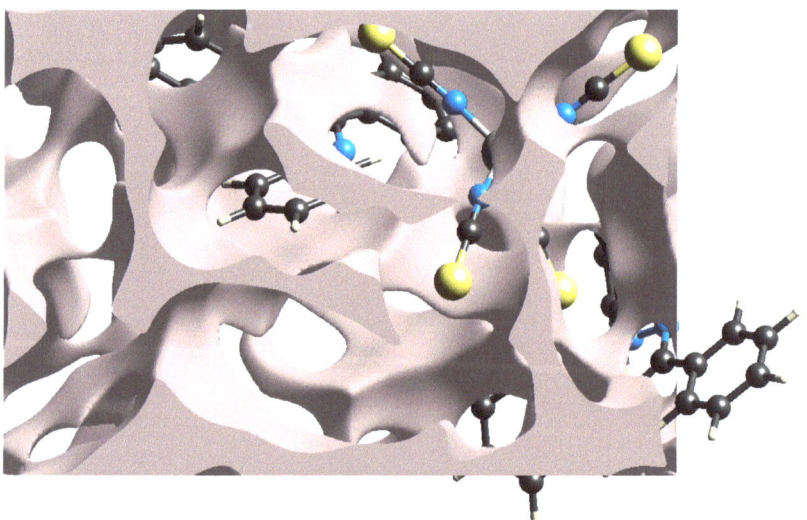

Figure 6. Graphical representation of voids in the crystal packing of (**1**).

3.5. Growth Morphology Prediction by the AE Method

The crystal morphologies predicted by the BFDH and growth morphology (GM) models (Figure 7) reveal the main shape of as-grown crystals to be a trigonal plate shape, with a small deviation in the GM model. Both models expect a trigonal plate morphology, with (001) facets having the most morphological importance (MI) followed by (010) facets as the second most MI. However, the (GM) model indicates the (111) facet would be the third dominant facet, while the BFDH model predicts the (100) facet would be the third largest exposed face.

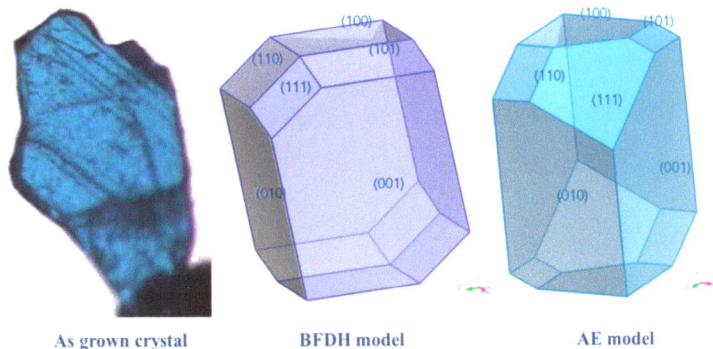

Figure 7. As grown graph of synthetic crystal, growth morphology prediction by the AE method and BFDH model morphological predictions showing the major faces that are predicted by the models.

The surface structures of all important facets of (**1**) given by the GM model were studied. The (GM) model assumes the growth of the crystals to take place in a vacuum and at a low driving force. The significance of the (001) and (010) morphological facets, which account for almost 73% of the crystal surface, is apparent due to their limited interactions characterized by a small number of polar groups. The attachment energies of (001) and (010) are −42.3605 and −45.7329 kJ mol^{-1}, respectively. The additional morphologically significant aspects, denoted as (100), (101), and (110), indicate the existence of more pronounced interactions, specifically the presence of polar functional groups. The estimates of attachment energy were conducted to elucidate the potential energetic interactions that occur during the process of crystal formation. According to the literature [35], when the attachment energy is at its minimum in a specific direction, the facet that plays a crucial role in morphology and bounds that development direction will exhibit the slowest rate of growth and hence have the smallest size. So, according to the E_{att} values (Table 6), the (1 0 1), and (110) facets present smaller attachment energies and therefore slower growth rates and lower morphological importance (Figure 8).

Table 6. Crystal facets and related parameters of (**1**) predicted by the BFDH and GM models.

(h k l)	Multiplicity	d_{hkl} (Å)		% of TFA
		BFDH		
(0 0 1)	2	18.42		44.62
(0 1 0)	2	12.45		27.81
(1 0 0)	2	9.22		11.04
(1 0 1)	2	8.92		6.55
(1 1 0)	2	8.52		6.28
(1 1 1)	2	8.28		3.69

(h k l)	Multiplicity	d_{hkl} (Å)	E_{att} (Total) (kcal·mol^{-1})	% of TFA
		Growth Morphology		
(0 0 1)	2	18.42	−42.3605	38.93
(0 1 0)	2	12.45	−45.7329	34.03
(1 0 0)	2	9.22	−79.0048	10.73
(1 0 1)	2	8.92	−87.8669	1.97
(1 1 0)	2	8.53	−80.935	3.22
(1 1 1)	2	8.28	−71.4886	11.12

(h k l) is the crystal plane; d_{hkl} is the distance between the planes; TFA is the total facet area; and Eatt (Tot) is the total potential energy.

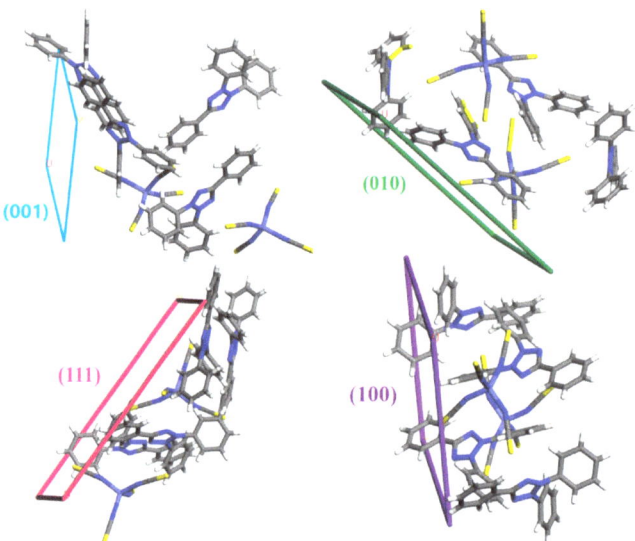

Figure 8. The molecular arrangement of different crystal facets in (**1**).

3.6. Computational Results

3.6.1. Geometry Optimization

The geometry parameters and their error relative to the experimental values (Table 7) prove the method used in our study reproduces well crystallographic geometry, where the largest error is around 3.2% in Co1-N10 bond.

Table 7. Selected theoretical and experimental equilibrium geometry parameters of (**1**) in Å.

	Cal	Exp	Error%
Co1-N12	1.962	1.946 (4)	0.8%
Co1-N9	1.968	1.947 (3)	1.1%
Co1-N11	1.972	1.950 (3)	1.1%
Co1-N10	2.013	1.950 (3)	3.2%
S3-C42	1.58	1.618 (4)	2.3%
S1-C40	1.57	1.605 (4)	2.2%
S2-C41	1.587	1.600 (4)	0.8%

3.6.2. Chemical Descriptors

The energy gap (E_g) between HOMO and LUMO plays a pivotal role in elucidating both the bioactivity and the intermolecular charge transfer processes [45]. The calculated energy gap of our crystal holds paramount importance in elucidating its electronic properties and potential applications. This modest E_g value of 0.13 eV categorizes the material as a semiconductor, providing strong evidence for the occurrence of intramolecular charge transfer (ICT) (Table S1). This noteworthy ICT phenomenon plays a central role in reinforcing and confirming the inherent antioxidant capabilities of the compound under examination. This particular E_g value falls within the range typically associated with certain types of semiconductors, making it particularly suitable not only for electronic and optoelectronic applications but also for use in biomedical sensing applications, where it can serve as a sensitive element for detecting biomolecules or environmental changes. The calculated electrophilicity (ω) value of the studied crystal (Table 8) suggests the crystal is highly electrophilic. This high electrophilicity value typically indicates the crystal has a strong tendency to accept electrons and engage in chemical reactions where it acts as an electron acceptor. This reactivity can make the crystal prone to forming chemical bonds or undergoing reactions with other molecules or ions that can donate electrons.

Table 8. HOMO-LUMO energy and global reactivity descriptor values of (1) in (eV).

E_{SOMO}	−6.35
E_{LUMO}	−6.22
Eg	0.13
Ionis	6.35
Elec. Aff	6.22
Hardness	0.07
Chem. pot	−6.28
Electrophilicity	295.0
Electronegativity	6.28

The contour surfaces of the frontier molecular orbitals (FMOs) calculations are shown in Figure 9. As depicted, the spatially occupied molecular orbital SOMO such as half-filled α-HOMO for the open shell system, SOMO-1 (α-HOMO-1) and LUMO (α-LUMO) predominantly reside within the inorganic component of the studied crystal. This observation highlights a significant and intriguing facet of the compound's electronic structure, signifying that the inorganic portion plays a paramount role in facilitating intramolecular charge transfer (ICT) within the crystal.

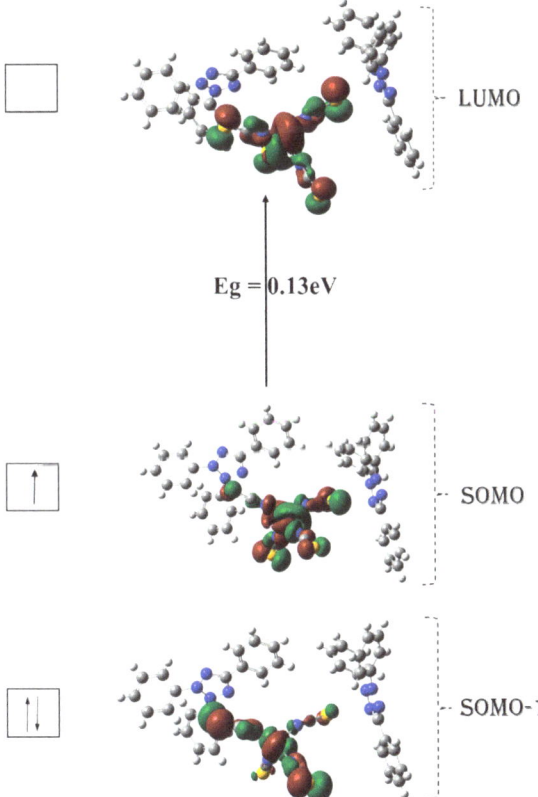

Figure 9. Frontier molecular orbitals and the energy gap (Eg in electron volt) of (1) (Isosurface 0.05 e/Å^3).

4. Conclusions

The crystal structure is determined by single crystal X-ray diffraction analysis. The molecular structure is composed of one tetrathiocyanatecobaltate [Co^{II}(NCS)$_4$]$^{2-}$ anionic complex and two triphenyltetrazolium (TPT)$^+$ organic cations. The supramolecular assembly is mainly stabilized by C-H$\cdots$N hydrogen bonding and S$\cdots$π non-covalent interactions. The crystal packing of compound (1) was determined to be primarily influenced by C$\cdots$H, H$\cdots$H, S$\cdots$H, and N$\cdots$H contacts, as indicated by the analysis of Hirshfeld surface and enrichment ratio estimates. The study of the void revealed the absence of a significant cavity, indicating the compound is anticipated to possess favorable mechanical characteristics. The final crystal morphology of (1) was obtained using BFDH and growth morphology models. The results show that (1) crystals simulated in vacuum have six habit faces, of which the (0 0 1) and (0 1 0) faces are morphologically more important. Finally, in terms of electronic properties, the low energy gap (E$_g$ = 0.13 eV), implies stability, high kinetics, and low chemical reactivity, as it is energetically unfavorable to add electrons to the LUMO or to extract electrons from the HOMO.

Supplementary Materials: The following supporting information can be downloaded at: https://www.mdpi.com/article/10.3390/cryst13111598/s1; Figure S1: The crystal packing of (1) along *a* axis; Table S1: Calculated UV-Vis spectrum, oscillator strength and the major contribution calculated using B3LYP functional.

Author Contributions: Formula analysis, investigation and writing—reviewer editing, E.A.A.; Investigation, software, visualization, data curation, writing the origin draft, and writing—reviewer editing, R.B.; Methodology, investigation and writing—reviewer editing, R.A.-S.; Investigation, methodology, A.S.M.A.-J.; Conceptualization, data curation, investigation and writing—reviewer editing, M.W.A. and G.A.E.M. All authors have read and agreed to the published version of the manuscript.

Funding: The research work was funded by researchers supporting project number (RSPD2023R1000), King Saud University, Riyadh, Saudi Arabia.

Data Availability Statement: All data are available within the manuscript.

Acknowledgments: The authors extend their appreciation to the researchers supporting project number (RSPD2023R1000), King Saud University, Riyadh, Saudi Arabia, for financial support.

Conflicts of Interest: The authors declare no conflict of interest.

References

1. Peppel, T.; Hinz, A.; Thiele, P.; Geppert-Rybczyńska, M.; Lehmann, J.K.; Köckerling, M. Synthesis, Properties, and Structures of Low-Melting Tetraisocyanatocobaltate (II)-Based Ionic Liquids. *Eur. J. Inorg. Chem.* **2017**, *2017*, 885–893. [CrossRef]
2. Triki, H.; Nagy, B.; Overgaard, J.; Jensen, F.; Kamoun, S. Structure, DFT Based Investigations on Vibrational and Nonlinear Optical Behavior of a New Guanidinium Cobalt Thiocyanate Complex. *Struct. Chem.* **2020**, *31*, 103–114. [CrossRef]
3. Burmeister, J.L.; Al-Janabi, M.Y. Selenocyanate Complexes of Cobalt (III), Palladium (II), and Platinum (II). *Inorg. Chem.* **1965**, *4*, 962–965. [CrossRef]
4. Poddar, R.K.; Parashad, R.; Agarwala, U. Linkage Isomerism of NCS– Group in Ruthenium Complexes. *J. Inorg. Nucl. Chem.* **1980**, *42*, 837–838. [CrossRef]
5. Ferchichi, A.; Makhlouf, J.; El Bakri, Y.; Saravanan, K.; Valkonen, A.; Hashem, H.E.; Ahmad, S.; Smirani, W. Self-Assembly of New Cobalt Complexes Based on [Co (SCN)$_4$], Synthesis, Empirical, Antioxidant Activity, and Quantum Theory Investigations. *Sci. Rep.* **2022**, *12*, 15828. [CrossRef]
6. Golovanov, D.G.; Perekalin, D.S.; Yakovenko, A.A.; Antipin, M.Y.; Lyssenko, K.A. The Remarkable Stability of the Cl–$\cdots$(π-System) Contacts in 2, 3, 5-Triphenyltetrazolium Chloride. *Mendeleev Commun.* **2005**, *15*, 237–239. [CrossRef]
7. Gjikaj, M.; Xie, T.; Brockner, W. Uncommon Compounds in Antimony Pentachloride–Ionic Liquid Systems: Synthesis, Crystal Structure and Vibrational Spectra of the Complexes [TPT][SbCl6] and [Cl-EMIm][SbCl6]. *Z. Anorg. Allg. Chem.* **2009**, *635*, 1036–1040. [CrossRef]
8. Benon, H.J.B.; Grace, G.S.; Stanley, B. Reduction of Nitro Blue Tetrazolium by CO$_2$–and O$_2$–Radicals. *J. Phys. Chem.* **1980**, *84*, 830–833.
9. Předota, M.; Petříček, V.; Žák, Z.; Głowiak, T.; Novotný, J. Structure Du Perrhenate de Triphényl-2, 3, 5 Tétrazolium. *Acta Crystallogr. Sect. C Cryst. Struct. Commun.* **1991**, *47*, 738–740. [CrossRef]

10. Gavazov, K.B.; Dimitrov, A.N.; Lekova, V.D. The Use of Tetrazolium Salts in Inorganic Analysis. *Russ. Chem. Rev.* **2007**, *76*, 169. [CrossRef]
11. Mostafa, G.A.-H. PVC Matrix Membrane Sensor for Potentiometric Determination of Triphenyltetrazolium Chloride and Ascorbic Acid. *Ann. Chim.* **2007**, *97*, 1247–1256. [CrossRef]
12. Hassanien, M.M.; Abou-El-Sherbini, K.S.; Mostafa, G.A.E. A Novel Tetrachlorothallate (III)-PVC Membrane Sensor for the Potentiometric Determination of Thallium (III). *Talanta* **2003**, *59*, 383–392. [CrossRef] [PubMed]
13. Abbas, M.N.; Mostafa, G.A.E.; Homoda, A.M.A. PVC Membrane Ion Selective Electrode for the Determination of Pentachlorophenol in Water, Wood and Soil Using Tetrazolium Pentachlorophenolate. *Talanta* **2001**, *55*, 647–656. [CrossRef] [PubMed]
14. Bruker, A. *Bruker Advanced X-ray Solutions SAINT Software Reference Manual SAINT v8. 34A*; Bruker AXS Inc.: Madison, WI, USA, 2013.
15. Sheldrick, G.M. Crystal Structure Refinement with SHELXL. *Acta Crystallogr. Sect. C Struct. Chem.* **2015**, *71*, 3–8. [CrossRef] [PubMed]
16. Sheldrick, G.M. SHELXT–Integrated Space-Group and Crystal-Structure Determination. *Acta Crystallogr. Sect. A Found. Adv.* **2015**, *71*, 3–8. [CrossRef] [PubMed]
17. Macrae, C.F.; Sovago, I.; Cottrell, S.J.; Galek, P.T.A.; McCabe, P.; Pidcock, E.; Platings, M.; Shields, G.P.; Stevens, J.S.; Towler, M. Mercury 4.0: From Visualization to Analysis, Design and Prediction. *J. Appl. Crystallogr.* **2020**, *53*, 226–235. [PubMed]
18. Spackman, M.A.; Jayatilaka, D. Hirshfeld Surface Analysis. *CrystEngComm* **2009**, *11*, 19–32. [CrossRef]
19. Spackman, M.A.; McKinnon, J.J. Fingerprinting Intermolecular Interactions in Molecular Crystals. *CrystEngComm* **2002**, *4*, 378–392.
20. Spackman, P.R.; Turner, M.J.; McKinnon, J.J.; Wolff, S.K.; Grimwood, D.J.; Jayatilaka, D.; Spackman, M.A. CrystalExplorer: A Program for Hirshfeld Surface Analysis, Visualization and Quantitative Analysis of Molecular Crystals. *J. Appl. Crystallogr.* **2021**, *54*, 1006–1011. [CrossRef]
21. Bondi, A. Van Der Waals Volumes and Radii of Metals in Covalent Compounds. *J. Phys. Chem.* **1966**, *70*, 3006–3007. [CrossRef]
22. Jelsch, C.; Ejsmont, K.; Huder, L. The Enrichment Ratio of Atomic Contacts in Crystals, an Indicator Derived from the Hirshfeld Surface Analysis. *IUCrJ* **2014**, *1*, 119–128. [CrossRef] [PubMed]
23. Ferjani, H.; Chebbi, H.; Guesmi, A.; AlRuqi, O.S.; Al-Hussain, S.A. Two-Dimensional Hydrogen-Bonded Crystal Structure, Hirshfeld Surface Analysis and Morphology Prediction of a New Polymorph of 1H-Nicotineamidium Chloride Salt. *Crystals* **2019**, *9*, 571. [CrossRef]
24. Miglani Bhardwaj, R.; Ho, R.; Gui, Y.; Brackemeyer, P.; Schneider-Rauber, G.; Nordstrom, F.L.; Sheikh, A.Y. Origins and Implications of Extraordinarily Soft Crystals in a Fixed-Dose Combination Hepatitis C Regimen. *Cryst. Growth Des.* **2022**, *22*, 4250–4259. [CrossRef]
25. Studio, A.D. Accelrys Materials Studio 7.0. Available online: www.scientific-computing/com/press-release/accelrys-materials-Studio7.0 (accessed on 15 October 2023).
26. Ferjani, H. Structural, Hirshfeld Surface Analysis, Morphological Approach, and Spectroscopic Study of New Hybrid Iodobismuthate Containing Tetranuclear 0D Cluster $Bi_4I_{16} \cdot 4(C_6H_9N_2)$ $2(H_2O)$. *Crystals* **2020**, *10*, 397. [CrossRef]
27. Frisch, M.J.; Trucks, G.W.; Schlegel, H.B.; Scuseria, G.E.; Robb, M.A.; Cheeseman, J.R.; Scalmani, G.; Barone, V.; Petersson, G.A.; Nakatsuji, H. *Gaussian 16*; Gaussian, Inc.: Wallingford, CT, USA, 2016.
28. Becke, A.D. Density-functional exchange-energy approximation with correct asymptotic behavior. *Phys. Rev. A At. Mol. Opt. Phys.* **1988**, *38*, 3098–3100. [CrossRef]
29. Lee, C.; Yang, W.; Parr, R.G. Development of the Colle-Salvetti Correlation-Energy Formula into a Functional of the Electron Density. *Phys. Rev. B* **1988**, *37*, 785. [CrossRef]
30. Chiodo, S.; Russo, N.; Sicilia, E. LANL2DZ Basis Sets Recontracted in the Framework of Density Functional Theory. *J. Chem. Phys.* **2006**, *125*, 104107. [CrossRef]
31. Zárate, X.; Schott, E.; Carey, D.M.-L.; Bustos, C.; Arratia-Pérez, R. DFT Study on the Electronic Structure, Energetics and Spectral Properties of Several Bis (Organohydrazido (2−)) Molybdenum Complexes Containing Substituted Phosphines and Chloro Atoms as Ancillary Ligands. *J. Mol. Struct. THEOCHEM* **2010**, *957*, 126–132. [CrossRef]
32. Xu, Z.; Li, Y.; Zhang, W.; Yuan, S.; Hao, L.; Xu, T.; Lu, X. DFT/TD-DFT Study of Novel T Shaped Phenothiazine-Based Organic Dyes for Dye-Sensitized Solar Cells Applications. *Spectrochim. Acta Part A Mol. Biomol. Spectrosc.* **2019**, *212*, 272–280. [CrossRef]
33. Wei, J.; Song, P.; Ma, F.; Saputra, R.M.; Li, Y. Tunable Linear and Nonlinear Optical Properties of Chromophores Containing 3, 7-(Di) Vinylquinoxalinone Core by Modification of Receptors Moieties. *Opt. Mater.* **2020**, *99*, 109580. [CrossRef]
34. Wang, L.; Zhang, J.; Duan, Y.-C.; Pan, Q.-Q.; Wu, Y.; Geng, Y.; Su, Z.-M. Theoretical Insights on the Rigidified Dithiophene Effects on the Performance of Near-Infrared Cis-Squaraine-Based Dye-Sensitized Solar Cells with Panchromatic Absorption. *J. Photochem. Photobiol. A Chem.* **2019**, *369*, 150–158. [CrossRef]
35. Ferjani, H.; Bechaieb, R.; Dege, N.; Abd El-Fattah, W.; Elamin, N.Y.; Frigui, W. Stabilization of Supramolecular Network of Fluconazole Drug Polyiodide: Synthesis, Computational and Spectroscopic Studies. *J. Mol. Struct.* **2022**, *1263*, 133192. [CrossRef]
36. Ferjani, H.; Bechaieb, R.; Abd El-Fattah, W.; Fettouhi, M. Broad-Band Luminescence Involving Fluconazole Antifungal Drug in a Lead-Free Bismuth Iodide Perovskite: Combined Experimental and Computational Insights. *Spectrochim. Acta Part A Mol. Biomol. Spectrosc.* **2020**, *237*, 118354. [CrossRef] [PubMed]

37. Ferjani, H.; Bechaieb, R.; Alshammari, M.; Lemine, O.M.; Dege, N. New Organic–Inorganic Salt Based on Fluconazole Drug: TD-DFT Benchmark and Computational Insights into Halogen Substitution. *Int. J. Mol. Sci.* **2022**, *23*, 8765. [CrossRef]
38. Gümüş, H.P.; Tamer, Ö.; Avcı, D.; Atalay, Y. Quantum Chemical Calculations on the Geometrical, Conformational, Spectroscopic and Nonlinear Optical Parameters of 5-(2-Chloroethyl)-2, 4-Dichloro-6-Methylpyrimidine. *Spectrochim. Acta Part A Mol. Biomol. Spectrosc.* **2014**, *129*, 219–226. [CrossRef]
39. Tamer, Ö.; Bhatti, M.H.; Yunus, U.; Avcı, D.; Atalay, Y.; Nadeem, M.; Shah, S.R.; Helliwell, M. Structural, Spectroscopic, Nonlinear Optical and Electronic Properties of Calcium N-Phthaloylglycinate: A Combined Experimental and Theoretical Study. *J. Mol. Struct.* **2016**, *1125*, 315–322. [CrossRef]
40. Nakashima, K.; Kawame, N.; Kawamura, Y.; Tamada, O.; Yamauchi, J. Bis (2,3,5-Triphenyltetrazolium) Tetrathiocyanatocobaltate (II). *Acta Crystallogr. Sect. E Struct. Rep. Online* **2009**, *65*, m1406–m1407. [CrossRef]
41. Hsieh, C.-H.; Brothers, S.M.; Reibenspies, J.H.; Hall, M.B.; Popescu, C.V.; Darensbourg, M.Y. Ambidentate Thiocyanate and Cyanate Ligands in Dinitrosyl Iron Complexes. *Inorg. Chem.* **2013**, *52*, 2119–2124. [CrossRef]
42. Steiner, T. The Hydrogen Bond in the Solid State. *Angew. Chemie Int. Ed.* **2002**, *41*, 48–76. [CrossRef]
43. Turner, M.J.; McKinnon, J.J.; Jayatilaka, D.; Spackman, M.A. Visualisation and Characterisation of Voids in Crystalline Materials. *CrystEngComm* **2011**, *13*, 1804–1813. [CrossRef]
44. Setifi, Z.; Ferjani, H.; Smida, Y.B.; Jelsch, C.; Setifi, F.; Glidewell, C. A Novel CuII/8-Aminoquinoline Isomer Complex [Cu(H$_2$O)$_2$(C$_9$H$_8$N$_2$)$_2$]Cl$_2$: Solvothermal Synthesis, Molecular Structure, Hirshfeld Surface Analysis, and Computational Study. *Chem. Afr.* **2023**, *6*, 891–901. [CrossRef]
45. Hajji, M.; Kouraichi, C.; Guerfel, T. Modelling, Structural, Thermal, Optical and Vibrational Studies of a New Organic–Inorganic Hybrid Material (C$_5$H$_{16}$N$_2$)Cd$_{1.5}$Cl$_5$. *Bull. Mater. Sci.* **2017**, *40*, 55–66. [CrossRef]

Disclaimer/Publisher's Note: The statements, opinions and data contained in all publications are solely those of the individual author(s) and contributor(s) and not of MDPI and/or the editor(s). MDPI and/or the editor(s) disclaim responsibility for any injury to people or property resulting from any ideas, methods, instructions or products referred to in the content.

 *crystals*

Article

Intermolecular Interactions in Molecular Ferroelectric Zinc Complexes of Cinchonine

Marko Očić and Lidija Androš Dubraja *

Division of Materials Chemistry, Ruđer Bošković Institute, Bijenička Cesta 54, 10000 Zagreb, Croatia
* Correspondence: lidija.andros@irb.hr; Tel.: +385-1-456-1184

Abstract: The use of chiral organic ligands as linkers and metal ion nodes with specific coordination geometry is an effective strategy for creating homochiral structures with potential ferroelectric properties. Natural *Cinchona* alkaloids, e.g., quinine and cinchonine, as compounds with a polar quinuclidine fragment and aromatic quinoline ring, are suitable candidates for the construction of molecular ferroelectrics. In this work, the compounds [CnZnCl$_3$]·MeOH and [CnZnBr$_3$]·MeOH, which crystallize in the ferroelectric polar space group $P2_1$, were prepared by reacting the cinchoninium cation (Cn) with zinc(II) chloride or zinc(II) bromide. The structure of [CnZnBr$_3$]·MeOH was determined from single-crystal X-ray diffraction analysis and was isostructural with the previously reported chloride analog [CnZnCl$_3$]·MeOH. The compounds were characterized by infrared spectroscopy, and their thermal stability was determined by thermogravimetric analysis and temperature-modulated powder X-ray diffraction experiments. The intermolecular interactions of the different cinchoninium halogenometalate complexes were evaluated and compared.

Keywords: quinoline; quinuclidine; cinchonine; zinc(II); halogenometalate; hydrogen bonds; ferroelectric; stacking interactions

Citation: Očić, M.; Androš Dubraja, L. Intermolecular Interactions in Molecular Ferroelectric Zinc Complexes of Cinchonine. *Crystals* 2024, 14, 978. https://doi.org/10.3390/cryst14110978

Academic Editor: Peng Shi

Received: 14 October 2024
Revised: 8 November 2024
Accepted: 11 November 2024
Published: 13 November 2024

Copyright: © 2024 by the authors. Licensee MDPI, Basel, Switzerland. This article is an open access article distributed under the terms and conditions of the Creative Commons Attribution (CC BY) license (https://creativecommons.org/licenses/by/4.0/).

1. Introduction

Ferroelectrics are an important class of materials that are of great interest from both a fundamental and an applied point of view, e.g., in the electronics and medical industries [1]. Their main characteristic is the occurrence of a permanent and spontaneous polarization that can be altered by applying an external electric field. Although various inorganic ferroelectrics have been discovered to date, the most common are lead-based oxides, which are now becoming an increasing environmental concern due to the toxicity of lead, the scarcity of elemental resources, and the high cost of producing oxide materials [2,3]. Recent research has shown that there is great potential for soft materials based on organic and inorganic–organic molecules [4–6]. Such materials are produced using relatively simple processes at low temperatures and according to the principles of green chemistry. A prerequisite for the existence of a permanent dipole moment is that the material crystallizes in a space group with a unique axis of rotation and without a center of symmetry. The advantage of these soft materials is that the crystal packing can be influenced by the careful selection of the molecular fragments, relying on the intermolecular contacts they will achieve in the solid state, but also by some external stimuli such as crystallizing solvents, pressure, and heat [7]. For example, polar spherical molecules such as quinuclidine can be very easily reoriented in an electric field and thus influence the occurrence of ferroelectric polarization [8]. In addition, the chirality of certain molecules is important for the design of polar structures, which is essential for ferroelectrics. The use of naturally occurring chiral ligands can be an effective strategy for the preparation of molecular ferroelectrics [9,10]. Many small organic molecules are asymmetric and can retain their asymmetry upon crystallization. However, this does not necessarily mean that asymmetric molecules will consistently crystallize in an asymmetric or polar structure. A polar asymmetric molecule can interact with a neighboring molecule and form a supramolecular synthon that forms a crystal structure with an inversion center, which is undesirable for ferroelectrics [11]. For this reason, the understanding and prediction of intermolecular interactions is crucial for the design of supramolecular ferroelectrics [3]. However, the literature reports on the ferroelectric properties of homochiral molecules are not very extensive, and among the best studied are those of organic salts such as R-3-hydroxyquinuclidinium halides [12], three-dimensional metal-free perovskites of (3-ammonioquinuclidinium)NH$_4$Br$_3$ [13], and bis (imidazolium)-L-tartrate [14]. Recently, a pair of homochiral organic simple-component ferroelectrics based on a heterocyclic derivative of spirooxazacamphorsultam was reported to exhibit well-defined ferroelectricity with spontaneous polarization of 2.2 µC cm^{-2} at a coercive field of ~50 kV cm^{-1} [15]. Homochiral organic molecules are also responsible for ferroelectricity in semi-crystalline solid materials with polar symmetry, i.e., liquid crystals. For example, the ferroelectric chiral cholesterol derivatives exhibit a spontaneous polarization switching of ~4 µC cm^{-2} at a coercive field of ~50 kV cm^{-1} [16]. There are far fewer reports of ferroelectric metal–organic complexes constructed from a chiral ligand. The presence of transition metal atoms can impart additional physical properties to the material, as in the case of the plastic hybrid compound R-3-hydroxyquinuclidium tetrachloroferatte, which exhibits both ferroelectricity and long-range magnetic ordering [5], or N,N'-dimethyl-1,4-diazoniabicyclo [2.2.2]octonium tetrachlorocuprate, in which thermochromism is observed due to a change in coordination geometry around the metal center [4]. One strategy to prepare molecular ferroelectrics is to introduce the chiral solvent, i.e., R-1,2-propanediol or S-1,2-propanediol, into the crystal structure of the metal–organic complex, which successfully achieved ferroelectricity in the compound Cu(1,10-phenanthroline)$_2$SeO$_4$ [17]. In all the systems mentioned, the values of the coercive fields and the magnitude of the polarization are similar, indicating that the ferroelectric polarization occurs as a result of energetically similar processes, i.e., similar ferroelectric switching mechanisms.

In the search for suitable organic ligands for the construction of polar crystal structures, chiral alkaloids have proven to be excellent candidates, which is of crucial importance for ferroelectrics. The development of ligands based on 4-quinolones has made considerable progress, and to date more than 10,000 analogs have been prepared by various modifications of the quinoline ring system [18], providing a platform of nearly 200 biologically active alkaloids for material design. Among the best known are those that can be isolated from the *Cinchona* plant, in particular quinine, which has been used for many years to treat malaria [19]. In addition to quinine, its quasi-enantiomer quinidine and its analogs without a methoxy group in the quinoline ring, cinchonidine and cinchonine, are also known [20]. In addition to quinoline fragments, these alkaloids have a polar quinuclidine fragment which is responsible for the ferroelectric properties of organic and organic–inorganic compounds [5,8]. Nevertheless, reports on the use of these alkaloids for molecular ferroelectrics are rather scarce, and only two organic–inorganic compounds have been reported to exhibit ferroelectricity, namely, the quinine–copper(II) complex (H$_2$-quinine)$_2$Cu$_5$Cl$_9$ [21] and the quinine–copper(I) coordination polymer (H-quinine)$_2$Cu$_8$Cl$_{10}$ [22], which achieve a relatively low value of remanent polarization of about 0.1 µC cm^{-2} at a coercive field of 10 kV cm^{-1}. These two examples show that quinine and related alkaloids can be used to tune the dimensionality of metal–organic systems. Polymeric species are formed when only one nitrogen is protonated, whereas isolated complexes are formed when both nitrogen atoms are protonated and the quinine molecule appears as a dication. Our motivation was to investigate whether other alkaloids from the *Cinchona* family are suitable for the preparation of molecular ferroelectrics.

In this work, we selected the cinchoninium cation [Cn, (C$_{19}$H$_{23}$N$_2$O)$^+$], also a member of the *Cinchona* alkaloids, to prepare the metal complexes with zinc(II) chloride and zinc(II) bromide. The properties of the prepared [CnZnCl$_3$]·MeOH and [CnZnBr$_3$]·MeOH complexes were investigated by FTIR-ATR spectroscopy, powder and single-crystal X-ray diffraction, thermal analysis and measurements of polarization as a function of applied voltage. The structure of the prepared cinchoninium–trihalogenozinc(II) complexes was compared with similar compounds in the literature, and it was investigated how intermolecular interactions in these systems influence the formation of polar (ferroelectric) structures.

2. Materials and Methods

2.1. Syntesis of [CnZnX$_3$] MeOH, X = Cl, Br

Cinchoninium chloride dihydrate (85%), CnCl·2H$_2$O, zinc(II) chloride (98%), ZnCl$_2$, and zinc(II) bromide (99%), ZnBr$_2$, were purchased from Sigma Aldrich. The compounds [CnZnCl$_3$]·MeOH and [CnZnBr$_3$]·MeOH were prepared using a solvent-layering technique. Methanolic solution (2 mL; 0.105 M) of cinchoninium chloride dihydrate was covered with acetonitrile solution (2 mL; 0.095 M) of zinc(II) halide. After a few days, rod-shaped crystals of the compound [CnZnX$_3$]·MeOH (X = Cl, Br) formed in a closed test tube. The rod-shaped crystals were separated and briefly dried in air (70% yield).

2.2. Spectroscopic Measreumtns

Attenuated total reflectance Fourier transform infrared (ATR-FTIR) spectra were recorded in the 4000–400 cm^{-1} range using a PerkinElmer FT-IR Frontier spectrometer.

2.3. Thermal Analyis

Thermal analysis was performed with a Shimadzu DTG-60H analyser, in the range from 290 to 1000 K, in a stream of synthetic air at a heating rate of 10 K min^{-1}.

2.4. Single-Crystal and Powder X-Ray Diffraction

The single-crystal X-ray diffraction data for compound [CnZnBr$_3$]·MeOH were collected by ω-scans using Cu-Kα radiation (λ = 1.54179 Å, microfocus tube, mirror monochromator) on a Rigaku XtaLAB Synergy S diffractometer at 293 K. The crystal data, experimental conditions, and final refinement parameters are summarized in Table 1. Data reduction, including the multiscan absorption correction, was performed with the CrysAlisPRO software package (version 1.171.42.62a). The molecular and crystal structures were solved by direct methods using the program SIR2019 [23] and refined by the full-matrix least-squares method based on F^2 with anisotropic displacement parameters for all non-hydrogen atoms (SHELXL-2014/7) [24]. Both programs were operating under the WinGX program package [25]. The positions of the hydrogen atoms attached to the carbon and nitrogen of the cinchoninium cation were found in the electron density map, but were placed in idealized positions. The hydrogen atoms of the methanol molecule were also identified based on a difference Fourier map [O–H distances were restrained to a target value of 0.85 (2) Å]. Geometrical calculations were carried out with PLATON [26] and the figures were generated using the CCDC-Mercury program [27].

Table 1. Crystallographic data and structural refinement details for the compound [CnZnBr$_3$]·MeOH.

Empirical formula	C$_{20}$H$_{27}$Br$_3$N$_2$O$_2$Zn	ρ_{calcd}/g cm^{-3}	1.84
Crystal color, habit	Colorless, rod-like	μ/mm^{-1}	7.784
Formula weight/g mol^{-1}	632.53	θ range/°	4.73–79.62
Crystal system	monoclinic	No. of measured reflections	8820
Space group	$P2_1$	No. of independent reflections	4059
a/Å	9.3262(1)	No. of observed reflections	3948
b/Å	13.1230(2)	No. of parameters, restraints	264, 6
c/Å	9.3436(1)	R_{int}	0.0409
α/°	90	R, wR [I > 2σ(I)]	0.0833, 0.2575
β/°	92.709(1)	R, wR [all data]	0.0842, 0.2623
γ/°	90	Flack parameter	0.005(5)
V/Å^3	1142.26(4)	Goodness of fit	1.266
Z	2	$\Delta\rho_{max}$, $\Delta\rho_{min}$/e Å^{-3}	4.124, −1.778

The Hirshfeld surfaces and 2D fingerprints of the Hirshfeld surface were calculated using the program Crystal Explorer [28]. The normalized contact distances d_{norm} were mapped onto the generated Hirshfeld surface, with red regions indicating close intermolecular contacts (negative d_{norm}), blue regions indicating longer contacts (positive d_{norm}), and white regions with intermolecular contacts corresponding to the van der Waals radii of the atoms in contact (d_{norm} = 0).

The powder X-ray diffraction data (PXRD) were collected in reflection mode with Cu-Kα radiation (λ = 1.54060 Å) on a Malvern Panalytical Empyrean diffractometer using a step size of 0.013° in the 2θ range between 5° and 50°. For temperature-modulated PXRD measurements, a high temperature camera was used.

2.5. Polarization Measurements

Ferroelectric tests based on the positive-up–negative-down method [29,30] at room temperature were measured at a frequency of 10 Hz and under a voltage of 450 V using a ferroelectric analyser TF1000 from AixACCT (Aachen, Germany).

3. Results and Discussion

3.1. Synthesis and Spectroscopic Characterization

In the study carried out, the compound [CnZnCl$_3$]·MeOH was prepared in the form of single crystals through a modified literature method by layering a methanol solution of cinchoninium chloride with an acetonitrile solution of zinc(II) chloride. This method allows the preparation of high-quality single crystals in very high yield. The choice of solvent for dissolving the starting compounds played a decisive role in crystallization as well as in the chemical composition and crystal structure of the products obtained. According to the PXRD analysis, the prepared cinchoninium–trichlorozinc(II) compound corresponds to the structure deposited in the Cambridge Structural Database (CSD) under the reference code JORQIQ (Supplementary Figure S3) [31]. By replacing zinc(II) chloride with zinc(II) bromide in the reaction with cinchoninium chloride (CnCl), a new mononuclear coordination complex of the formula [CnZnBr$_3$]·MeOH was obtained. The compound crystallized as a solvate with a methanol molecule in the crystal structure and was isostructural with the compound [CnZnCl$_3$]·MeOH [32].

The spectrum of cinchoninium chloride dihydrate and the complex compounds [CnZnX$_3$]·MeOH (X = Cl, Br) shows bands at similar wavenumbers (Supplementary Figures S1 and S2). The band at 3490 cm^{-1} is related to the stretching vibration of the O–H bond of the hydroxyl group of cinchonine and methanol [ν(O–H)], while the band at 3134 cm^{-1} corresponds to the stretching of the N–H bond of the protonated quinuclidine nitrogen [ν(N–H)] [33]. The most intense band in the spectrum appears at 778 cm^{-1} and is related to the deformation of the quinolone group [34].

3.2. Thermal Stability

The thermal stability of the compound [CnZnCl$_3$]·MeOH was investigated by TG/DTA analysis (Supplementary Figure S4). In the first stage of decomposition, which starts at 323 K and ends at 387 K, the methanol molecule leaves the crystal structure (mass loss for CH$_3$OH: w_{calc} = 6.41%; w_{exp} = 6.43%). The next step, which corresponds to the cleavage of the ethylene group on the quinuclidine fragment of the cinchonine, begins at 550 K and ends at 600 K (mass loss for C$_2$H$_4$: w_{calc} = 5.61%; w_{exp} = 4.77%). Further heating leads to the complete decomposition of the organic part of the molecule, and apparently the inorganic part also decomposes with the formation of volatile products, so that no residue remains after heat treatment at 1000 K. Compared to traditional ferroelectrics such as Pb(Zr,Ti)O$_3$ and LiNbO$_3$, which are stable at high temperatures above 650 K [35], the thermal stability of [CnZnCl$_3$]·MeOH up to 323 K limits its potential applicability. So far, the highest Curie temperature of ~521 K has been reported for a purely organic ferroelectric crystal based on a phenanthroimidazole derivative [36].

In addition, temperature-modulated PXRD experiments were performed to reveal the structural changes that the original structure of [CnZnBr$_3$]·MeOH undergoes when heated from room temperature to 463 K (Figure 1). Even at a slight heating to 323 K, the PXRD pattern changes, and in addition to the peaks corresponding to the [CnZnBr$_3$]·MeOH phase, additional peaks belonging to a new phase are detected. The correlation of these observations with the TG/DTA experiment on the isostructural compound [CnZnCl$_3$]·MeOH suggests that the structural transformation is related to the removal of the solvent molecule (methanol) from the crystal structure. According to PXRD, the solvent-free form is stable up to 463 K.

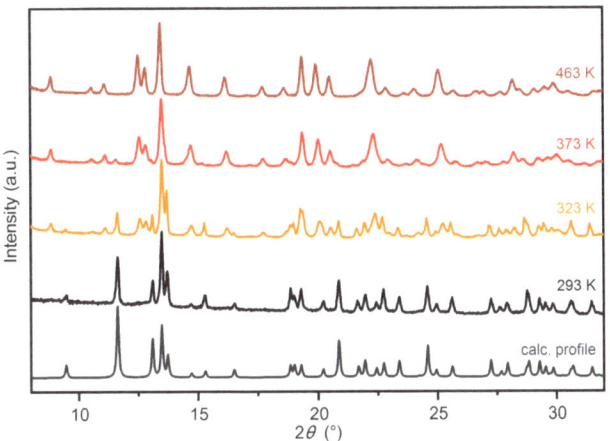

Figure 1. Temperature-modulated PXRD experiments on the initial sample [CnZnBr$_3$]·MeOH (black line) measured at 293 K. The diffractogram simulated from the single-crystal XRD data is given for comparison (gray line).

3.3. Crystal Structures of [CnZnX$_3$]·MeOH, X = Cl, Br

At room temperature, the compounds [CnZnX$_3$]·MeOH, where X = Cl, Br, crystallized in the monoclinic space group $P2_1$. The coordination of the Zn(II) center is a tetrahedron with a nitrogen atom from the quinolone fragment and three halide ligands (Figure 2). In the [CnZnBr$_3$]·MeOH complex, the Zn–N bond length is 2.096(8) Å, which is typical for complexes with a similar coordination polyhedron according to the CSD [31] (values found in the CSD: average 2.063 Å, range 2.010–2.114 Å). The Zn–Br bond lengths are also uniform and lie in the range of 2.318–2.365 Å. Similar values were found in other tetrahedral zinc(II) complexes with nitrogen and bromine atoms in the coordination sphere (average 2.371 Å, range 2.309–2.410 Å). Details of the coordination geometry around zinc are given in Supplementary Tables S1 and S2. Calculation of the continuous symmetry measures (CSM) using the CoSyM calculator [37] shows that the deviation of the geometry from the ideal tetrahedron is 0.40 in the [CnZnBr$_3$]·MeOH complex and only slightly smaller, about 0.34, in [CnZnCl$_3$]·MeOH, indicating that both compounds exhibit some degree of distortion from the ideal tetrahedral geometry.

The intermolecular interactions were analyzed by generating Hirshfeld surfaces with normalized contact distance (d_{norm}) and two-dimensional (d_i vs. d_e) fingerprint plots for the compounds [CnZnBr$_3$]·MeOH and [CnZnCl$_3$]·MeOH. The calculation of the Hirshfeld surface without solvent clearly shows a short interaction between the methanol molecule and the halogen atom of the [CnZnX$_3$] complex (Figure 3a,d). The red regions on the Hirshfeld surface of [CnZnBr$_3$]·MeOH were significantly smaller than those of [CnZnCl$_3$]·MeOH (Figure 3b,e). These red regions represent areas with high electron density and strong interactions, which is consistent with the fingerprint plot analysis. The interaction between the H atoms in the organic component and the halide atoms in the inorganic component was calculated and analyzed from these fingerprint plots (Figure 3c,f). It was found that the H···Br interactions in [CnZnBr$_3$]·MeOH accounted for 36.5%, slightly more than the 34.7% for H···Cl interactions in [CnZnCl$_3$]·MeOH. The contributions of other contact types were similar in both compounds, with the exception of H···H interactions, which were associated with 42.8% and 45.8% of the surface area of [CnZnBr$_3$]·MeOH and [CnZnCl$_3$]·MeOH compounds, respectively.

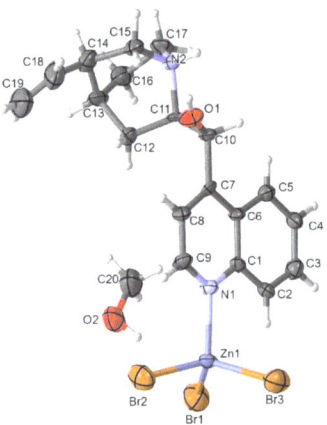

Figure 2. Asymmetric unit in [CnZnBr$_3$]·MeOH with the atom numbering scheme. Displacement ellipsoids are drawn for a probability of 50% and hydrogen atoms are shown as spheres of arbitrary radii.

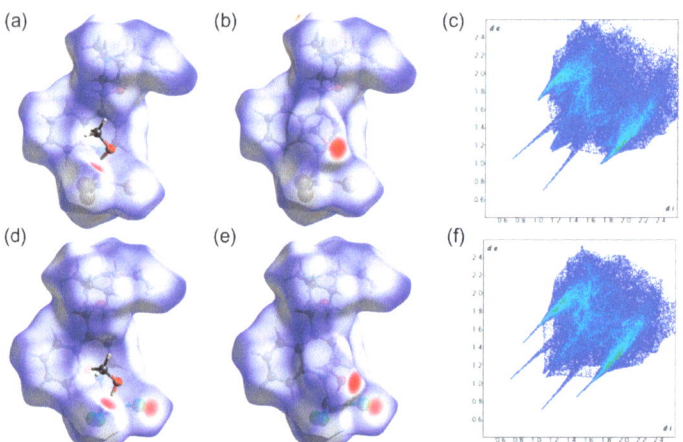

Figure 3. Hirshfeld surface mapped with normalized contact distance of (**a**) complex [CnZnBr$_3$], (**b**) [CnZnBr$_3$]·MeOH; (**d**) complex [CnZnCl$_3$]; (**e**) [CnZnBr$_3$]·MeOH. Fingerprint plots for all contacts in (**c**) [CnZnBr$_3$]·MeOH; (**f**) [CnZnCl$_3$]·MeOH.

The crystal packing of [CnZnBr$_3$]·MeOH is determined by hydrogen bonds between the halide atoms and the hydrogen atoms of the hydroxyl group and the protonated quinuclidine group of the cinchoninium. The methanol molecule mediates the hydrogen bonds between two [CnZnBr$_3$] complexes, and this type of interaction forms a cooperative hydrogen bonding chain along the *c*-axis (Figure 4a). Besides participating as a proton donor and acceptor in the cooperative hydrogen bond, methanol serves as an additional proton donor for the C–H···Br contact propagating along the *a*-axis. Along the polar *b*-axis, there is a hydrogen bonding chain between the two [CnZnBr$_3$] complexes (Figure 4b), which results from a hydroxyl-O–H···Br contact. In addition to these contacts, the two C–H···π interactions also stabilize the crystal packing along the polar axis. Details of these interactions can be found in Supplementary Tables S3 and S4. The intermolecular potentials calculated in Mercury using the UNI force field calculation [38,39] are in good agreement with the Hirshfeld surface analysis [28], which predicts stronger contacts for the [CnZnCl$_3$]·MeOH compound. Each [CnZnCl$_3$] complex forms two contacts of -40.5 kJ mol^{-1} and -33.9 kJ mol^{-1} with two neighboring [CnZnCl$_3$] complexes and one contact of -29.1 kJ mol^{-1} with the methanol molecule. In the [CnZnBr$_3$]·MeOH compound, the mentioned contacts reached energy levels of -38.2 kJ mol^{-1}, -32.2 kJ mol^{-1} and -28.4 kJ mol^{-1}, respectively.

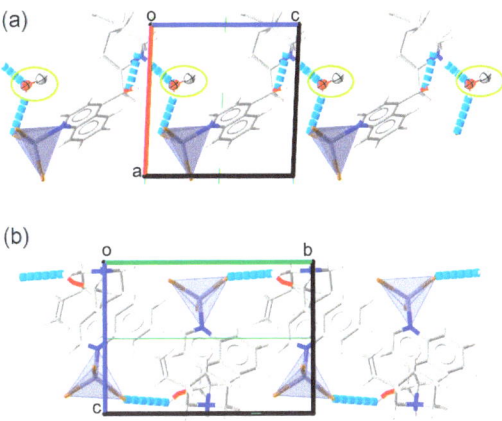

Figure 4. Hydrogen bonding in [CnZnBr$_3$]·MeOH: (**a**) cooperative hydrogen bond chain between methanol and [CnZnBr$_3$]; (**b**) hydrogen bonding along the direction of the polar axis. Hydrogen contacts are shown as blue dashes; the coordination sphere around zinc is shown as a gray tetrahedron. The green line represents the two-fold screw axis.

3.4. Ferroelectric Properties

The measurements of the dependence of the polarization on the voltage at room temperature (298 K) confirmed the ferroelectric polarization in the compound [CnZnCl$_3$]·MeOH. A typical hysteresis loop describing the polarization as a function of the applied voltage is shown in Figure 5 together with the voltage-dependent maxima of the electric current, confirming the macroscopic ferroelectric response due to intrinsic spontaneous polarization. Since the measurements were performed on thin pressed pellets of the compound, the saturation value of the spontaneous polarization under the above conditions is very low and is about 2 nC cm^{-2}. The coercive field for this compound is about 50 kV cm^{-1}.

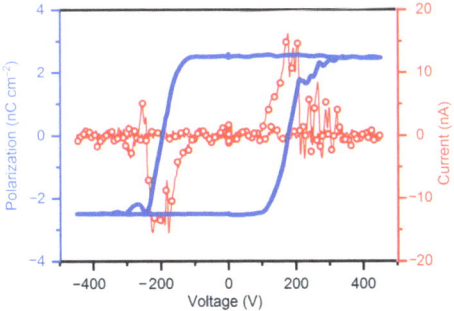

Figure 5. Polarization–voltage loop measured on a [CnZnCl$_3$]·MeOH 50 µm thick pellet sample.

4. Discussion

The results presented in this work include the preparation and characterization of metal–organic compounds based on natural alkaloids from the *Cinchona* group, a cinchoninium cation, and halogenometalates, trichlorozinc(II), and tribromozinc(II). Two compounds were prepared: the cinchoninium–trichlorozinc(II)–methanol complex, which has already been reported in the literature [32], and a new compound, cinchoninum–tribromozinc(II)–methanol. The compounds produced are neutral mononuclear units whose molecular structures are essentially the same. The zinc atom is in a tetrahedral geometry in which one coordination site is occupied by the quinolone nitrogen atom and the other three coordination sites are occupied by halogen anions (Cl– or Br–). The N–H groups of the protonated quinuclidine fragment and the O–H hydroxyl groups are donors of hydrogen bonds to the halogen anions of the zinc(II) tetrahedron. The compounds [CnZnBr$_3$]·MeOH and [CnZnCl$_3$]·MeOH (Ref. code in CSD JORQIQ) [32] are isostructural and crystallize in the polar space group $P2_1$. In addition to these compounds, there are five other structures in the CSD that contain a combination of protonated cinchoninium molecule and halogenometalate [31]. Three of them contain a doubly protonated cinchoninium cation and a tetrachlorometalate anion (M = Cd, Cu) [40–42]. These structures crystallize in the non-polar space group $P2_12_12_1$ (ref. codes in CSD CINCDC [40], FACFEU [41], and WATFUT [42]). Trichlorocobalt(II) complexes with a cinchoninium cation are also deposited in the CSD, namely as a non-solvent complex (ref. code in CSD WUXQIP [43]) and as an ethanol–solvent compound (ref. code in CDS WUXQOV [43]), both crystallizing in the polar monoclinic space group $P2_1$.

In orthorhombic structures with higher symmetry (CINCDC [40], FACFEU [41], WATFUT [42]), aromatic stacking interactions have a stabilizing and directing effect on the crystal packing in addition to hydrogen bonds (see Figure 6a,b). In polar crystal structures (JORQIQ [32], WUXQIP [43], [CnZnBr$_3$]·MeOH), these types of interactions are absent (Figure 6c). Another observation is that in non-polar structures, cinchoninium molecules appear as doubly protonated and isolated cations, whereas in all polar structures with cinchoninium, the quinolone nitrogen is coordinated to the metal center and the organic fragment is part of the complex. The cinchoninium molecules probably have more freedom of movement in the structures in which they occur as isolated cations, and their packing is determined by stacking interactions. In the structures where cinchoninium is part of the metal complex, the molecule is more rigid and other types of interactions, especially hydrogen bonds, are more pronounced. If the influence of the inorganic moiety is taken into account, structures with trihalogenometalate contribute more to the overall dipole moment of the complex, while the tetrahalogenometalate anions are non-polar in ideal tetrahedron geometry.

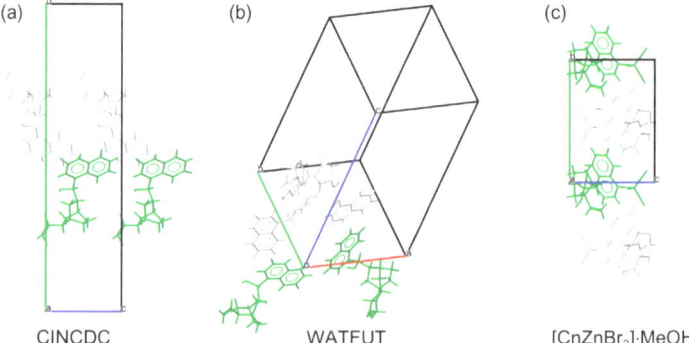

Figure 6. Interactions between cinchoninium molecules related by two-fold screw axes in the structures of (**a**) cinchoninium tetrachlorocadmium(II) dihydrate (CINCDC) [40]; (**b**) bis(cinchoninium) tetrachlorocadmium(II) tetrachlorocopper(II) (WAFFUT) [42]; and (**c**) cinchoninium tribromozinc(II) methanol.

Temperature-modulated experiments corroborating the TG/DTA analysis indicate that the loss of the methanol molecule from the crystal structure causes a structural transformation in the solvent-free structure, which is stable up to 460 K. The structure of this solvent-free phase could be related to that of the cinchoninium–trichlorocobalt(II) complex (ref. code WUXQIP [43]) which crystalizes in a monoclinic $P2_1$ structure with the following unit cell parameters: a = 8.414(1) Å, b = 12.768(2) Å, c = 10.058(2) Å, β = 97.41(2)°, V = 1071.51 Å^3. The intermolecular potentials of the cinchoninium–trichlorocobalt(II) complex calculated in Mercury using the UNI force field calculation [38,39] show two strong interactions with an energy of −41.2 kJ mol^{-1} and two with −33.6 kJ mol^{-1}, of similar value to those found in the structures of [CnZnCl$_3$]·MeOH and [CnZnBr$_3$]·MeOH. The next strongest interaction with 13.8 kJ mol^{-1} is only half as large as the corresponding contacts in [CnZnCl$_3$]·MeOH and [CnZnBr$_3$]·MeOH. These intermolecular potential values indicate that the methanol molecule significantly influences the stability of the crystal packing through intermolecular interactions in the [CnZnCl$_3$]·MeOH and [CnZnBr$_3$]·MeOH complexes.

In supramolecular compounds, the crystal symmetry and thus the physical properties are largely influenced by intermolecular interactions. For example, the choice of solvents with different polarities can trigger specific structural rearrangements. This was observed for the mononuclear iron(III) complex [Fe(sap)(acac)(solvent)] (H$_2$sap = 2-salicylideneaminophenol; acac = acetylacetate; solvent = MeOH, pyridine, DMSO), where the presence of the highly polar DMSO molecule triggered crystallization into the polar crystal structure and ferroelectric properties, while the solvates of pyridine and methanol were non-polar [44]. The compound [CnZnCl$_3$]·MeOH also crystallizes in the polar space group, and a relatively small but detectable ferroelectric polarization value was measured based on ferroelectric positive-up–negative-down tests performed on the pressed bulk sample. A possible explanation for the observed ferroelectricity could be related to the presence of permanent dipoles originating from the solvent molecule (methanol) or the quinuclidine part of the [CnZnCl$_3$] complex, both of which have some freedom of movement in the solid state [33].

Given the low value of remanent polarization, there are several issues that can lead to its underestimation. First, it is not possible to apply the electric field exactly along the polar axis because the measurement was performed on a pressed pellet in which different crystal and domain orientations were present. Secondly, such bulk samples are always accompanied by various defects and depolarization fields. To improve polarization, sample preparation needs to be optimized to obtain a more suitable morphology for ferroelectric testing, e.g., by growing defect-free single crystals or producing thin films for better control of the electric field along the polar axis. Another problem is the search for compatible electrical contacts on the surfaces of crystals or thin films. While conductive pastes allow the deposition of contacts under ambient conditions, they can be destructive to soft metal–organic materials due to the presence of organic solvents. On the other hand, sputtering or thermal evaporation of metal on crystals is difficult to apply, and the associated heating of the surface can also cause undesirable processes such as the destruction of samples or short circuits.

Overall, the advantage of metal–organic compounds over conventional ferroelectrics is that they definitely offer more possibilities to tune the structural properties and can be prepared by simple synthesis protocols at low temperatures. However, it is very difficult to maintain ferroelectric polarization in these materials. Therefore, additional efforts need to be invested in the fabrication of functional devices based on such soft molecular materials.

5. Conclusions

In summary, we describe the preparation of *Cinchona*-based materials in the form of single crystals by a solvent-layering technique. This investigation has shown that the structural changes in the prepared [CnZnCl$_3$]·MeOH and [CnZnBr$_3$]·MeOH complexes already start at moderately low temperatures (323 K), but the metal–organic material is crystalline and stable up to 463 K. The room temperature phase of these isostructural compounds belongs to the ferroelectric polar space group $P2_1$. Macroscopic ferroelectricity, based on polarization–voltage measurements at room temperature, was indeed detected in the [CnZnCl$_3$]·MeOH complex, confirming that the natural alkaloids of the *Cinchona* family are suitable candidates for the design of molecular ferroelectrics. The study of intermolecular interactions in a small group of structures with cinchoninium cations and halogenometallates deposited in CSD shows that the formation of polar structures occurs with monoprotonated cinchoninium cations coordinating the metal center, in contrast to structures with double protonated cinchoninium cations. Furthermore, a detailed insight into the crystal structures of these compounds and calculations of the intermolecular potentials revealed the significant influence of solvent molecules (methanol) in the supramolecular arrangement, mediated by hydrogen bonding.

Supplementary Materials: The following supporting information can be downloaded at: https://www.mdpi.com/article/10.3390/cryst14110978/s1, Figures S1 and S2: IR spectra; Figure S3: PXRD; Figure S4: TG/DTA; Figure S5: Hirshfeld surface analysis; Tables S1 and S2: selected bond lengths and angles; Table S3 and S4: intra- and intermolecular interaction analysis.

Author Contributions: Conceptualization, L.A.D.; methodology, M.O.; formal analysis, M.O.; investigation, M.O.; resources, L.A.D.; data curation, M.O.; writing—original draft preparation, L.A.D.; supervision, L.A.D.; funding acquisition, L.A.D. All authors have read and agreed to the published version of the manuscript.

Funding: This research was funded by the Croatian Science Foundation, grant number UIP-2019-04-7433.

Data Availability Statement: The deposition number CCDC 2389956 contains the supplementary crystallographic data for this article, including structure factors. Other data are available upon reasonable request.

Conflicts of Interest: The authors declare no conflicts of interest.

References

1. Nuraje, N.; Su, K. Perovskite ferroelectric nanomaterials. *Nanoscale* **2013**, *5*, 8752–8780. [CrossRef] [PubMed]
2. Lallart, M. *Ferroelectrics—Applications*; IntechOpen: Rijeka, Croatia, 2011.
3. Tayi, A.S.; Kaeser, A.; Matsumoto, M.; Aida, T.; Stupp, S.I. Supramolecular ferroelectrics. *Nat. Chem.* **2015**, *7*, 281–294. [CrossRef] [PubMed]
4. Liu, J.-C.; Liao, W.-Q.; Li, P.-F.; Tang, Y.-Y.; Chen, X.-G.; Song, X.-J.; Zhang, H.-Y.; Zhang, Y.; You, Y.-M.; Xiong, R.-G. A Molecular Thermochromic Ferroelectric. *Angew. Chem. Int. Ed.* **2020**, *59*, 3495–3499. [CrossRef] [PubMed]
5. González-Izquierdo, P.; Fabelo, O.; Cañadillas-Delgado, L.; Beobide, G.; Vallcorba, O.; Salgado-Beceiro, J.; Sánchez-Andújar, M.; Martin, C.; Ruiz-Fuentes, J.; García, J.E.; et al. ((R)-(−)-3-Hydroxyquinuclidium)[FeCl$_4$]; a plastic hybrid compound with chirality, ferroelectricity and long range magnetic ordering. *J. Mater. Chem. C* **2021**, *9*, 4453–4465. [CrossRef]
6. Li, J.-Y.; Xu, Q.-L.; Ye, S.-Y.; Tong, L.; Chena, X.; Chen, L.-Z. A multiaxial molecular ferroelectric with record high T$_C$ designed by intermolecular interaction modulation. *Chem. Commun.* **2021**, *57*, 943–946. [CrossRef]
7. Xu, L.; Zhang, Y.; Jiang, H.-H.; Zhang, N.; Xiong, R.-G.; Zhang, H.-Y. Solvent Selective Effect Occurs in Iodinated Adamantanone Ferroelectrics. *Adv. Sci.* **2022**, *9*, 2201702. [CrossRef]
8. Harada, J.; Shimojo, T.; Oyamaguchi, H.; Hasegawa, H.; Takahashi, Y.; Satomi, K.; Suzuki, Y.; Kawamata, J.; Inabe, T. Directionally tunable and mechanically deformable ferroelectric crystals from rotating polar globular ionic molecules. *Nat. Chem.* **2016**, *8*, 946–952. [CrossRef]
9. Puškarić, A.; Dunatov, M.; Jerić, I.; Sabljić, I.; Androš Dubraja, L. Room temperature ferroelectric copper(ii) coordination polymers based on amino acid hydrazide ligands. *New J. Chem.* **2022**, *46*, 3504–3511. [CrossRef]
10. Dunatov, M.; Puškarić, A.; Androš Dubraja, L. Multi-Stimuli Responsive (L-tartrato)oxovanadium(V) Complex Salt with Ferroelectric Switching and Thermistor Properties. *J. Mater. Chem. C* **2023**, *11*, 2880–2888. [CrossRef]
11. Centore, R.; Fusco, S.; Capone, F.; Causà, M. Competition between Polar and Centrosymmetric Packings in Molecular Crystals: Analysis of Actual and Virtual Structures. *Cryst. Growth Des.* **2016**, *16*, 2260–2265. [CrossRef]
12. Li, P.-F.; Tang, Y.-Y.; Wang, Z.-X.; Ye, H.-Y.; You, Y.-M.; Xiong, R.-G. Anomalously rotary polarization discovered in homochiral organic ferroelectrics. *Nat. Commun.* **2016**, *7*, 13535. [CrossRef] [PubMed]
13. Ye, H.-Y.; Tang, Y.-Y.; Li, P.-F.; Liao, W.-Q.; Gao, J.-X.; Hua, X.-N.; Cai, H.; Shi, P.-P.; You, Y.-M.; Xiong, R.-G. Metal-free three-dimensional perovskite ferroelectrics. *Science* **2018**, *361*, 151–155. [CrossRef] [PubMed]
14. Sun, Z.; Chen, T.; Luo, J.; Hong, M. Bis(imidazolium) L-Tartrate: A Hydrogen-Bonded Displacive-Type Molecular Ferroelectric Material. *Angew. Chem. Int. Ed.* **2012**, *51*, 3871–3876. [CrossRef]

15. Song, X.-J.; Tang, S.-Y.; Chen, X.-G.; Ai, Y. Chemical design of homochiral heterocyclic organic ferroelectric crystals. *Chem. Commun.* **2022**, *58*, 10361. [CrossRef]
16. Song, X.-J.; Chen, X.-G.; Liu, J.-C.; Liu, Q.; Zeng, Y.-P.; Tang, Y.-Y.; Li, P.-F.; Xiong, R.-G.; Liao, W.-Q. Biferroelectricity of a homochiral organic molecule in both solid crystal and liquid crystal phases. *Nat. Commun.* **2022**, *13*, 6150. [CrossRef]
17. Liu, Y.-L.; Ge, J.-Z.; Wang, Z.-X.; Xiong, R.-G. Metal–organic ferroelectric complexes: Enantiomer directional induction achieved above-room-temperature homochiral molecular ferroelectrics. *Inorg. Chem. Front.* **2020**, *7*, 128–133. [CrossRef]
18. Banerjee, S.; Prasad, P.; Hussain, A.; Khan, I.; Kondaiah, P.; Chakravarty, A.R. Remarkable photocytotoxicity of curcumin in HeLa cells in visible light and arresting its degradation on oxovanadium(iv) complex formation. *Chem. Commun.* **2012**, *48*, 7702–7704. [CrossRef]
19. Castillo-Blum, S.E.; Barba-Behrens, N. Coordination chemistry of some biologically active ligands. *Coord. Chem. Rev.* **2000**, *196*, 3–30. [CrossRef]
20. Hoffmann, H.M.R.; Frackenpohl, J. Recent Advances in Cinchona Alkaloid Chemistry. *Eur. J. Org. Chem.* **2004**, *2004*, 4293–4312. [CrossRef]
21. Zhao, H.; Qu, Z.-R.; Ye, Q.; Abrahams, B.F.; Wang, Y.-P.; Liu, Z.-G.; Xue, Z.; Xiong, R.-G.; You, X.-Z. Ferroelectric Copper Quinine Complexes. *Chem. Mater.* **2003**, *15*, 4166–4168. [CrossRef]
22. Qu, Z.-R.; Chen, Z.-F.; Zhang, J.; Xiong, R.-G.; Abrahams, B.F.; Xue, Z.-L. The First Highly Stable Homochiral Olefin−Copper(I) 2D Coordination Polymer Grid Based on Quinine as a Building Block. *Organometallics* **2003**, *22*, 2814–2816. [CrossRef]
23. Burla, M.C.; Caliandro, R.; Carrozzini, B.; Cascarano, G.L.; Cuocci, C.; Giacovazzo, C.; Mallamo, M.; Mazzone, A.; Polidori, G. Crystal Structure Determination and Refinement via SIR2014. *J. Appl. Crystallogr.* **2015**, *48*, 306–309. [CrossRef]
24. Sheldrick, G.M. Crystal Structure Refinement with SHELXL. *Acta Crystallogr. Sect. C Struct. Chem.* **2015**, *C71*, 3–8. [CrossRef] [PubMed]
25. Farrugia, L.J. WinGX and ORTEP for Windows: An Update. *J. Appl. Crystallogr.* **2012**, *45*, 849–854. [CrossRef]
26. Spek, A.L. Structure Validation in Chemical Crystallography. *Acta Crystallogr. Sect. D Biol. Crystallogr.* **2009**, *D65*, 148–155. [CrossRef]
27. Macrae, F.; Edgington, P.R.; McCabe, P.; Pidcock, E.; Shields, G.P.; Taylor, R.; Towler, M.; van de Streek, J. Mercury: Visualization and Analysis of Crystal Structures. *J. Appl. Crystallogr.* **2006**, *39*, 453–457. [CrossRef]
28. Spackman, P.R.; Turner, M.J.; McKinnon, J.J.; Wolff, S.K.; Grimwood, D.J.; Jayatilaka, D.; Spackman, M.A. CrystalExplorer: A program for Hirshfeld surface analysis, visualization and quantitative analysis of molecular crystals. *J. Appl. Cryst.* **2021**, *54*, 1006–1011. [CrossRef]
29. Sawaguchi, E. Ferroelectricity versus Antiferroelectricity in the Solid Solutions of $PbZrO_3$ and $PbTiO_3$. *J. Phys. Soc. Jpn.* **1953**, *8*, 615–629. [CrossRef]
30. Suzuki, E.; Shiozaki, Y. Ferroelectric displacement of atoms in Rochelle salt. *Phys. Rev. B Condens. Matter Mater. Phys.* **1996**, *53*, 5217. [CrossRef]
31. Groom, C.R.; Bruno, I.J.; Lightfoot, M.P.; Ward, S.C. The Cambridge Structural Database. *Acta Cryst. B* **2016**, *72*, 171–179. [CrossRef]
32. Hubel, R.; Polborn, K.; Beck, W. Cinchona Alkaloids as Versatile Ambivalent Ligands—Coordination of Transition Metals to the Four Potential Donor Sites of Quinine. *Eur. J. Inorg. Chem.* **1999**, *1999*, 471–482. [CrossRef]
33. Dunatov, M.; Puškarić, A.; Pavić, L.; Štefanić, Z.; Androš Dubraja, L. Electrically Responsive Structural Transformations Triggered by Vapour and Temperature in a Series of Pleochroic Bis(oxalato)chromium(iii) Complex Salts. *J. Mater. Chem. C* **2022**, *10*, 8024–8033. [CrossRef]
34. Özel, A.E.; Büyükmurat, Y., Akyüz, S. Molecular structure and vibrational assignment of 2-,4-,6-methylquinoline by density functional theory (DFT) and ab initio Hartree-Fock (HF) calculations. *Vib. Spectrsc.* **2006**, *42*, 325–332. [CrossRef]
35. Mikolajick, T.; Slesazeck, S.; Mulaosmanovic, H.; Park, M.H.; Fichtner, S.; Lomenzo, P.D.; Hoffmann, M.; Schroeder, U. Next generation ferroelectric materials for semiconductor process integration and their applications. *J. Appl. Phys.* **2021**, *129*, 100901. [CrossRef]
36. Dutta, S.; Vikas; Yadav, A.; Boomishankar, R.; Bala, A.; Kumar, V.; Chakraborty, T.; Elizabeth, S.; Munshi, P. Record-high thermal stability achieved in a novel single-component all-organic ferroelectric crystal exhibiting polymorphism. *Chem. Commun.* **2019**, *55*, 9610–9613. [CrossRef]
37. Pinsky, M.; Avnir, D. Continuous symmetry measures. 5. The classical polyhedra. *Inorg. Chem.* **1998**, *37*, 5575–5582. [CrossRef]
38. Gavezzotti, A. Are Crystal Structures Predictable? *Acc. Chem. Res.* **1994**, *27*, 309–314. [CrossRef]
39. Gavezzotti, A.; Filippini, G. Geometry of the Intermolecular X-H.cntdot..cntdot..cntdot.Y (X, Y = N, O) Hydrogen Bond and the Calibration of Empirical Hydrogen-Bond Potentials. *J. Phys. Chem.* **1994**, *98*, 4831–4837. [CrossRef]
40. Oleksyn, B.J.; Stadnicka, K.M.; Hodorowicz, S.A. The crystal structure and absolute configuration of cinchoninium tetrachlorocadmate(II) dihydrate. *Acta Crystallogr. Sect. B Struct. Crystallogr. Cryst. Chem.* **1978**, *34*, 811–816. [CrossRef]
41. Weselucha-Birczynska, A.; Oleksyn, B.J.; Hoffmann, S.K.; Sliwinski, J.; Borzecka-Prokop, B.; Goslar, J.; Hilczer, W. Flexibility of $CuCl_4$-tetrahedra in Bis[Cinchoninium Tetrachlorocuprate(II)]trihydrate Single Crystals. X-ray Diffraction and EPR Studies. *Inorg. Chem.* **2001**, *40*, 4526–4533. [CrossRef]

42. Weselucha-Birczynska, A.; Oleksyn, B.J.; Sliwinski, J.; Goslar, J.; Hilczer, W.; Hoffmann, S.K. Crystal structure and EPR studies of (cinchonineH$_2$)$_2$(CdCl$_4$)(Cd/CuCl$_4$) crystals with thermochromic and Jahn–Teller effect. *J. Mol. Struct.* **2005**, *751*, 109–120. [CrossRef]
43. Skorska, A.; Oleksyn, B.J.; Sliwinski, J. Cobalt Complex of Cinchonine: Intermolecular Interactions in Two Crystalline Modifications. *Enantiomer* **2002**, *7*, 295–303. [CrossRef] [PubMed]
44. Kobayashi, F.; Akiyoshi, R.; Kosumi, D.; Nakamura, M.; Lindoy, L.F.; Hayami, S. Solvent vapor-induced polarity and ferroelectricity switching. *Chem. Commun.* **2020**, *56*, 10509–10512. [CrossRef] [PubMed]

Disclaimer/Publisher's Note: The statements, opinions and data contained in all publications are solely those of the individual author(s) and contributor(s) and not of MDPI and/or the editor(s). MDPI and/or the editor(s) disclaim responsibility for any injury to people or property resulting from any ideas, methods, instructions or products referred to in the content.

Article

CO$_2$ Promoting Polymorphic Transformation of Clarithromycin: Polymorph Characterization, Pathway Design, and Mechanism Study

Lixin Hou [1], Dingding Jing [2], Yanfeng Wang [1] and Ying Bao [1,*]

[1] School of Chemical Engineering and Technology, Tianjin University, Tianjin 300072, China; hlx20000117@tju.edu.cn (L.H.); wyf0815@tju.edu.cn (Y.W.)
[2] Asymchem Life Science (Tianjin) Co., Ltd., Tianjin 300072, China; jingdingding@asymchem.com.cn
* Correspondence: yingbao@tju.edu.cn

Citation: Hou, L.; Jing, D.; Wang, Y.; Bao, Y. CO$_2$ Promoting Polymorphic Transformation of Clarithromycin: Polymorph Characterization, Pathway Design, and Mechanism Study. *Crystals* **2024**, *14*, 394. https://doi.org/10.3390/cryst14050394

Academic Editor: Sławomir Grabowski

Received: 3 April 2024
Revised: 17 April 2024
Accepted: 19 April 2024
Published: 24 April 2024

Copyright: © 2024 by the authors. Licensee MDPI, Basel, Switzerland. This article is an open access article distributed under the terms and conditions of the Creative Commons Attribution (CC BY) license (https://creativecommons.org/licenses/by/4.0/).

Abstract: Carbon dioxide (CO$_2$) has a wide range of uses such as food additives and raw materials for synthetic chemicals, while its application in the solid-state transformation of pharmaceutical crystals is rare. In this work, we report a case of using 1 atm CO$_2$ as an accelerator to promote the polymorphic transformation of clarithromycin (CLA). Initially, crystal structures of Form 0′ and three solvates were successfully determined by single crystal X-ray diffraction (SCXRD) analysis for the first time and found to be isomorphous. Powder X-ray diffraction (PXRD) and thermal analysis indicated that the solvate desolvates and transforms into the structurally similar non-solvated Form 0′ at room temperature to ~50 °C. Form 0′ and Form II are monotropically related polymorphs with Form II being the most stable. Subsequently, the effect of CO$_2$ on the transformation of CLA solvates to Form II was studied. The results show that CO$_2$ can significantly facilitate the transformation of Form 0′ to Form II, despite no significant effect on the desolvation process. Finally, the molecular mechanism of CO$_2$ promoting the polymorphic transformation was revealed by the combination of the measurement of adsorption capacity, theoretical calculations as well as crystal structure analysis. Based on the above results, a new pathway of preparing CLA Form II was designed: transform CLA solvates into Form 0′ in 1 atm air at 50 °C followed by the transformation of Form 0′ to Form II in 1 atm CO$_2$ at 50 °C. This work provides a new idea for promoting the phase transformation of pharmaceutical crystals as well as a new scenario for the utilization of CO$_2$.

Keywords: clarithromycin solvate; polymorphic transformation; CO$_2$; adsorption

1. Introduction

Polymorphism is a common phenomenon present in active pharmaceutical ingredients (APIs). To prevent the change in drug efficacy, bioavailability, and toxicity during a long period of storage, the most stable Form Is typically prioritized in the drug formulations on the market. Some APIs can only be produced by the desolvation of their solvates. Under non-solvent conditions, channel solvates may transform to either a structurally different stable polymorph after desolvation such as the solvated forms of azoxystrobin [1], Form 2, Form 5, Form 7, and Form 9, or a structurally similar metastable polymorph such as the solvates of methyl cholate [2]. In the latter case, further transformation to a stable polymorph is usually necessary for APIs. Polymorphic transformation under non-solvent conditions is often more difficult in comparison to solvent-mediated transformation because of the restricted molecular migration and high resistance when rearranging. As a result, some metastable polymorphs may remain a long period without transforming to a stable one such as the metastable Form II of isotactic polybutene-1, which cannot totally transForm Into the stable Form I even after a remarkably long annealing time [3]. Transformation from a metastable polymorph to a stable one requires overcoming an energy barrier. Providing energy or lowering the energy barrier will facilitate the process. Heating, mechanical force, seeding, and high-pressure CO_2 are commonly used to induce the transformation. For example, the gold(I) complex, $C_{24}H_{15}AuF_3N \cdot 2(CH_4O)$, exhibits a reversible solid-state polymorphic transformation by mechanical stimulation [4]. Hanna et al. [5] reported a single-crystal-to-single-crystal phase transition of the 2D uranium MOF NU-1302 after supercritical CO_2 activation. They hypothesized that CO_2 pressure induced the adjacent sheets in the crystal structure to shift from their closed conformation to a more open stacking arrangement. Yu et al. [6] found that the plasticization effect of supercritical CO_2 resulted in a decreased energy barrier of phase transition, which promoted the transformation from the α Form to β Form of syndiotactic polystyrene. Several green processes affected by CO_2 are reported such as the CO_2-solvated liquefaction of polyethylene glycol at low temperatures [7] and the CO_2-induced glassification of sucrose octaacetate and its implications in the spontaneous release of the drug from drug-excipient composites [8]. Furthermore, it was found that by adjusting the annealing time in supercritical CO_2, mixtures of polymorphs Form I and Form I' of isotactic polybutene-1 can be obtained with different crystal phase ratios [9]. The solid-state transformation of APIs induced by external factors, especially atmospheric pressure CO_2, has rarely been reported, and an in-depth understanding of its molecular mechanism is still lacking.

Clarithromycin (CLA, $C_{38}H_{69}NO_{13}$, Figure 1) is a common macrolide antibiotic that is mainly used in the treatment of upper respiratory tract infections, lower respiratory tract infections, and skin and soft tissue infections caused by bacteria. CLA is known to exist in three polymorphs, termed Form 0' [10], Form I [11], and Form II [12] as well as many solvates such as hydrate, methanol solvate, ethanol solvate (CLA-Eth solvate), acetonitrile solvate, isopropyl acetate solvate, and tetrahydrofuran solvate, etc. [12–17]. Drugs currently on the market are formulated from the thermodynamically more stable Form II. Form I is crystallized in orthorhombic space group $P2_12_12$, while Form II is in orthorhombic space group $P2_12_12_1$. The two polymorphs are monotropically related, with Form II the most stable [18]. Form I converts to Form II at 130 °C~132 °C [11,13,15]. Form II melts at 227 °C~233 °C [11,15,19,20]. Form II and the methanol solvate are isomorphic [12]. No reports in the literature have been found on the crystal structure of Form 0' or the stability relationship between Form 0' and Form II. It has been reported that under non-solvent conditions, Form II is obtained directly by the transformation of solvates or by undergoing an intermediate Form I [18,21]. Usually, the process takes about 18 h under vacuum and/or high temperature (70 °C~110 °C). Tian et al. [12] proposed that high pressure CO_2 could promote the solid-state transformation of the CLA-Eth solvate at room temperature. Their experiments showed that complete transformation directly to Form II took about 6 days at 6.8 atm CO_2 atmosphere but only 4 h at 23.8 atm. They speculated that CO_2 molecules cause CLA molecules to slide over one another, resulting in the break of the intermolecular hydrogen bonds between ethanol and CLA, forcing the CLA molecules to move closer together to form Form II. Additionally, their experiments at 1 atm CO_2 showed that the CLA-Eth solvate did not exhibit phase change, while Form I occasionally displayed the onset of phase transformation over a period of ~24 h. They also pointed out that prior to reaching Form II, Form I first transformed to the isomorphic Form 0' of the CLA-Eth solvate.

Currently reported methods of the preparation of Form II from CLA solvates require high temperature, high pressure, or vacuum, which means a large amount of energy consumption. Moreover, the mechanism of the phase transformation induced by CO_2 is still not clear. On the basis of the experimental research in this work, we propose a method for the transformation of CLA solvates to Form II under 1 atm CO_2. Through SCXRD and PXRD characterizations, thermal analysis, measurements of adsorption capacity, and theoretical calculations, the mechanism of the phase transformation promoted by CO_2 is revealed. This work provides an energy-saving, time-saving, and environmentally friendly route for the preparation of CLA Form II, expands the perspective of promoting the polymorphic transformation of APIs, and opens up a new avenue for the resource utilization of CO_2.

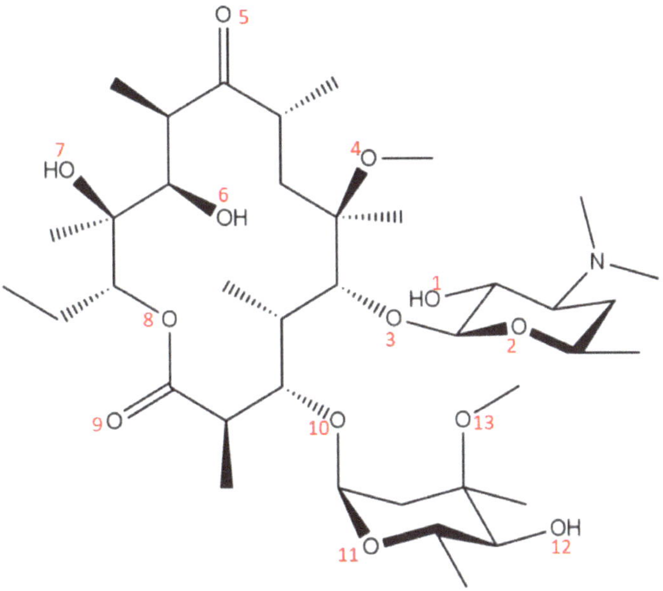

Figure 1. Molecular structure of CLA. Oxygen at 13 different locations in CLA was labeled in red.

2. Experimental Section

2.1. Materials

CLA Form II (purity > 98%) was purchased from Heowns Technology Co. Ltd. (Tianjin, China). N-Propyl acetate (PAC), isopropyl acetate (IPA), ethyl butyrate (EB), and dimethyl carbonate (DMC) (purity > 99%) were purchased from Chemart Chemical Technology Co. Ltd. (Tianjin, China) and used without further purification. CO_2 (purity > 99.999%) was purchased from Bolimin Technology Co. Ltd. (Tianjin, China).

2.2. Preparation of Solvates and Form 0′

A total of 3.0 g of CLA raw material was dissolved in 50 mL PAC, 40 mL IPA, and 30 mL EB, respectively, at 80 °C. The obtained solutions were cooled to 15 °C in 3 h, and the clarithromycin n-propyl acetate (CLA-PAC) solvate, clarithromycin isopropyl acetate (CLA-IPA) solvate, and clarithromycin ethyl butyrate (CLA-EB) were crystallized, respectively. The resulting slurry was processed in three ways. (1) Removing the crystals with a spoon and placing them on a piece of filter paper. After 5–10 min, there was no obvious solvent on the crystal surface. (2) Filtered under vacuum at 20 °C for 0.5 h. (3) The crystals obtained by process (2) was dried at 50 °C under vacuum or atmospheric pressure for 9~19 h. The obtained products were immediately sampled and characterized by PXRD and thermal analysis. The results showed that the products proceeded by the three methods were the corresponding CLA solvates, a mixture of the corresponding CLA solvate and Form 0′, and pure Form 0′, respectively.

2.3. Single Crystals Preparation of Solvates and Form 0′

Single crystals of CLA solvates were obtained through slowly cooling crystallization. A total of 3.0 g CLA raw material and 50 mL PAC were dissolved at 80 °C. Then, the solution was slowly cooled to 15 °C in 6 h to obtain large single crystals of the CLA-PAC solvate. Similarly, single crystals of the CLA-IPA and CLA-EB solvates were obtained by slowly cooling IPA and EB solutions of CLA, respectively. A saturated DMC solution of CLA was placed at room temperature and atmospheric pressure for slow evaporation. About three weeks later, high-quality single crystals of CLA Form 0′ were obtained. In order to avoid desolvation, the obtained solvate single crystals were immediately subjected to single crystal X-ray diffraction analysis after being removed from solution.

2.4. Phase Transformation Experiments

In this section, the CLA-PAC solvate was selected as a representative of the three CLA solvates to perform the experiments on the phase transformation of the solvates and was prepared by method (2) in Section 2.2.

2.4.1. Phase Transformation Experiments in Air and Vacuum

The phase transformation experiment of the solvate was performed under vacuum at 50 °C. Crystalline phases of the crystals as a function of time were monitored using DSC measurements by sampling at 2 h, 3 h, 4 h, 7 h, 8 h, and 9 h.

The phase transformation experiment of the solvate was also performed in 1 atm air at 50 °C. Crystalline phases of the crystals as a function of time were monitored through TGA measurements. Samples of the crystals were taken at a time interval of 2 h.

2.4.2. Phase Transformation Experiments in CO_2 Atmosphere

The phase transformation experiment of the solvate was carried out in 1 atm CO_2 at 50 °C. The CLA-PAC solvate was placed in a custom built micro-vacuum box. The micro-vacuum box was first vacuumed for 2 min using a vacuum line followed by the introduction of dry CO_2 until the pressure reached 1 atm. Afterward, the micro-vacuum box was placed in a thermostat at 50 °C. Crystalline phases of the crystals as a function of time were monitored using TGA measurements by sampling at a time interval of 2 h. After each sampling, the micro-vacuum box was vacuumed and CO_2 was introduced again to restore the CO_2 atmosphere of 1 atm.

The phase transformation experiment of Form 0' was also carried out in 1 atm CO_2 at 50 °C. Crystalline phases of the crystals as a function of time were monitored using PXRD analysis by sampling at a time interval of 1 h. Form 0' used in the experiment was prepared by method (2) in Section 2.2.

The custom built micro-vacuum box used in the experiment is shown in Figure S1. Its outer diameter, inner diameter, inner height, and outer height were 50 mm, 30 mm, 40 mm, and 100 mm, respectively. The micro-vacuum box was equipped with an intake valve, an outlet valve, a pressure gauge, an intake port, an outlet port, and a sample cell. The intake port was connected to the CO_2 cylinder through a soft rubber tube. In the experiment, powdered samples were placed on a piece of plastic wrap and then put in the sample cell. Please note that the powder was lightly covered with plastic wrap to prevent the powder from splashing during decompression.

2.5. Characterization

2.5.1. Single Crystal X-ray Diffraction (SCXRD)

Suitable single crystals were selected and analyzed using a Rigaku mm007 Saturn70 diffractometer. The diffraction data of single crystals of the three solvates (CLA-PAC, CLA-IPA, and CLA-EB solvates) and Form 0' were collected at −165 °C and 25 °C, respectively. The structures were resolved using Olex2 [22] and SHELXT [23], and refined by the least squares methods using SHELXT [24]. The absolute configuration of all structures were determined through comparison with the structure of clarithromycin Form II (deposited at the CCDC, deposition number 780856 [12]).

2.5.2. Powder X-ray Diffraction (PXRD)

All PXRD data were collected on a Rigaku D/max-2500 diffractometer using Cu Kα radiation (λ = 1.54056 Å, 40.0 kV, and 200 mA) with a scanning rate of 8°·min^{-1} between 2° and 35° (2θ). For the variable temperature PXRD (VT-PXRD) trials, the sample was heated from 30 °C to 120 °C with a heating rate of 10 °C·min^{-1} and stabilized for 5 min before the measurements were taken.

2.5.3. Thermal Analysis

Differential scanning calorimetry (DSC) analysis was performed using DSC 1/500 (Mettler Toledo, Greifensee, Switzerland). A quantity of 5~10 mg of powder was added to an aluminum pan and heated at a rate of 10 °C·min^{-1} with a nitrogen flux of 50 mL·min^{-1}. Thermogravimetric analysis (TGA) was carried out under a nitrogen flow of 20 mL·min^{-1} using a TGA/DSC STARe (Mettler Toledo, Greifensee, Switzerland). Then, 5~10 mg of powder was placed in a ceramic crucible and heated at a rate of 10 °C·min^{-1}.

2.5.4. Hot Stage Microscopy (HSM)

The thermal behaviors of the CLA-PAC solvate were observed on a Kofler hot stage microscope (Reichert Thermovar, Vienna, Austria) under an optical microscope mounted with a charge-coupled device (CCD) camera. The samples were placed on the sample stage and heated from room temperature to 190 °C at a rate of 5 °C·min^{-1}. The temperature of the hot stage was monitored with a central processor (TMS 94, Linkam Scientific Instruments Ltd., Surrey, UK). Morphological changes during heating were recorded with the CCD camera.

2.5.5. Measurement of the Physical Adsorption Capacity

CO_2 adsorption at 25 °C and 50 °C was conducted using an ASAP 2020 PLUS HD88 surface area analyzer (Micromeritics, Norcross, GA, USA) to obtain the adsorption capacity of Form 0′ and Form II. All samples were pretreated at 60 °C for 3 h.

2.5.6. Measurement of Chemical Adsorption Capacity

The temperature-programmed desorption (TPD) of CO_2 was measured by using a Micromeritics AutoChem 2920 equipped with a thermal conductivity detector (TCD). The sample was pretreated in a helium flow for 3 h at 60 °C, then cooled to 50 °C to conduct CO_2 adsorption. A mixture of CO_2 and helium (40 mL/min, CO_2 volume fraction 90%) was vented for 1 h until saturation. Afterward, the helium flow was switched (30–50 mL/min) to purge 1 h to remove the weakly physically adsorbed CO_2 on the surface of the sample, then the temperature was increased to 200 °C at a rate of 10 °C·min^{-1} in a helium flow to desorb CO_2. It should be noted that Form II was used in the TPD test since Form 0′ would undergo polymorphic transformation during the test temperature range of 50 °C~200 °C, causing unstable signals.

2.6. Computational Method

2.6.1. Molecular Electrostatic Potential Surface (MEPS)

Geometry optimization and the wave function computation of the CLA molecule were carried out using both density functional theory (DFT) and B3LYP/6-311G** methods using Gaussian 09 [25]. Furthermore, Multiwfn 3.7 [26] was used to calculate the MEPS to 0.001 Bohr^{-3} electron density equivalence surface for analysis. Finally, Visual Molecular Dynamics (VMD) 1.9.3 software [27] was used to visualize the results.

2.6.2. Adsorption Energy Calculation

In order to obtain an adsorption model of CO_2 on Form 0′ and calculate the adsorption energy, the software package Gaussian 09 [25] was used for all calculations. The adsorption model consisted of two CLA molecules extracted from a 'Z' shape structure in Form 0′ (as shown in Figure 10a) and one CO_2 molecule. The CO_2 molecule was initially placed at O7 of the CLA molecule. The B3LYP [28,29] function and D3BJ [28] dispersion correction were used. Considering the calculation time due to a large number of atoms (245 atoms in all), the 6-31G* basis set was used to determine the optimized adsorption model and the vibration frequencies of the molecules [30]. Grid data generation was performed using Multiwfn Version 3.8 (development version) [26,31]. The adsorption energy was calculated by the following equation:

$$E_{adsorb} = E_{total} - E_{CLA} - E_{CO2} \qquad (1)$$

where E_{adsorb} is the adsorption energy of the CO_2 on the CLA molecule, E_{total} is the system's total energy once adsorption is completed; E_{CLA} and E_{CO2} are the energy of the CLA molecule and CO_2 molecule before adsorption, respectively. A negative adsorption energy indicates that the adsorption process can take place.

2.6.3. Crystal Habit Prediction

Universal force field package COMPASS [32] in the Discover module was used as the initialization force field in Material Studio (MS) 8.0. Smart Minimizer was selected for cell configuration optimization. After optimization, cell parameters changed within 5% compared with those before optimization. The BFDH [33] method in the Morphology module was used in predicting the crystal habit of Form 0′.

3. Results and Discussion

3.1. Crystal Structure Analysis of Form 0′ and Solvates

High-quality single crystals of a CLA non-solvated form and three CLA solvates, namely the CLA-PAC solvate, CLA-IPA solvate, and CLA-EB solvate, were prepared and their crystal structures were successfully resolved for the first time. The crystallographic data and structural refinement parameters are given in detail in Table 1. It can be seen that the non-solvated form and the three solvates were all in the orthorhombic system with a space group of $P2_12_12_1$, possessed the same formula units per cell (Z = 4) as well as similar unit cell parameters. All of these properties indicate that they are isomorphic. In the crystal structures of the three CLA solvates, the stoichiometric ratios of CLA and solvent were all determined to be 1:1.

Table 1. Crystallographic data of CLA Form 0′ and three solvates.

Phase	CLA-Form 0′	CLA-PAC Solvate	CLA-IPA Solvate	CLA-EB Solvate
Empirical formula	$C_{38}H_{69}NO_{13}$	$C_{38}H_{69}NO_{13} \cdot C_5H_{10}O_2$	$C_{38}H_{69}NO_{13} \cdot C_5H_{10}O_2$	$C_{38}H_{69}NO_{13} \cdot C_6H_{12}O_2$
Formula weight	747.94	850.07	850.07	864.09
Crystal system	orthorhombic	orthorhombic	orthorhombic	orthorhombic
Space group	$P2_12_12_1$	$P2_12_12_1$	$P2_12_12_1$	$P2_12_12_1$
Temperature (°C)	25	−165	−165	−165
a (Å)	8.7700 (4)	8.7114 (2)	8.6827 (2)	8.7063 (3)
b (Å)	14.5393 (6)	14.5078 (3)	14.4474 (3)	14.5448 (5)
c (Å)	38.5455 (16)	37.6484 (12)	37.9694 (8)	37.8502 (11)
α (°)	90	90	90	90
β (°)	90	90	90	90
γ (°)	90	90	90	90
Cell volume (Å^3)	4914.90	4758.10	4762.97	4793.02
ρ, kg·m^3	1.011×10^3	1.187×10^3	1.185×10^3	1.197×10^3
Z	4	4	4	4
Rint	0.0801	0.0683	0.0566	0.0853
R1 (I > 2σ(I))	0.0671	0.0526	0.0490	0.0614
wR$_2$	0.1640	0.1035	0.1040	0.1326
CCDC NO.	2,339,166	2,339,163	2,339,164	2,339,165

Comparing the non-solvated form with the CLA-Eth solvate (deposited at the CCDC, deposition number 700729, Table S1 [15]), it can be seen that the crystal system and space group of the two forms were the same, and the CLA molecular conformation (see Figure S2 cyan and orange) and molecular packing pattern (see Figure 2a and Figure S3) of the two forms were similar. Form 0′ has been reported as a desolvated isostructural form of the CLA-Eth solvate by Tian et al. [12]. Therefore, it is reasonable to infer that the non-solvated form crystallized in DMC was Form 0′. In addition, the molecular conformations of CLA in Form 0′ and Form II (deposited at the CCDC, deposition number 780856 [12], colored in red in Figure 2) were basically identical, suggesting that the two forms are packing polymorphs.

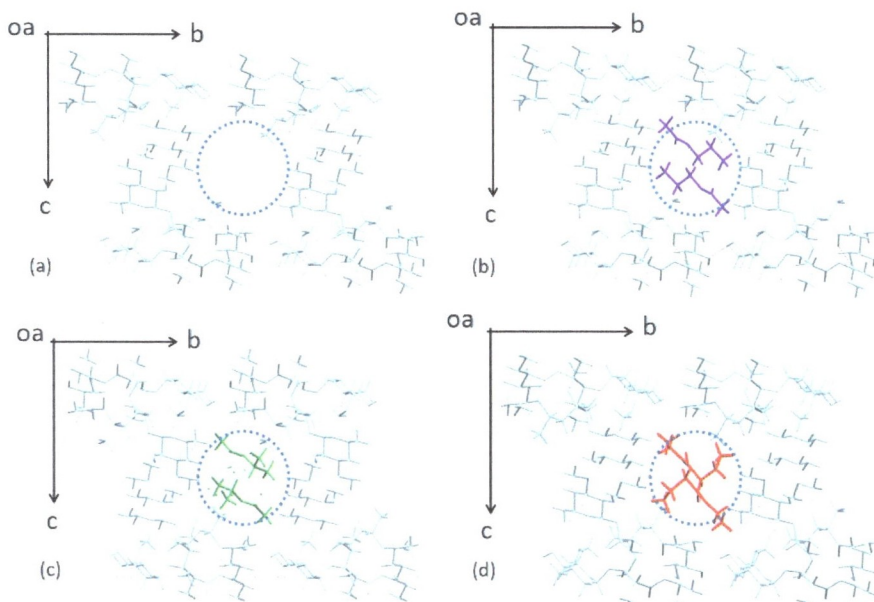

Figure 2. Molecular arrangements of Form 0′ and three solvates. (**a**) Form 0′, (**b**) CLA-PAC solvate, (**c**) CLA-IPA solvate, (**d**) CLA-EB solvate, where the blue dashed circles represent the channel structure; CLA, PAC, IPA, and EB molecules are colored in light blue, purple, green, and red, respectively.

The molecular arrangements in the crystal structures of the three solvates are shown in Figure 2b–d. It shows that the stacking patterns of the molecules in the three solvates were the same as that in Form 0′. Solvent molecules, as a guest, enter the channels formed in the framework stacked by CLA molecules. The solvent molecules do not form hydrogen bonds with the CLA molecules, only weak van der Waals forces. This means that the solvent molecules can easily escape from the channel. In contrast, as shown in Figure S3, in the crystal structure of the CLA-Eth solvate, the ethanol molecules are connected not only with CLA molecules by O60-H60...O12, but are also connected to each other by hydrogen bonds, indicating a greater difficulty of solvent removal.

3.2. Thermal Analysis of Form 0′ and Solvates

DSC and TGA analyses were performed to gain information about the thermal behaviors of the crystals prepared in Experiment 2.2 via method (3). As shown in Figure 3a, the weight loss of the sample in the temperature range of 250 °C~350 °C on the TGA curve corresponded to the decomposition of CLA, which is basically in agreement with that (250 °C~340 °C) reported by Li Wei [34]. There was no weight loss before decomposition on the TGA curve, indicating that the crystal was a non-solvated form. Combined with the PXRD analysis in Section 3.3, it can be proven that the crystals prepared in Experiment 2.2 by method (3) are Form 0′. More discussion can be seen in Section 3.3. In the DSC curve of Form 0′ shown in Figure 3b, Form 0′ exhibited an exotherm at 114.5 °C, and an endotherm at 228.6 °C during heating. It has been reported that the melting point of Form II is between approximately 227 °C and 233 °C. The DSC curve of Form II was measured in this work and showed that Form II melted at 228.1 °C (see Figure S4). These indicate that the endothermic DSC signal of Form 0′ represents the melting process of Form II, while the exothermic event represents the transformation of Form 0′ to Form II. The exothermic transition demonstrates that Form 0′ and Form II bear a monotropic relationship, and that

Form 0′ is less stable than Form II. Moreover, the density of Form 0′ (1.011 × 10³ kg·m⁻³) being smaller than that of Form II (1.164 × 10³ kg·m⁻³) also supports this conclusion.

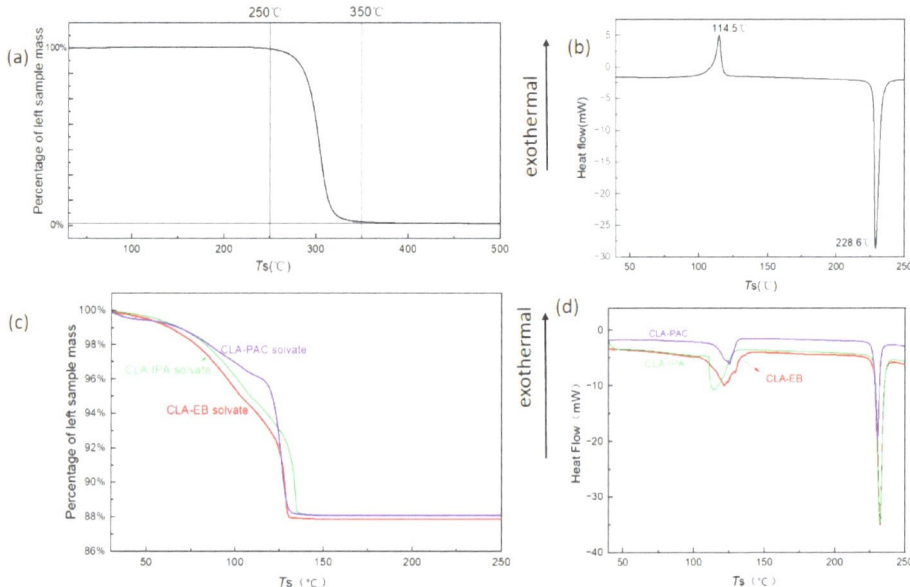

Figure 3. (**a**) TGA curve. (**b**) DSC curve of Form 0′. (**c**) TGA curve. (**d**) DSC curve of the three solvates.

The TGA and DSC profiles of the three crystals prepared in Experiment 2.2 via method (1) are shown in Figure 3c,d. All TGA profiles presented significant weight losses, which corresponded to wide endothermic peaks in the DSC profiles in the same temperature ranges (109.3 °C~134.9 °C, 109.7 °C~130.8 °C, and 99.3 °C~133.8 °C, respectively, for the three solvates), indicating solvent removal. The weight losses of the three samples were 11.88%, 12.10%, and 12.15%, respectively, which basically matched the solvent stoichiometries determined by SCXRD (12.01%, 12.01%, and 13.44%, respectively), indicating that the obtained crystals were the CLA-PAC solvate, CLA-IPA solvate, and CLA-EB solvate. The weight loss of the sample of CLA-EB solvate was slightly lower than the crystallographic data, which was due to the escape of the solvent molecules during the test. The CLA-IPA solvate has been published by Liang's work [12]; here, we determined its crystal structure. At the same time, the solvates of CLA-PAC and CLA-EB obtained in this work were reported for the first time. This second endothermic peaks in the DSC curves of the three solvates appeared at approximate temperatures of 230.3 °C, 231.7 °C, and 231.7 °C, respectively. There were no weight losses at these temperature ranges in the TGA curves, thus we speculate that the endothermic peaks on the DSC curves should belong to the melting process of Form II.

To confirm the above speculation, the phase transformation process of the solvates was visualized by HSM using the CLA-PAC solvate as an example. A fresh block-shaped single crystal was used in the observation experiment, and its morphologies during heating are depicted in Figure S5. In the microscopic field of view, at the beginning, the crystal could be seen brightly, and the dark periphery was due to a certain thickness of the crystal. When heated to 110 °C, the middle of the crystal in the field of view gradually became dark, which corresponded to the first desolvation endothermic process observed in the DSC curve of the CLA-PAC solvate. When heated to 138 °C, the crystal in the field of vision became completely dark, while its boundary remained unchanged. After heated to 190 °C, the crystal was removed to perform the DSC measurement. The DSC curve (see Figure S6) showed an endothermic peak at 229.3 °C, corresponding to the melting point of Form II. Therefore, it can be concluded that the second endothermic peaks in the DSC curves of the three CLA solvates represent the melting process of Form II, and that the first endothermic peaks represent the desolvation and transformation to Form II of the solvates. Moreover, no melting behavior and distinct shape change of the single crystal were observed during the whole temperature range investigated. This means that the transition of the CLA solvate to Form II belongs to a solid-state transformation process.

The thermal behaviors of the crystals prepared in Experiment 2.2 by method (2) was characterized by DSC measurements. As shown in Figure 4a, all DSC curves showed exothermic peaks followed closely by endothermic peaks. Combined with the DSC curves of the three solvates and Form $0'$ (Figure 3b,d), it can be speculated that the exothermic peak in Figure 4a represents the polymorphic transformation of Form $0'$ to Form II, while the endothermic one shows the desolvation of the solvates. To confirm this speculation, a phase transformation experiment of the solvates was performed under vacuum using the CLA-PAC solvate as a representative sample. DSC measurements were conducted after the CLA-PAC solvate experienced different durations. As seen in Figure 4b, after 2 h, the DSC curve rose slightly near 91.0 °C, followed by an endothermic peak at 101.6 °C corresponding to the desolvation of the solvate. Four hours later, a clear exothermic peak appeared at 100.0 °C, followed by an endothermic peak at 106.63 °C. With the increase in duration, both the peak temperatures of the exotherm and the endotherm on the DSC curve increased. After 9 h, the DSC curve only had one sharp exothermic peak at 114.2 °C, corresponding to the transformation of Form $0'$ to Form II. Figure 4c shows the correlation between the exothermic peak temperature and heating duration. It is obvious that the longer the duration, the higher the peak value, which can be attributed to the greater solvent removal and transformation of the solvate to Form $0'$. When heated for up to 9 h, the solvate completely desolvated and transformed to Form $0'$. The analysis above demonstrates that when the solvate is placed at a reduced pressure, solvent removal will occur, resulting in the formation of a mixture of Form $0'$ and the solvate. This rationalizes the thermal events where the exothermic peaks are followed by the endothermic peaks in the DSC curves in Figure 4a.

3.3. PXRD Analysis of Form $0'$ and Solvates

The PXRD of the solvent-free form prepared in Experiment 2.2 by method (3) as well as the PXRDs calculated from the single crystal structures of forms $0'$, I, and II are shown in Figure 5a. The PXRD of the solvent-free form (red curve) was consistent with the calculated PXRD of Form $0'$, confirming it was Form $0'$. The experimental and calculated PXRDs of the CLA-PAC solvate, CLA-IPA solvate, and CLA-EB solvate are also depicted in Figure 5a. It is obvious that the PXRDs of the three solvates are identical and basically consistent with that of Form $0'$. This indicates that the three solvates and Form $0'$ are isomorphic, in agreement with the results determined by SCXRD.

The VT-PXRD of Form $0'$ in the temperature range of 30 °C~120 °C is shown in Figure 5b. In comparison to the PXRD at 100 °C, three characteristic peaks of Form $0'$ at 2θ = 4.601°, 6.492°, and 7.607° disappeared (marked as red circle) on equilibration at 114 °C. At the same time, two characteristic peaks (marked as yellow triangle) belonging to Form II at 10.748° and 11.356° formed. It can be seen that Form $0'$ transforms into Form II at about 114 °C, which is consistent with that determined by the thermal analysis.

The VT-PXRD of the CLA-PAC solvate in the temperature range 30 °C~120 °C is shown in Figure 5c. On equilibration at 110 °C, characteristic peaks belonging to Form II emerged at 2θ = 8.445°, 9.363°, 10.748°, and 11.356° (marked as an yellow triangle). At the same time, the characteristic peaks of 2θ = 4.601° and 7.607° (green circle marks) belonging to the CLA-PAC solvate disappeared. Upon a further rise to 120 °C, the characteristic diffraction peaks of 2θ = 6.492° and 10.106° belonging to the CLA-PAC solvate disappeared (black circle marks). It can be seen that the CLA-PAC solvate begins to transForm Into Form II at about 110 °C, which is in agreement with the previous results of the thermal analysis. The VT-PXRDs of the CLA-IPA solvate and CLA-EB solvate could obtain similar results, as shown in Figures S7 and S8.

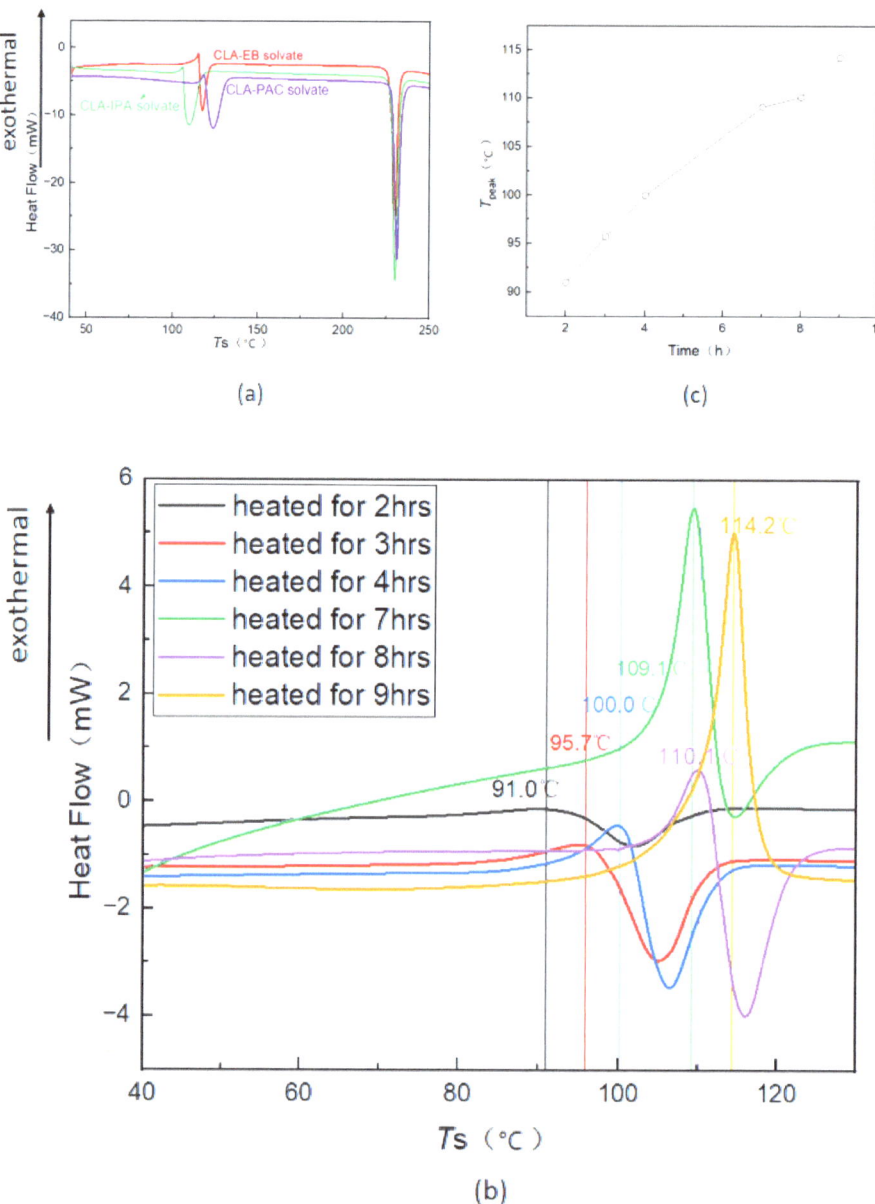

Figure 4. (**a**) DSC of the mixture obtained after the solvate was processed by method (2). (**b**) DSC of heating the CLA-PAC solvate processed by method (1) for different hours. (**c**) Temperature of the exothermic peak changing with the heating duration.

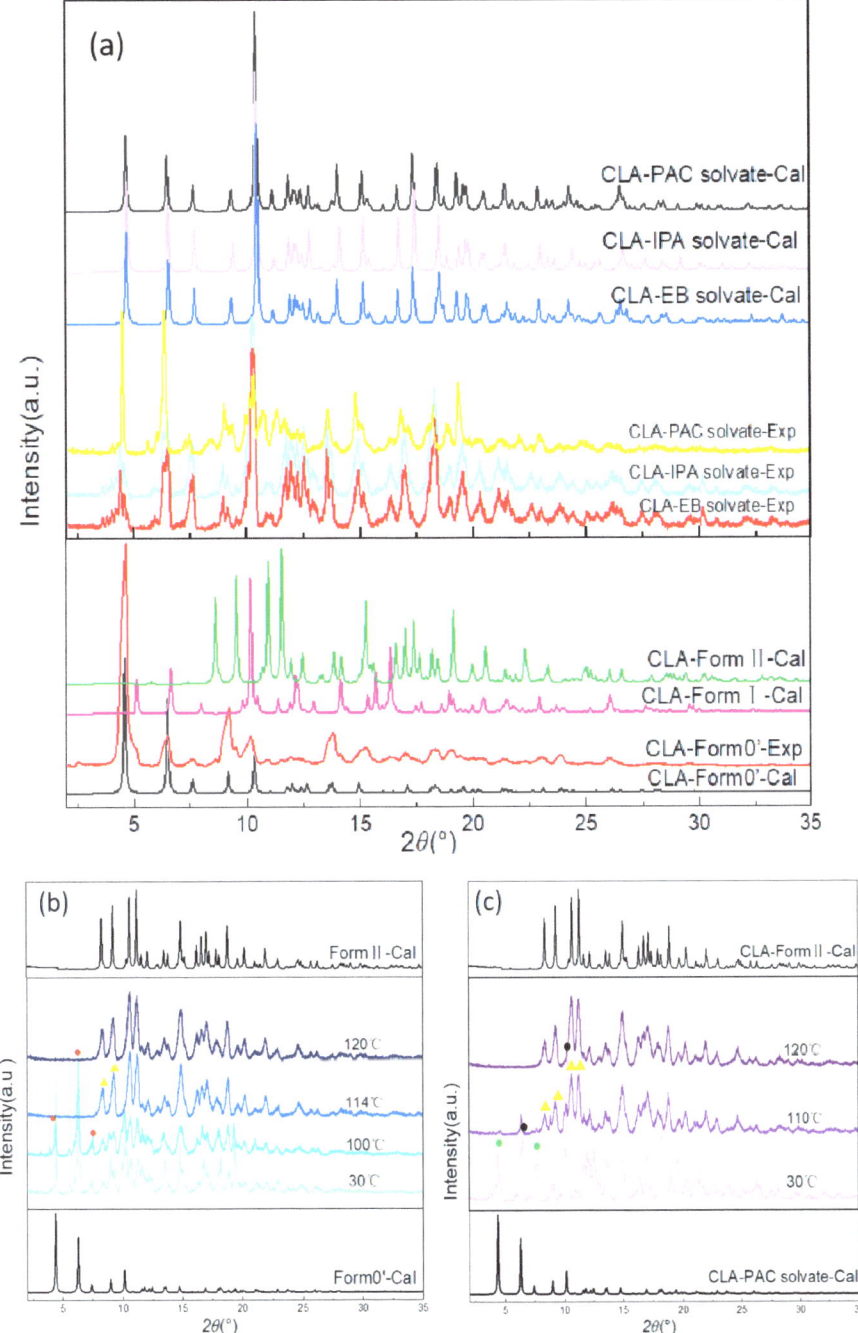

Figure 5. (**a**) Experimental and calculated PXRD patterns of the solvates and Form 0′. (**b**) VT-PXRD patterns of CLA Form 0′ (In (**b**), the red circle represents the disappeared peaks, while the yellow triangle marks represent the peaks belong to the formed crystal). (**c**) VT-PXRD patterns of the CLA-PAC solvate (In (**c**), the yellow triangle marks represent the peaks belong to the formed crystal, while the green and the black circle marks represent the disappeared peaks).

3.4. Phase Transformation of Solvates and Form 0′

From the analyses above, it can be seen that the three solvates, namely the CLA-PAC solvate, CLA-IPA solvate, and CLA-EB solvate, can transform directly to Form II at 100 °C~110 °C or first to Form 0′ at room temperature to ~50 °C, and then further to Form II at near 114 °C, as shown in Figure 6. The effects of atmospheric pressure CO_2 on the desolvation of the solvates and polymorphic transformation of Form 0′ were investigated. Firstly, the solvate desolvation experiment in 1 atm CO_2 was carried out with CLA-PAC as the representative, and the solvent residual in the solvate was monitored by TGA analysis (shown in Figure S9). As shown in Figure S11, it took about 19 h for the complete desolvation of the CLA-PAC solvate in 1 atm CO_2 at 50 °C (marked as red circle). In an air atmosphere under the same temperature and pressure (marked as a black square in Figure S11, TGA curves shown in Figure S10), the solvent removal of CLA-PAC also took about 19 h. The result of the comparative experiment indicates that CO_2 has no effect on the solvent removal of the CLA solvates.

(a) Solvates $\xrightarrow{100\sim110\,°C}$ FormII

(b) Solvates $\xrightarrow{\text{Room temperature}\sim50\,°C}$ Form 0′ $\xrightarrow{114\,°C}$ FormII

(c) Solvates $\xrightarrow{\text{Room temperature}\sim50\,°C}$ Form 0′ $\xrightarrow[\text{1 atm }CO_2]{50\,°C}$ FormII

Figure 6. Three pathways of transforming the solvates to Form II (For pathway a: the solvates are directly transformed to Form II by heated in 100~110 °C; for pathway b: the solvates have the chance to desolvate at room temperature~50 °C, forming Form 0′. And then by being heated in 114 °C, Form 0′ transform to Form II; for pathway c: the solvates have the chance to desolvate at room temperature~50 °C, forming Form 0′. With Form 0′ being heated in 50 °C under 1 atm CO_2, Form II is formed).

Next, the effect of the 1 atm CO_2 atmosphere on the polymorphic transformation of Form 0′ to Form II was investigated using crystalline phase monitoring by PXRD. The PXRDs of the samples at 1 h, 2 h, 3 h, and 4 h are shown in Figure 7a. After 3 h (green curve), the intensities of the characteristic peaks of Form 0′ at 2θ = 4.601°, 6.492°, and 10.106° decreased (marked as purple circle). At the same time, the characteristic peaks of Form II appeared at 2θ = 8.445°, 10.748°, and 11.356° (yellow triangle marks). At 4 h, the PXRD was in agreement with that of Form II, indicating that Form 0′ completely transforms into Form II at this time. The results of the comparative experiments in the air atmosphere at the same temperature and pressure are shown in Figure 7b. The PXRD pattern at 61 h indicates that the polymorph was Form 0′ at this time. At 88 h, the characteristic peaks of Form 0′ (2θ = 4.601°, 6.492°, and 10.106°) were still present, although their intensities became weak. Up to 108 h, no characteristic peaks of Form II were observed. The results show that CO_2 can significantly promote the polymorphic transformation of CLA Form 0′ to Form II. Based on the above results, a new pathway was designed for the preparation of Form II with the CLA solvates, as shown in Figure 6c: the solvate (either the CLA-PAC solvate, the CLA-IPA solvate. or the CLA-EB solvate) desolvates to form Form 0′ in the temperature range of room temperature to ~50 °C, and then Form 0′ transforms into Form II in a 1 atm CO_2 atmosphere at 50 °C. The crystals obtained by this method were characterized by DSC and PXRD. As shown in Figure S12, there was only one endothermic peak at 229.3 °C in the DSC curve, and the PXRD pattern was consistent with that of Form

II with no characteristic peaks of Form 0′, indicating that the crystals prepared by the new method were pure Form II.

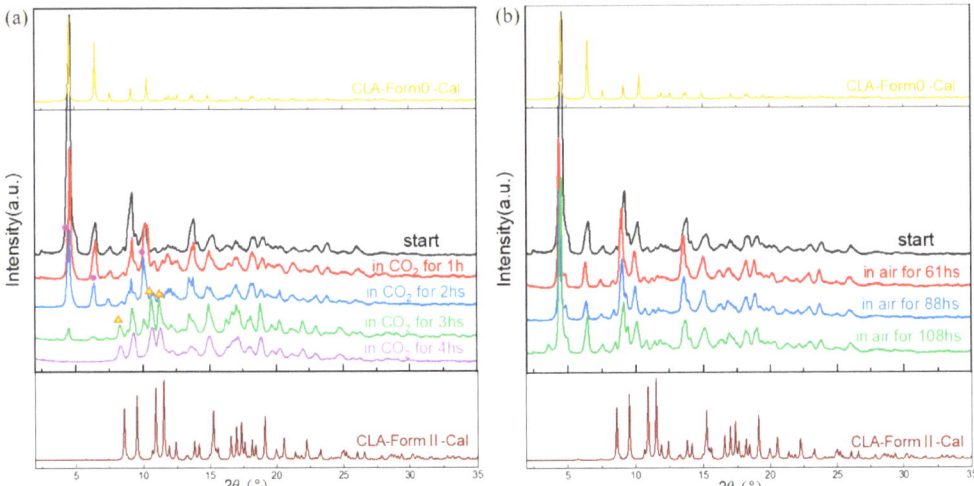

Figure 7. Polymorphic transformation of Form 0′ in (**a**) 1 atm CO_2; (**b**) 1 atm air.

3.5. Mechanism of CO_2 Promoting Polymorphic Transformation

The experiments show that it requires more than 108 h for the polymorphic transformation of Form 0′ into Form II in 1 atm air at 50 °C. However, at the same temperature and pressure, CO_2 can dramatically shorten the transformation duration of CLA from Form 0′ to Form II. Tian et al. [15] have reported that neither air, N_2, O_2, H_2, CH_4, C_2H_6, nor C_3H_8 can accelerate the transformation. It is speculated that CO_2 is likely to be chemically adsorbed on the CLA molecules, which promotes the polymorphic transformation. To confirm this hypothesis, the physical and chemical adsorption capacity of CLA for CO_2 were measured. As shown in Table 2, the physical adsorption capacity of Form 0′ and Form II for CO_2 was 0.31240 mmol·g^{-1} and 0.24877 mmol·g^{-1}, respectively, at 25 °C, and 0.12914 mmol·g^{-1} and 0.11153 mmol·g^{-1}, respectively, at 50 °C. The decrease in adsorption capacity with the increase in temperature reflects the character of physical adsorption. The adsorption capacity of Form 0′ being slightly larger than that of Form II may be related to the difference in the void structure of the two polymorphs.

The results of TPD measurement are shown in Figure 8. The results showed that two signals appeared at 63.81 °C and 196.53 °C, respectively, during the desorption process at higher temperature, indicating that there were two types of chemisorption sites (labeled L and R, respectively, in Figure 8). The desorption amounts of CO_2 at peaks L and R were 0.04813 mmol·g^{-1} and 0.9487 mmol·g^{-1}, respectively. The low desorption amount corresponded to the chemical adsorption of CO_2 by the groups with weak electronegativity, while the high desorption amount corresponded to the chemical adsorption of CO_2 by the groups with strong electronegativity. Considering the incomplete peak R caused by the sample melting when heated to 200 °C, the total capacity of chemisorption should be greater than 0.9968 mmol·g^{-1}. Obviously, the capacity of the chemical adsorption is significantly greater than that of the physical adsorption. Therefore, the chemisorption of CO_2 plays a major role in promoting the polymorphic transformation.

Table 2. Physical adsorption capacity at 25 °C and 50 °C.

Adsorption Capacity (mol·g^{-1})	25 (°C)	50 (°C)
Form 0′	0.31240	0.12914
Form II	0.24877	0.11153

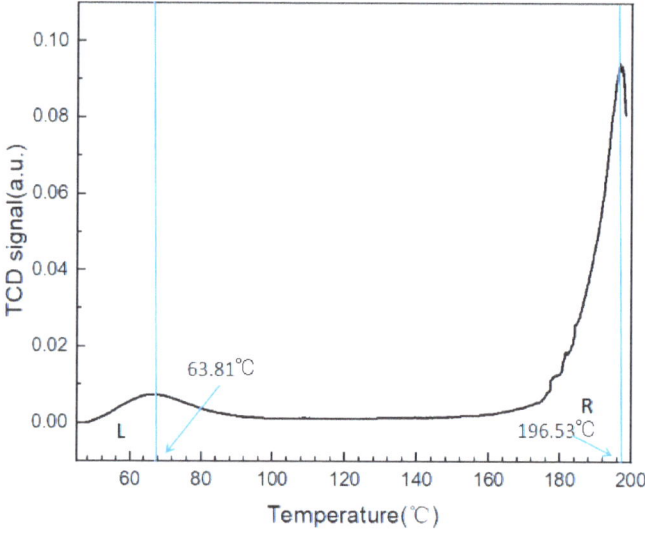

Figure 8. TPD results of CLA Form II.

In order to identify the adsorption site of CO_2, the MEPS of the CLA molecule in Form 0′ was calculated and is shown in Figure 9. Corresponding to the negative region displayed in the MEPS, the electronegativity of the groups in the molecular structure were ordered from strong to weak as: O7 (hydroxyl oxygen on the 14-membered lactone ring) > O1 (hydroxyl oxygen on the 6-membered desamine ring) > O9 (carbonyl oxygen on the 14-membered lactone ring) > O13 (oxygen on the 6-membered cladinose ring) > O6 (hydroxyl group on the 14-membered lactone ring) > O5 (carbonyl oxygen on the 14-membered lactone ring). Furthermore, the groups exposed on the crystal faces of Form 0′ were analyzed. There were four morphologically important crystal faces, (101), (002), (011), and (110) in the Form 0′ morphology predicted by the BFDH model (as shown in Figure S13). The groups exposed on the four crystal faces are shown in Figure S14. On faces (101) and (002), only the dimethylamino group was exposed. On face (011), O13 (EP = −42.51 kcal·mol^{-1}, where EP denotes electrostatic potential) on the 6-membered cladinose ring was exposed. On face (110), all oxygens on the 14-membered lactone ring were exposed, among which O7, O9, O6, and O5 exhibited strong electronegativity with EP values of −47.63 kcal·mol^{-1}, −42.99 kcal·mol^{-1}, −40.68 kcal·mol^{-1}, and −39.55 kcal·mol^{-1}, respectively. It can be inferred that chemisorption may occur at O7, O9, O13, O6, and O5 when Form 0′ is exposed in a CO_2 atmosphere. O7, which had the most negative EP value, was taken as the representative to construct the adsorption model (as shown in Figure S15), and the adsorption energy of CO_2 on O7 was calculated to be −8.43 kcal·mol^{-1}. The negative adsorption energy indicates that CO_2 can be stably adsorbed at the site of O7.

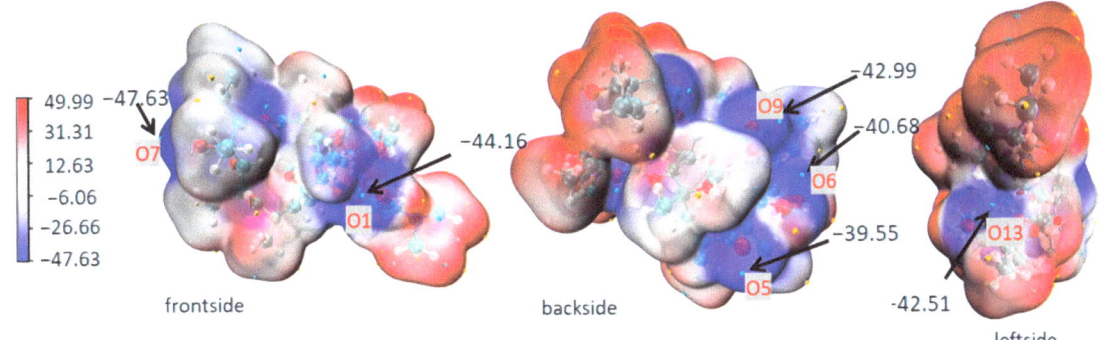

Figure 9. MEPS of the CLA molecule in Form 0′.

As shown in Figure 10a, in Form 0′, two adjacent CLA molecules along the c axis compose a Z-shaped structure by van der Waals forces. The adjacent Z-shaped structures along the b axis are connected by intermolecular hydrogen bond O12-H12...O7 to form a one-dimensional chain. O12-H12...O7 is the only type of intermolecular hydrogen bond in Form 0′, with O7 as the hydrogen bond acceptor and O12 on the 6-membered cladinose ring as the hydrogen bond donor. The two Z-shaped structures (e.g., gray 'Z' composed of molecules A (colored in light green) and A′ (colored in dark green), and pink 'Z' composed of molecules B (colored in orange) and B′ (colored in rose purple), respectively) in the up–down chains along the c axis orient oppositely and they are connected by van der Waals forces. As shown in Figure 10b, in Form II, two adjacent CLA molecules along the b axis form a Z-shaped structure in parallel through van der Waals forces. The adjacent Z-shaped structures along the b axis are connected to form a zigzag molecular chain by van der Waals forces. Along the c axis, the up–down molecular chains with opposite orientation are connected by intermolecular hydrogen bond O12...H7-O7, which is the only type of intermolecular hydrogen bond in Form II. Form 0′ is the metastable form with a void ratio of 21.6%, while Form II is the stable form with a void ratio of 5.4%. Molecular packing always tends to be closer and more stable. When Form 0′ is exposed in a CO_2 atmosphere, CO_2 will be adsorbed on O7, O9, O13, O6, and O5 of the CLA molecule. The adsorption of CO_2 changes the electronegativity of the region around O7, resulting in the break of the hydrogen bond O12-H12...O7, which connects the two 'Z' structures in a molecular chain extending along the b axis. In the boc plane, the two molecules A and A′ (colored in light green and dark green, respectively) composing the 'Z' structure in the upper molecular chain (gray) rotate counterclockwise, while the two molecules B and B′ (colored in orange and reddish-brown, respectively) in the lower molecular chain (pink) rotate clockwise. Molecule A and its upper adjacent molecule B form an intermolecular hydrogen bond O7-H7...O12 through the OH on the 14-membered ring of molecule A and the O on the 6-membered cladinose ring of molecule B. At the same time, molecule A and its lower adjacent molecule B form another hydrogen bond O7-H7...O12. Similarly, molecule A′ connects to its upper and lower adjacent molecule B′ through two hydrogen bonds O7-H7...O12. The molecules reassemble into a densely packed zigzag arrangement structure (see Figure 10b). In this way, Form 0′ transforms into Form II after CO_2 activation.

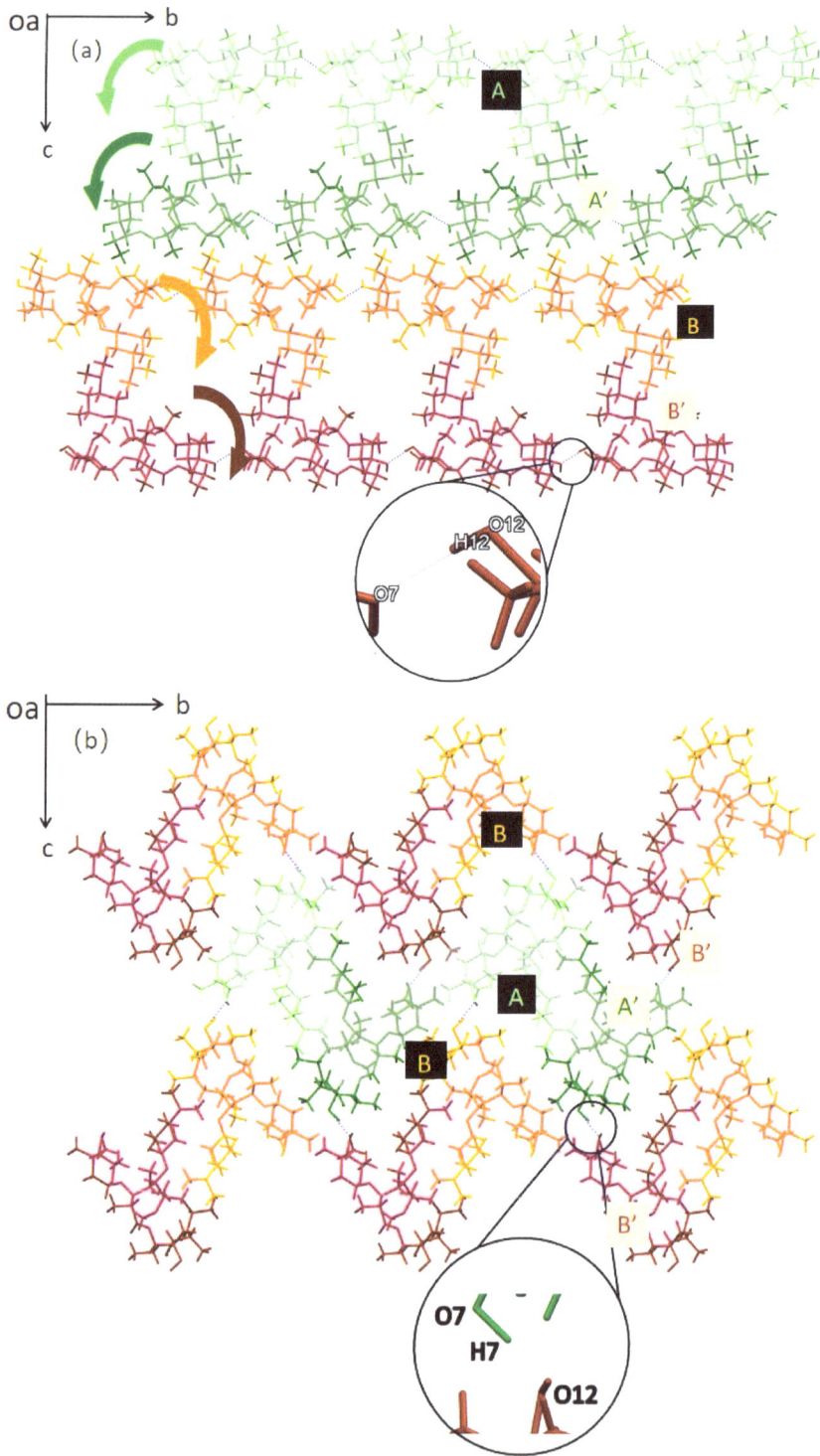

Figure 10. Molecular arrangements in Form 0′ (**a**) and Form II (**b**).

4. Conclusions

The currently reported methods of preparing Form II by CLA solvates require high temperature, high pressure, or vacuum, which means a large amount of energy cost. In this work, a new method for the conversion of CLA solvates to Form II promoted by 1 atm CO_2 was proposed. The mechanism of the effect of CO_2 was revealed through experiments and theoretical calculations.

First of all, three channel CLA solvates were prepared that are easy to remove solvents, namely the CLA-PAC solvate, CLA-IPA solvate, and CLA-EB solvate. The single crystal structures of the three solvates and Form 0′ were resolved for the first time by using SCXRD. The results showed that these are isomorphic. In the crystal structures of three solvates, the solvent molecules located in the channel did not form hydrogen bonds with the CLA molecules, only weak van der Waals forces. This is the reason why the solvent removal of the three solvates was easier than that of the CLA-Eth solvate. Thermal analysis and PXRD showed that the three solvates could desolvate in the temperature range of room temperature to 50 °C to Form Isostructural Form 0′. Form 0′ and Form II are monotropic, related to Form 0′ transforming into Form II at nearly 114 °C. Subsequently, the comparative experiment of the phase transformation in the air atmosphere and 1 atm CO_2 atmosphere showed that CO_2 had no obvious effect on the desolvation process of the solvates, while it could dramatically accelerate the polymorphic transformation of Form 0′ to Form II. In the 1 atm CO_2 atmosphere, the conversion time was only 4 h, which was much less time than that in the air atmosphere. In addition, through the combination of the measurement of adsorption capacity, MEPS calculation, and crystal structure analysis, the mechanism of the CO_2 effect on the polymorphic transformation was uncovered. The chemisorption of CO_2 on O5, O6, O7, O9, and O13 of the CLA molecule with stronger electronegativity exposed on the crystal surfaces of Form 0′ changed the electronegativity of the region near O7, resulting in the break of the only type of hydrogen bond, O12-H12...O7, in Form 0′, and the formation of the hydrogen bond O7-H7...O12 to reassemble the molecules into Form II. Finally, a new pathway for the preparation of Form II from the CLA solvates was designed. Step 1: The solvates desolvate and transForm Into the structurally similar Form 0′ in 1 atm air. Step 2: Form 0′ transforms into the more stable Form II in 1 atm CO_2. This work provides a successful case for the application of CO_2 as an accelerator or alternative energy in the phase transformation of APIs, and also provides a new idea for the study of the mechanisms of polymorphic transformation.

Supplementary Materials: The following supporting information can be downloaded at: https://www.mdpi.com/article/10.3390/cryst14050394/s1, Figure S1: The custom built micro-vacuum box; Figure S2: The conformation overlays diagram of CLA Form 0′ (cyan), Form II (red) and CLA-Eth solvate (orange); Figure S3: Molecular arrangement of CLA-Eth solvate crystal structure; Figure S4: DSC curve of CLA raw material; Figure S5: HSM snapshots of CLA-PAC solvate; Figure S6: DSC curve of the sample after the observation experiment by HSM; Figure S7: VT-PXRD of CLA-IPA solvate; Figure S8: VT-PXRD of CLA-EB solvate; Figure S9: The solvent residual in the sample after heating at 50 °C and 1atm CO_2 for different duration; Figure S10: The solvent residual in the sample after heating at 50 °C and 1atm air for different duration; Figure S11: The change of solvent residual in CLA-PAC solvate with time; Figure S12: (a) DSC curve, (b) PXRD of the sample prepared by pathway; Figure S13: The morphology of CLA Form 0′ predicted by BFDH model; Figure S14: Molecular arrangements on crystal faces (101), (002), (011) and (110) of Form 0′; Figure S15: The model of the adsorption of CO_2 on Form 0′ (Purple represents for N atoms, red for oxygen atoms, light pink for hydrogen atoms, and brown for carbon atoms); Table S1: Crystallographic data of CLA-Eth solvate.

Author Contributions: Data curation, L.H.; Formal analysis, Y.W.; Methodology, L.H.; Resources, D.J.; Software, L.H.; Writing—original draft, L.H., Y.W. and Y.B.; Writing—review and editing, L.H. and Y.B. All authors have read and agreed to the published version of the manuscript.

Funding: This research received no external funding.

Data Availability Statement: Crystallographic information files are available from the Cambridge Crystallographic Data Center (CCDC) upon request (http://www.ccdc.cam.ac.uk, CCDC deposition numbers 2339163–2339166).

Conflicts of Interest: D.J. is employed by the company Asymchem Life Science (Tianjin) Co., Ltd., and remaining authors declare that no conflicts of interest.

References

1. Du, D.; Shi, Z.P.; Ren, G.B.; Qi, M.H.; Li, Z.; Xu, X.Y. Preparation and characterization of several azoxystrobin channel solvates. *J. Mol. Struct.* **2019**, *1189*, 40–50. [CrossRef]
2. Bērziņš, A.; Trimdale, A.; Kons, A.; Zvanina, D. On the Formation and Desolvation Mechanism of Organic Molecule Solvates: A Structural Study of Methyl Cholate Solvates. *Cryst. Growth Des.* **2017**, *17*, 5712–5724. [CrossRef]
3. Yuan, W.; Yu, C.; Xu, S.; Ni, L.; Xu, W.; Shan, G.; Bao, Y.; Pan, P. Self-evolving materials based on metastable-tostable crystal transition of a polymorphic. *Mater. Horiz.* **2022**, *9*, 756–763. [CrossRef]
4. Jin, M. Mechanical-Stimulation-Triggered and Solvent-Vapor-Induced Reverse Single-Crystal-to-Single-Crystal Phase Transitions with Alterations of the Luminescence Color. *J. Am. Chem. Soc.* **2018**, *140*, 2875–2879. [CrossRef]
5. Hanna, S.L.; Zhang, X.; Otake, K.I.; Drout, R.J.; Li, P.; Islamoglu, T.; Farha, O.K. Guest-Dependent Single-Crystal-to-Single-Crystal Phase Transitions in a Two-Dimensional Uranyl-Based Metal−Organic Framework. *Cryst. Growth Des.* **2019**, *19*, 506–512. [CrossRef]
6. Liao, X.; He, J.; Yu, J. Process analysis of phase transformation of a to b-form crystal of syndiotactic polystyrene investigated in supercritical CO_2. *Polymer* **2005**, *46*, 5789–5796. [CrossRef]
7. Ramachandran, J.P.; Antony, A.; Ramakrishnan, R.M.; Wallen, S.L.; Raveendran, P. CO_2-solvated liquefaction of polyethylene glycol (PEG): A novel, green process for the preparation of drug-excipient composites at low temperatures. *J. CO_2 Util.* **2022**, *59*, 101971.
8. Ramachandran, J.P.; Kottammal, A.P.; Antony, A.; Ramakrishnan, R.M.; Wallen, S.L.; Raveendran, P. Green processing: CO_2-induced glassification of sucrose octaacetate and its implications in the spontaneous release of drug from drug-excipient composites. *J. CO_2 Util.* **2021**, *47*, 101472. [CrossRef]
9. Hu, D.; Li, W.; Wu, K.; Cui, L.; Xu, Z.; Zhao, L. Utilization of supercritical CO_2 for controlling the crystal phase transition and cell morphology of isotactic polybutene-1 foams. *J. CO_2 Util.* **2022**, *66*, 102265. [CrossRef]
10. Tian, J.; Thallapally, P.K.; Dalgarno, S.J.; Atwood, J.L. Free Transport of Water and CO_2 in Nonporous Hydrophobic Clarithromycin Form II Crystals. *J. Am. Chem. Soc.* **2009**, *131*, 13216–13217. [CrossRef]
11. Sohn, Y.-T.; Rhee, J.-K.; Im, W.-B. Polymorphism of Clarithromycin. *Pharmacol. Toxicol. Pharm.* **2000**, *4*, 381–384. [CrossRef] [PubMed]
12. Liang, J.H.; Yao, G.W. A New Crystal Structure of Clarithromycin. *J. Chem. Crystallogr.* **2008**, *38*, 61–64. [CrossRef]
13. Ito, M.; Shiba, R.; Watanabe, M.; Iwao, Y.; Itai, S.; Noguchi, S. Phase transitions of antibiotic clarithromycin forms I, IV and new form VII crystals. *Int. J. Pharm.* **2018**, *547*, 258–264. [CrossRef] [PubMed]
14. Iwasaki, H.; Sugawara, Y.; Adachi, T.; Morimoto, S.; Watanabe, Y. Structure of 6-O-methylerythromycin A (clarithromycin). *Acta Crystallogr. Sect. C Cryst. Struct. Commun.* **1993**, *49*, 1227–1230. [CrossRef]
15. Tian, J.; Dalgarno, S.J.; Atwood, J.L. A New Strategy of Transforming Pharmaceutical Crystal Forms. *J. Am. Chem. Soc.* **2011**, *133*, 1399–1404. [CrossRef] [PubMed]
16. Henry, R.; Zhang, G.G. Crystallographic characterization of several erythromycin A solvates: The environment of the solvent molecules in the crystal lattice. *J. Pharm. Sci.* **2007**, *96*, 1251–1257. [CrossRef] [PubMed]
17. Baronsky, J.; Preu, M.; Traeubel, M.; Urbanetz, N.A. Perfusion calorimetry in the characterization of solvates forming isomorphic desolvates. *Eur. J. Pharm. Sci.* **2011**, *44*, 74–82. [CrossRef] [PubMed]
18. Kuncham, S.; Shete, G.; Bansal, A.K. Quantification of clarithromycin polymorphs in presence of tablet excipients. *J. Excip. Food Chem.* **2000**, *5*, 65–78.
19. Miñambres, G.G.; Aiassa, V.; Longhi, M.R.; Chattah, A.K.; Garnero, C. Insights into the ethanol solvate form of clarithromycin. *J. Mol. Struct.* **2022**, *1264*, 133170. [CrossRef]
20. Rajbhar, P.; Sahu, A.K.; Gautam, S.S.; Prasad, R.K.; Singh, V.; Nair, S.K. Formulation and Evaluation of Clarithromycin CoCrystals Tablets Dosage Forms to Enhance the Bioavailability. *Pharma Innov. J.* **2016**, *5*, 5–13.
21. Liu, J.-H.; Riley, D.A. Herstellung von Kristallinen Form II von Clarithromycin. WO1998004574A1, 5 February 1998.
22. Gildea, R.J.; Bourhis, L.J.; Dolomanov, O.V.; Grosse-Kunstleve, R.W.; Puschmann, H.; Adams, P.D.; Howard, J.A. iotbx.cif: A comprehensive CIF toolbox. *J. Appl. Crystallogr.* **2011**, *44 Pt 6*, 1259–1263. [CrossRef] [PubMed]
23. Sheldrick, G.M. Crystal structure refinement with SHELXL. *Acta Crystallogr. C Struct. Chem.* **2015**, *71 Pt 1*, 3–8. [CrossRef] [PubMed]
24. Sheldrick, G.M. SHELXT—Integrated space-group and crystal-structure determination. *Acta Crystallogr. A Found. Adv.* **2015**, *71 Pt 1*, 3–8. [CrossRef] [PubMed]

25. Scalmani, G.; Frisch, M.J. Comment on "A smooth, nonsingular, and faithful discretization scheme for polarizable continuum models: The switching/gaussian approach" [J. Chem. Phys. 133, 244111 (2010)]. *J. Chem. Phys.* **2011**, *134*, 117101. [CrossRef] [PubMed]
26. Lu, T.; Chen, F. Multiwfn: A multifunctional wavefunction analyzer. *J. Comput. Chem.* **2012**, *33*, 580–592. [CrossRef]
27. Humphrey, W.; Dalke, A.; Schulten, K. VMD: Visual molecular dynamics. *J. Mol. Graph.* **1996**, *14*, 33–38. [CrossRef] [PubMed]
28. Grimme, S.; Antony, J.; Ehrlich, S.; Krieg, H. A consistent and accurate ab initio parametrization of density functional dispersion correction (DFT-D) for the 94 elements H-Pu. *J. Chem. Phys.* **2010**, *132*, 154104. [CrossRef]
29. Grimme, S.; Ehrlich, S.; Goerigk, L. Effect of the damping function in dispersion corrected density functional theory. *J. Comput. Chem.* **2011**, *32*, 1456–1465. [CrossRef] [PubMed]
30. Hehre, W.J.; Ditchfield, R.; Pople, J.A. Self—Consistent molecular orbital methods. XII. Further extensions of Gaussian—Type basis sets for use in molecular orbital studies of organic molecules. *J. Chem. Phys.* **1972**, *5*, 2257–2261. [CrossRef]
31. Zhang, J.; Lu, T. Efficient evaluation of electrostatic potential with computerized optimized code. *Phys. Chem. Chem. Phys.* **2021**, *23*, 20323–20328. [CrossRef]
32. Sun, H. COMPASS: An ab initio force-field optimized for condensed-phase applications overview with details on alkane and benzene compounds. *J. Phys. Chem. B* **1998**, *38*, 7338–7364. [CrossRef]
33. Hartman, P.; Bennema, P. The attachment energy as a habit controlling factor: I. Theoretical considerations. *J. Cryst. Growth* **1980**, *1*, 145–156. [CrossRef]
34. Li, W. Studies on thermal decomposition mechanism of clarithromycin and determination of the kinetic parameters. *Chin. J. Antibiot.* **2009**, *34*, 419–421.

Disclaimer/Publisher's Note: The statements, opinions and data contained in all publications are solely those of the individual author(s) and contributor(s) and not of MDPI and/or the editor(s). MDPI and/or the editor(s) disclaim responsibility for any injury to people or property resulting from any ideas, methods, instructions or products referred to in the content.

MDPI AG
Grosspeteranlage 5
4052 Basel
Switzerland
Tel.: +41 61 683 77 34

Crystals Editorial Office
E-mail: crystals@mdpi.com
www.mdpi.com/journal/crystals

Disclaimer/Publisher's Note: The title and front matter of this reprint are at the discretion of the Guest Editors. The publisher is not responsible for their content or any associated concerns. The statements, opinions and data contained in all individual articles are solely those of the individual Editors and contributors and not of MDPI. MDPI disclaims responsibility for any injury to people or property resulting from any ideas, methods, instructions or products referred to in the content.

www.ingramcontent.com/pod-product-compliance
Lightning Source LLC
LaVergne TN
LVHW072355090526
838202LV00019B/2553